Future Prospects of Wireless Transmission Based on Engineering Electrical Electronic and Telecommunications

Future Prospects of Wireless Transmission Based on Engineering Electrical Electronic and Telecommunications

Guest Editors

Muhammad Usman Hadi
Muhammad Ikram Ashraf

MDPI

Basel • Beijing • Wuhan • Barcelona • Belgrade • Novi Sad • Cluj • Manchester

Guest Editors
Muhammad Usman Hadi
School of Engineering
Ulster University
Belfast
UK

Muhammad Ikram Ashraf
Nokia
Espoo
Finland

Editorial Office
MDPI AG
Grosspeteranlage 5
4052 Basel, Switzerland

This is a reprint of the Special Issue, published open access by the journal *Electronics* (ISSN 2079-9292), freely accessible at: https://www.mdpi.com/journal/electronics/special_issues/QGKXV8467B.

For citation purposes, cite each article independently as indicated on the article page online and as indicated below:

Lastname, A.A.; Lastname, B.B. Article Title. *Journal Name* **Year**, *Volume Number*, Page Range.

ISBN 978-3-7258-8027-0 (Hbk)
ISBN 978-3-7258-8028-7 (PDF)
https://doi.org/10.3390/books978-3-7258-8028-7

Cover image courtesy of Usman Hadi

Contents

About the Editors

Muhammad Usman Hadi

Muhammad Usman Hadi is an Assistant Professor in the School of Engineering at Ulster University, Belfast, United Kingdom. He received his MS and PhD degrees from the University of Bologna, Italy. He previously worked as a Postdoctoral Researcher at Aalborg University, Denmark, in close collaboration with Nokia, and has also held visiting researcher positions at ESIEE Paris, France, and Nokia Bell Laboratories. His research focuses on advanced wireless transmission systems, including machine learning for engineering applications, autonomous aerial vehicles, wireless fiber communications, microwave photonics, and emerging telecommunication devices. He has authored over sixty peer-reviewed journal articles and contributions in leading international journals. As a Principal Investigator, he has secured funding from programmes including EPSRC DTNet+, Innovate UK, British Council Going Global Partnerships, the Garfield Weston Trust, International Science Partnership Fund, and internal Research and Innovation schemes, and he has also contributed as a Co-Investigator to EPSRC, DfE, and ISPF-funded projects. He has been listed among the top two percent of researchers worldwide for four consecutive years since 2021. His work bridges theoretical modelling and experimental validation, with a strong emphasis on the real-world impact of next-generation wireless and intelligent communication systems.

Muhammad Ikram Ashraf

Muhammad Ikram Ashraf received his M.Sc. and Ph.D. in Telecommunication Systems and Communication Engineering, respectively, from the University of Oulu, Finland. He is currently working as a Senior Research Specialist on Advanced 5G at Nokia Bell Labs, Espoo Finland. Prior to that, he worked as an experienced researcher in Network Architecture and Protocols at Ericsson Research in Jorvas, Finland. He has contributed to several technical papers, invention disclosures, and 3GPP. His research interests include Advanced 5G, AI/ML, Industry 4.0, XR, positioning, IoT, V2X, UAV, URLLC, and TSN. He is serving as a Series Editor of **IEEE Communication Standards Magazine** on Ultra-Low-Latency, and Reliable Communications for Future Wireless Networks, an Associate Editor of **IET Quantum Communication**, and a Guest Editor of Elsevier's **Journal of Industrial Information Integration** Special Issue on Data/Information Integration Techniques in Industry 4.0/5.0. Before that, he served as an Editor of the **IEEE Communication Standards Magazine**, a Guest Editor of **IEEE Wireless Communications**, an Editor of **IEEE Transactions on Cognitive Communications and Networking**, and as a Guest Editor of **IEEE Network**.

Preface

Wireless transmission technologies are undergoing rapid transformation driven by growing demands for connectivity, intelligence, and reliability across diverse application domains. This Reprint was developed to capture these developments and to provide a focused platform for presenting recent scientific advances and forward-looking perspectives in wireless transmission systems.

The aim of this Reprint is to explore both foundational and emerging research in electrical, electronic, and telecommunications engineering, with emphasis on next-generation wireless networks, AI-enabled communication frameworks, and advanced transmission techniques. It reflects the motivation to bridge theory and practice by showcasing contributions that combine analytical models, experimental results, and practical deployment insights.

This Reprint is intended for researchers, engineers, and industry professionals working in wireless communications, as well as early-career researchers seeking a consolidated overview of current trends and open challenges. By assembling contributions from diverse research communities, the Guest Editors aim to stimulate discussion, collaboration, and innovation that will help shape the future of wireless transmission technologies.

Muhammad Usman Hadi and Muhammad Ikram Ashraf
Guest Editors

electronics

MDPI

Article

Tri-Band Inverted-F Antenna for Wi-Fi 7 Laptops with Reduced Ground Plane Support

Yu-Kai Huang [1], Kuan-Hsueh Tseng [2] and Yen-Sheng Chen [1,*]

[1] Department of Electronic Engineering, National Taipei University of Technology, Taipei 10608, Taiwan
[2] Antenna Design Integration Department, Micro-Star International Co., Ltd., New Taipei City 23500, Taiwan
* Correspondence: yschen@ntut.edu.tw

Abstract

In modern laptops, antenna design for Wi-Fi 7 is constrained by limited space and reduced ground plane size, conditions under which many compact designs exhibit degraded bandwidth or efficiency or require large device grounds. This paper presents a miniaturized tri-band inverted-F antenna (IFA) that supports the 2.4, 5, and 6 GHz Wi-Fi 7 bands within a radiator area of $20 \times 5 \times 0.8$ mm^3 and a ground plane of 60×40 mm^2. The proposed design achieves wideband impedance matching and stable radiation efficiency under intentionally reduced grounding conditions, addressing a scenario rarely considered in prior studies where both radiator and ground plane miniaturization must be satisfied. Measurements confirm efficiencies of 74–81% at 2.4 GHz and 64–90% across 5–7 GHz, with performance in the lower band exceeding that of many compact designs and upper-band coverage comparable to structures requiring larger footprints. By demonstrating tri-band operation under simultaneous radiator and ground reduction, this work provides a practical antenna solution for next-generation Wi-Fi 7 laptop integration.

Keywords: mobile antennas; monopole antennas; multifrequency antennas; planar antennas; wireless LAN

1. Introduction

The evolution of wireless local area network (WLAN) technology has been driven by increasing requirements for data throughput, spectrum efficiency, and low-latency connectivity in congested environments. Early Wi-Fi standards, formalized under IEEE 802.11 in the late 1990s, operated exclusively in the 2.4 GHz band [1], providing adequate coverage and compatibility for initial consumer applications. With the rapid growth of network traffic, the limitations of the 2.4 GHz spectrum prompted adoption of the 5 GHz band [2], which enabled wider channel bandwidths and improved interference isolation. This transition was realized through standards such as IEEE 802.11a, 802.11n, and 802.11ac [3]. Continued expansion of connected devices, video streaming, and latency-sensitive applications eventually made dual-band operation insufficient. IEEE 802.11ax, branded as Wi-Fi 6E, introduced support for the 6 GHz band, adding 1.2 GHz of unlicensed spectrum [4]. Building on this foundation, IEEE 802.11be (Wi-Fi 7) consolidates simultaneous operation across 2.4, 5, and 6 GHz bands and incorporates new features such as 320 MHz channelization, 4096-QAM, and multi-link operation to support high-throughput and low-latency communication in dense device environments [5].

To meet these requirements, laptop-integrated antennas must provide consistent impedance matching across all three Wi-Fi 7 bands, sufficient radiation efficiency in compact enclosures, and limited sensitivity to surrounding metallic structures and noise sources. Early laptop antennas were typically single-band monopole, dipole, or loop structures designed for the 2.4 GHz band [6]. As dual-band operation became standard under IEEE 802.11n and 802.11ac, inverted-F antennas (IFAs) and folded structures were introduced to cover both 2.4 GHz and 5 GHz within narrow edge spaces [7–13]. With the extension to the 6 GHz band under Wi-Fi 6E and Wi-Fi 7, broader impedance bandwidth and enhanced frequency selectivity are required. Most reported solutions achieve radiator miniaturization but rely on relatively large ground planes to sustain acceptable radiation performance. In practice, thin and component-dense laptops often restrict available ground area, making it necessary to evaluate antenna behavior when both the radiator and the ground plane are constrained.

Recent designs for Wi-Fi 6E and Wi-Fi 7 antennas address multiband operation through a variety of strategies. Bandwidth enhancement has been achieved using metamaterial-inspired elements with interdigital capacitors and shorting strips to excite zeroth-order resonance modes [14] or modified inverted-L radiators with coupled-fed networks for compact dual-band operation [15]. Hybrid topologies integrating slot, dipole, and monopole elements have been used to extend coverage across all three bands [16]. Other studies emphasize geometric adaptability, such as folded IFAs optimized for bezel-limited regions [17] or monopoles miniaturized with chip inductors [18]. For multi-antenna platforms, conjoined layouts [19] and asymmetric isolation-enhanced structures [20] have been proposed to improve inter-element performance. Multi-feed configurations have also been explored to achieve continuous tri-band coverage in narrow bezels spanning 2.4–7.125 GHz [21]. More recent reports address Wi-Fi 6E/7 laptop and terminal antennas through display-edge closed-slot/T-feed arrays, metamaterial/zeroth-order-resonance miniaturization, hinge-region layouts for metal housings, and wideband/circularly polarized and wearable implementations spanning 2.4/5/6 GHz [22–26]; however, most assume standard or extended grounds or emphasize array layout, whereas concurrent radiator and ground-plane miniaturization remains less examined and is specifically addressed here.

This study proposes a tri-band antenna specifically designed for Wi-Fi 7 laptops, covering the 2.40–2.48 GHz, 5.15–5.85 GHz, and 5.925–7.125 GHz bands, corresponding to bandwidths of 80 MHz, 700 MHz, and 1.2 GHz, respectively. Unlike previous designs that assume extensive ground support, the proposed antenna is optimized to sustain impedance matching and radiation efficiency under constrained grounding. A systematic parameter analysis identifies the resonant mechanisms and guides geometric tuning. The design is fabricated and validated through full-wave simulation and experimental measurements, including reflection coefficient, radiation efficiency, and gain. Results confirm consistent tri-band operation without requiring full-sized ground or auxiliary isolation structures, addressing the gap between idealized radiator studies and practical laptop integration.

2. Antenna Structure and Design Methodology

This work develops a tri-band antenna structure tailored for Wi-Fi 7 laptop integration under constrained radiator and ground plane conditions. The design targets the IEEE 802.11be frequency allocations: 2.40–2.48 GHz, 5.15–5.85 GHz, and 5.925–7.125 GHz, with the 6 GHz band spanning a wide 1.2 GHz. To ensure consistent coverage, the impedance matching criterion is set to -7.4 dB, corresponding to a maximum input reflection of 25%, which is commonly adopted in compact device antenna evaluations. For radiation efficiency, thresholds follow commercial laptop RF design guidelines: $\geq$31.62% (-5 dB) at 2.4 GHz and $\geq$28.18% (-5.5 dB) across 5.15–7.125 GHz, providing margin for fabrication tolerances and packaging losses. These criteria frame the antenna's optimization under the dual constraint of radiator and ground plane miniaturization.

2.1. Antenna Geometry

The proposed structure adopts an IFA configuration, selected for its suitability in confined layouts while enabling quarter-wavelength resonance with a shorting element. Figure 1 illustrates the geometry. The antenna is realized on a single-layer FR4 substrate (dielectric constant $\varepsilon_r = 4.4$, loss tangent $\tan\delta = 0.02$, thickness = 0.8 mm), representing a low-cost PCB technology compatible with laptop integration. The radiator footprint is limited to 20 mm × 5 mm, while the ground plane is constrained to dimensions of $GND_y \times GND_x$, where $GND_y = 60$ mm and $GND_x = 40$ mm, both intentionally minimized to replicate space-restricted device conditions. A sideview schematic of the antenna geometry, including the substrate thickness of 0.8 mm, is shown in Figure 1c.

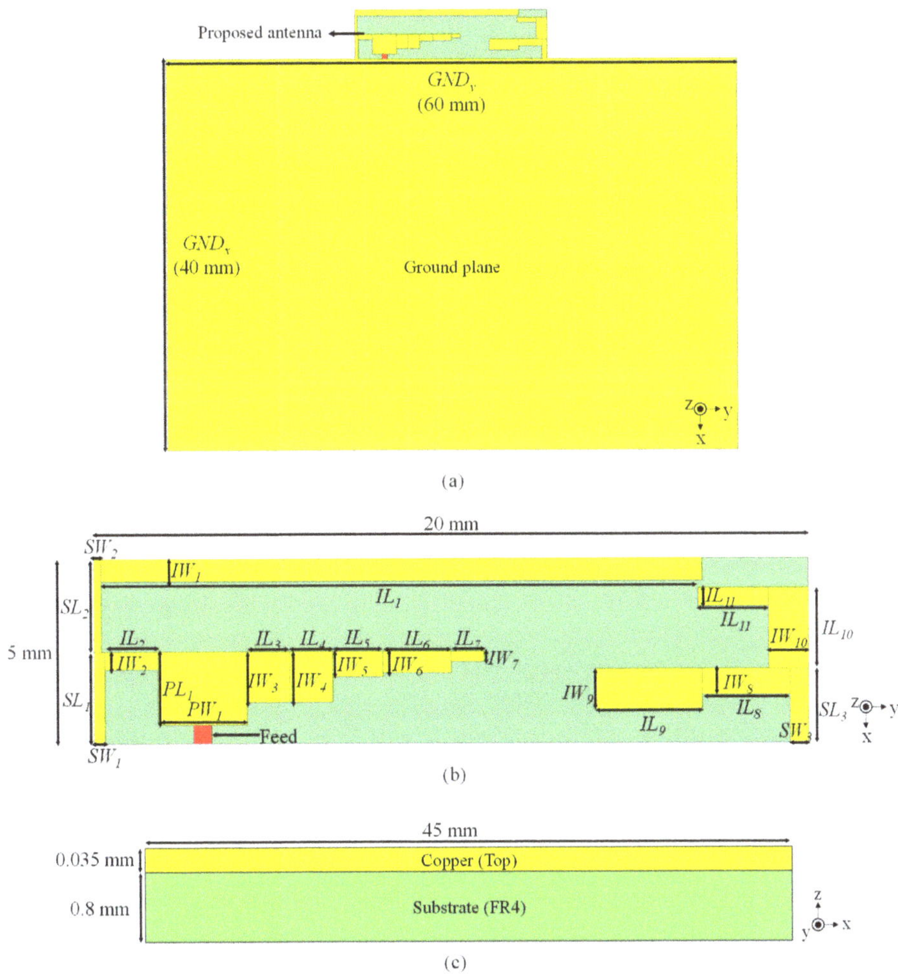

Figure 1. Geometry of the proposed antenna: (**a**) radiator and reduced ground plane; (**b**) segment labeling for feed, shorting, and radiating arm; (**c**) sideview of the proposed antenna structure.

The antenna consists of three functional regions: the feed patch, the shorting path, and the radiating arm. The feed patch, defined by parameters PW_1 and PL_1, introduces a 50 Ω excitation at the lower-left substrate corner. Adjacent to it, a vertical shorting strip connects directly to the ground plane, establishing the low-impedance return path that enforces quarter-wavelength resonance.

The radiating arm is implemented as a stepped meandered line, combining horizontal (IL_1–IL_{11}) and vertical (IW_1–IW_{10}) segments within the 5 mm height constraint. The central folded section (IL_1–IL_7, IW_1–IW_7) generates the 2.4 GHz and 5 GHz resonances. Local vertical stubs (e.g., IW_4–IW_6) perturb the impedance locally, improving bandwidth and modal selectivity. The terminal section (IL_8–IL_{10}, IW_8–IW_9) produces additional higher-order modes that enable 6 GHz coverage, with offsets IL_8 and IW_8 providing capacitive coupling and secondary loops that reinforce broadband behavior.

SW_1–SW_3 (horizontal) and SL_1–SL_3 (vertical) denote the geometric parameters of the physical conductors that connect the radiator to the ground plane. These shorting elements provide current return paths and influence the modal tuning across the operating bands. The final structure achieves full tri-band capability within the reduced 20 mm × 5 mm radiator area without vias or multilayer routing, addressing the ground-plane-constrained scenario.

2.2. Operating Principles

To clarify the physical mechanisms underlying the tri-band response, simulated surface current distributions at 2.4, 5.5, and 6.5 GHz are shown in Figure 2.

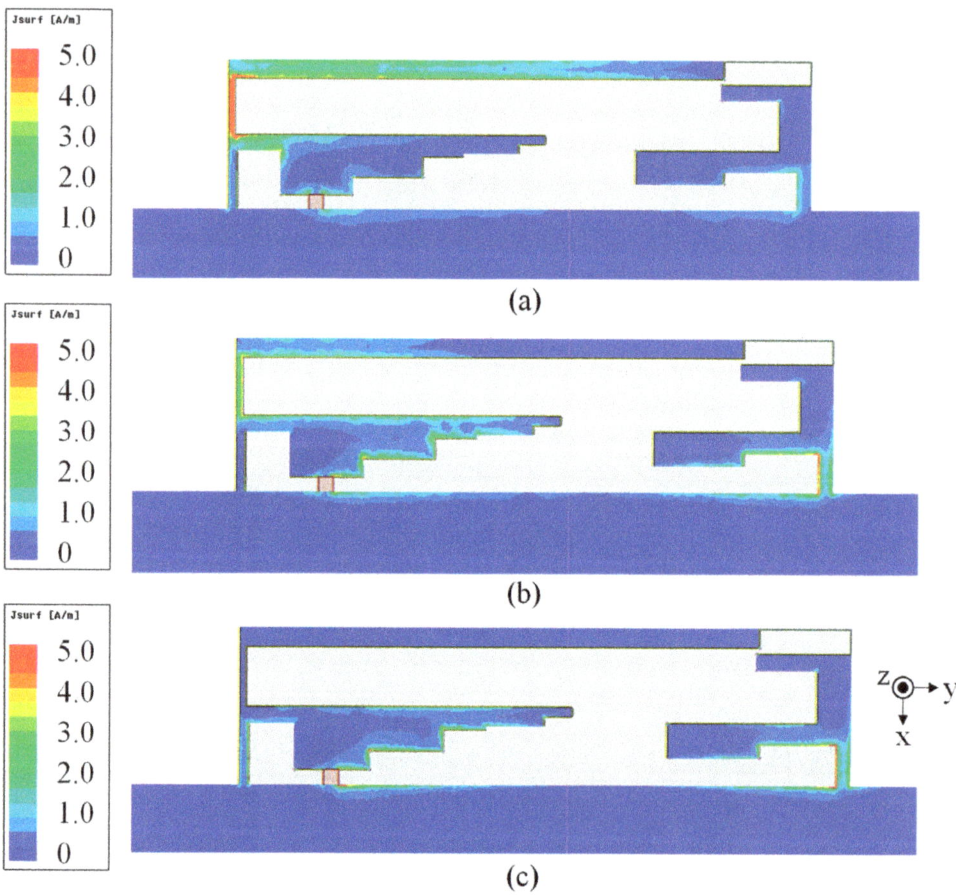

Figure 2. Simulated surface current distributions of the proposed IFA at (**a**) 2.45 GHz, (**b**) 5.5 GHz, and (**c**) 6.5 GHz.

At 2.4 GHz, the dominant resonance is established by a quarter-wavelength loop extending from the feed patch through the shorting strip into the initial horizontal segments (IL_1–IL_3), with strong current density between the feed and the shorting point. This confirms the fundamental IFA resonance under limited ground size.

At 5.5 GHz, resonance occurs through the extended meandered midsection (IL_3–IL_7), which elongates the current path without enlarging the footprint. The stub SL_2 interrupts current continuity, acting as a localized impedance transformer that generates multiple standing-wave lobes. This introduces additional resonance while preserving compactness.

At 6.5 GHz, the terminal lines (IL_8–IL_{10}, IW_8–IW_9) form the primary path, with shortened lengths and angular folds introducing localized high-impedance zones. These establish the boundary conditions for stable excitation of higher-order modes within the upper band, as shown in Figure 2c. The distinct segmentation of resonant regions confirms that the antenna achieves tri-band operation by distributing modal control across separate radiator sections.

2.3. Parametric Studies

To quantify the contribution of key geometric parameters, a set of five variables was analyzed: feed arm width (PW_1), front radiator length (IL_1), shorting width (SW_1), midsection length (IL_3), and terminal shorting width (SW_3). Each was varied independently, and the reflection coefficient response was recorded.

Figure 3 shows that increasing PW_1 from 1.5 mm to 3.5 mm shifts the 5 GHz resonance upward and reduces the depth of the matching dip. Narrower PW_1 enhances coupling, producing deeper but lower-frequency resonance, while wider PW_1 diffuses current and weakens excitation. Its effect is localized to the mid-band, with negligible influence on the 2.4 and 6.5 GHz resonances.

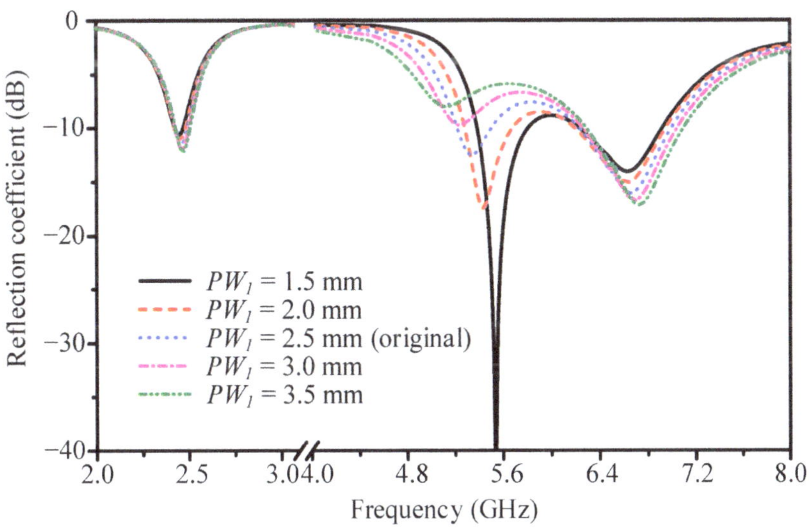

Figure 3. Simulated reflection coefficient variations with parameter PW_1.

Figure 4 demonstrates that extending IL_1 from 15.2 mm to 18.4 mm shifts the 2.4 GHz resonance downward, consistent with the longer current path of the fundamental mode. Secondary effects occur at 5 GHz, where IL_1 contributes to higher-order resonances. IL_1 is therefore the primary tuning parameter for the low band.

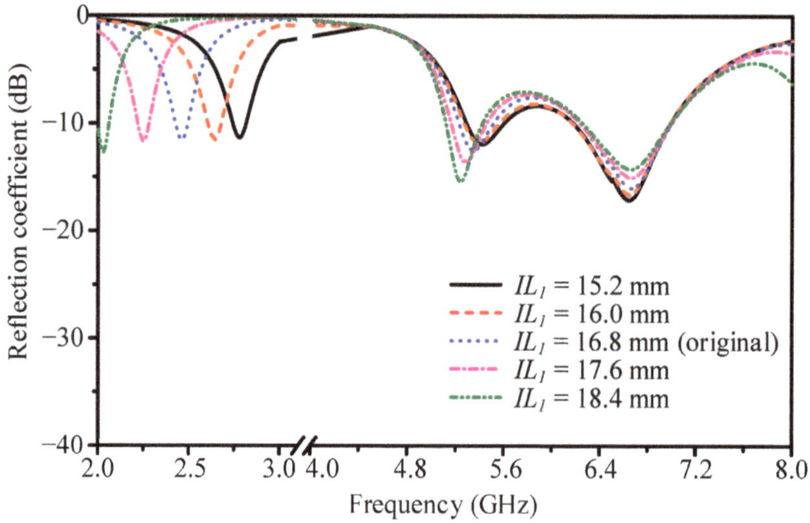

Figure 4. Simulated reflection coefficient variations with parameter IL_1.

As shown in Figure 5, increasing SW_1 from 0.3 mm to 2.3 mm degrades the 6.5 GHz resonance, reducing matching depth and shifting frequency. Excessive width enlarges the current injection region, disrupting field confinement at high frequency. Effects on 5 GHz are moderate, while at 2.4 GHz a slight deterioration in impedance matching can also be observed. Therefore, SW_1 acts mainly as a stabilizer for grounding and high-band control, but its influence is not completely negligible at the low band.

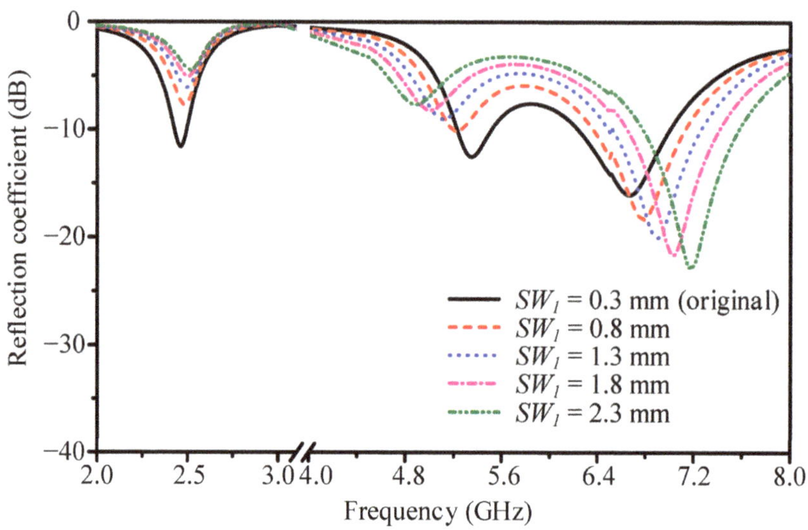

Figure 5. Simulated reflection coefficient variations with parameter SW_1.

Figure 6 shows that increasing IL_3 from 0.8 mm to 1.6 mm shifts the 5 GHz resonance downward, extending the electrical length. Shorter IL_3 configurations suppress resonance due to abrupt truncation. IL_3 is confirmed as the main tuning parameter for mid-band operation.

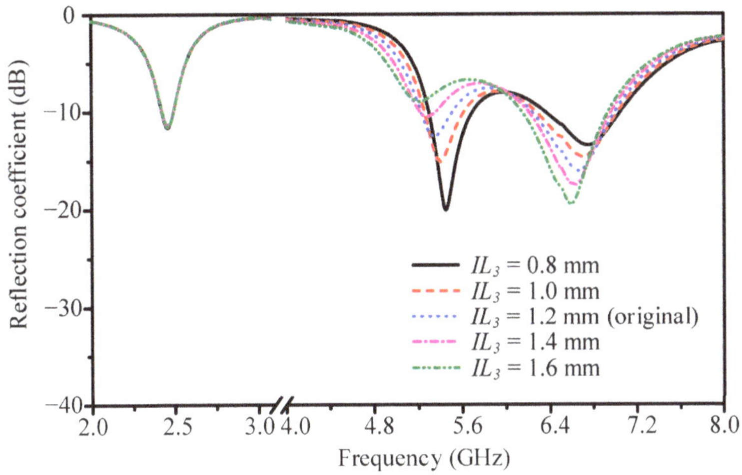

Figure 6. Simulated reflection coefficient variations with parameter IL_3.

Lastly, Figure 7 indicates that increasing SW_3 slightly reduces resonance depth at 6.5 GHz, while frequency shift remains small. Broader SW_3 disperses current, lowering coupling strength at high frequency. SW_3 is therefore a secondary but useful fine-tuning element for high-band matching in dense layouts.

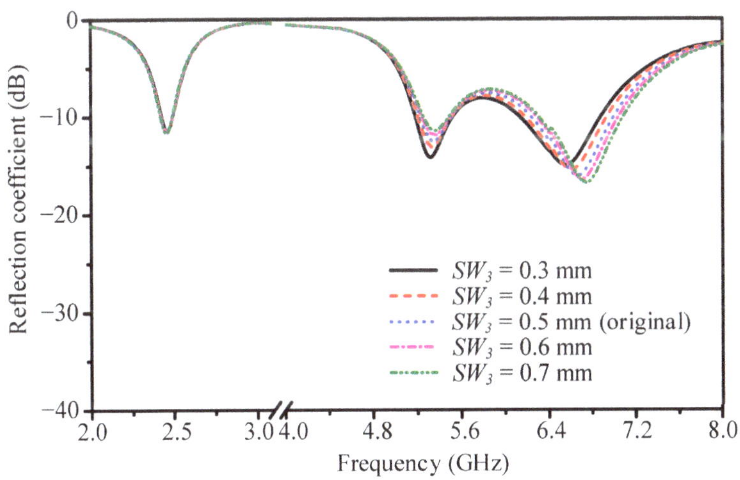

Figure 7. Simulated reflection coefficient variations with parameter SW_3.

Figure 8 shows the variation in antenna performance with respect to GND_y. At the 2.4 GHz band, the resonance frequency remains stable as the ground width is reduced, although the matching depth becomes slightly weaker. This behavior arises because a narrower ground increases the effective inductance in the return path, which elevates the input impedance seen by the fundamental mode. While this degradation is modest, it indicates that excessive reduction of GND_y gradually diminishes the low-band matching. In the 5 GHz band, the effect becomes more pronounced: as GND_y decreases, the counterpoise support from the ground is weakened, causing the second-order resonance to lose depth and bandwidth. The high-frequency band around 6–7 GHz is the most sensitive to ground width. A smaller GND_y not only reduces the depth of the resonance but also shifts its frequency upward, as the higher-order modes are strongly dependent on edge fringing fields at the ground boundary. Truncating the ground along the y direction shortens the effective resonant path and pulls the mode away from the intended value. These results confirm that a ground width of 60 mm is the smallest value that still preserves continuous tri-band coverage under the −7.4 dB criterion. Any further reduction leads to unacceptable deterioration, particularly in the mid- and high-band responses.

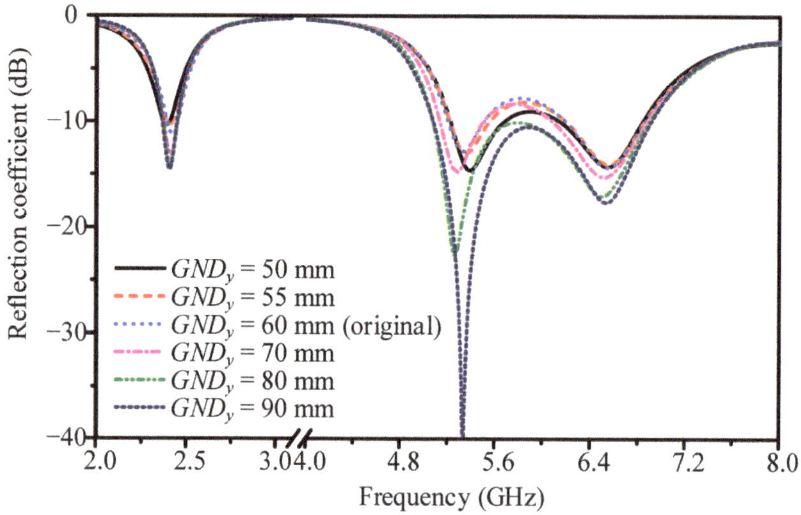

Figure 8. Simulated reflection coefficient variations with parameter GND_y.

Figure 9 illustrates the effect of varying GND_x. In contrast to GND_y, the influence of ground length is much weaker. At 2.4 GHz, the resonance frequency and matching depth are essentially unchanged when GND_x is varied, as the fundamental quarter-wavelength mode is dominated by vertical return currents close to the shorting strip rather than long horizontal extensions. At 5 GHz, only minor variations are observed, with the matching dip remaining below −7.4 dB across the tested range. At 6–7 GHz, there are small changes in matching depth, which can be attributed to weak interactions between higher-order modes and the ground edge capacitance along the x direction. However, these effects are secondary compared with the strong dependence on GND_y. Overall, the antenna is robust to variations in GND_x, meaning the ground length can be kept as short as 40 mm without compromising tri-band operation. This compactness along the x direction is especially important in laptop layouts where space is heavily constrained by other components.

Together, the parametric studies confirm that IL_1 and IL_3 directly define the low- and mid-band resonances, respectively, while PW_1 controls feed coupling that shapes mid-band impedance. SW_1 stabilizes grounding and SW_3 adjusts field confinement at the radiator end, both influencing high-frequency operation. In addition, the ground dimensions play a critical role: GND_y determines whether sufficient bandwidth and matching can be maintained in the 5–7 GHz region, while GND_x shows only weak influence across all bands, allowing the ground length to be kept compact without compromising performance. Effective tri-band performance therefore requires coordinated tuning of path-defining lengths, coupling elements, and the reduced ground size, especially under reduced ground conditions where modal interactions are more sensitive.

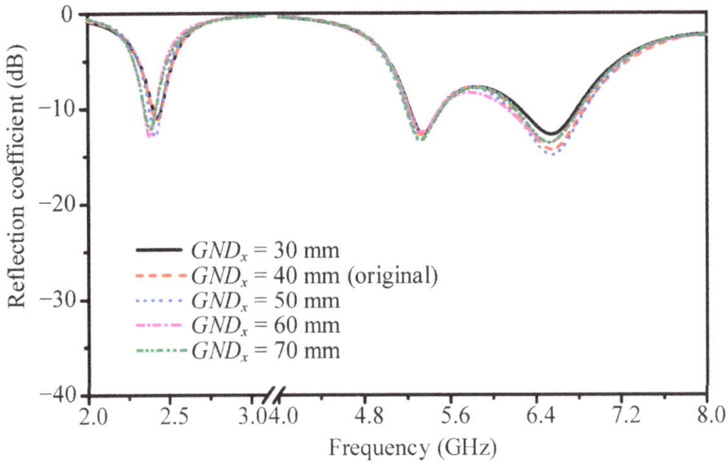

Figure 9. Simulated reflection coefficient variations with parameter GND_X.

3. Antenna Performance

To verify the impedance matching and tri-band operation of the proposed IFA, both simulations and measurements were conducted. The prototype was fabricated following the same dimensions used in simulation. The evaluation criterion for acceptable matching was a reflection coefficient below −7.4 dB (VSWR = 2.5:1), a commonly applied threshold in compact wireless antenna systems.

The fabricated prototype is shown in Figure 10. A polystyrene block provided structural support, while copper foil formed the radiator and ground plane. A 9 GHz-rated IPEX-I(M)-SMA(F) connector and a 10 cm low-loss coaxial cable were used for feeding. During measurements, the prototype was placed on a polystyrene pad to minimize external coupling. The reflection coefficient was recorded using a vector network analyzer (VNA) from 2 to 7 GHz.

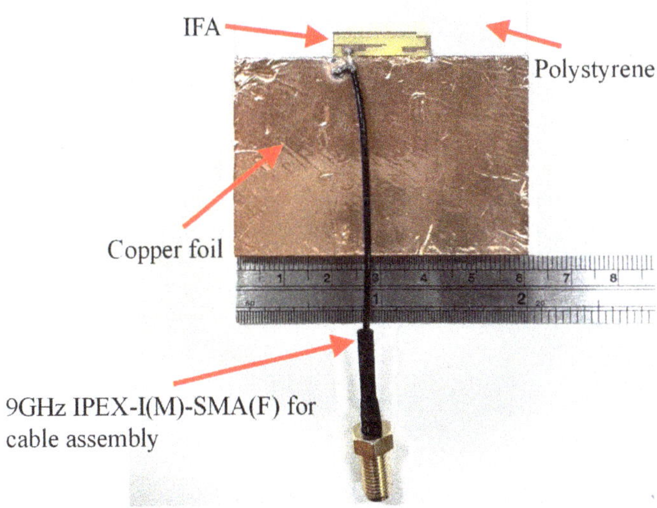

Figure 10. Fabricated prototype of the proposed IFA. The copper foil ground plane has a length of 60 mm, consistent with the design model. The polystyrene block serves only as mechanical support and does not extend the electrical ground plane.

The measured and simulated reflection coefficients are compared in Figure 11. Three resonant modes were observed near 2.4 GHz, 5.5 GHz, and 6.5 GHz. Simulated minima reached −11. 6 dB, −12.5 dB, and −16.1 dB, respectively. The measured data preserved all three resonances, with reflection coefficients below −6 dB at 2.4 GHz and below −7.4 dB at 5.5 GHz and 6.5 GHz.

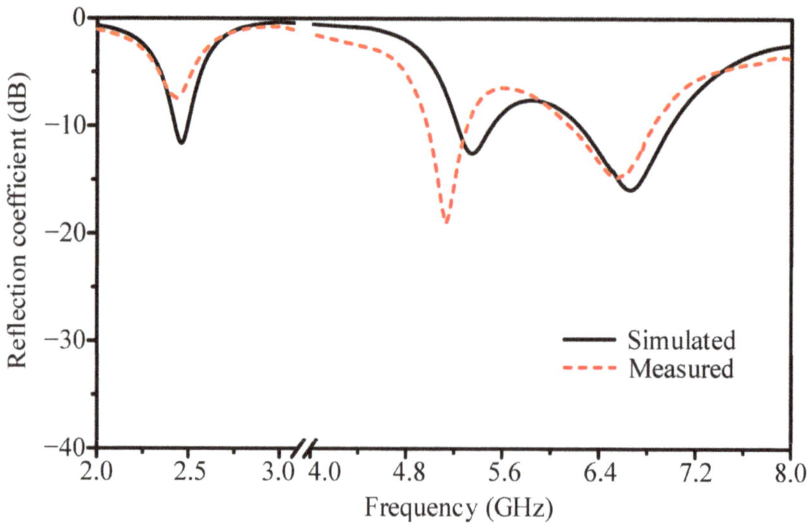

Figure 11. Simulated and measured reflection coefficient of the proposed antenna.

A downward frequency shift of about 100 MHz was observed in all measured resonances. Likely causes include geometric variation from the copper foil and substrate thickness, parasitic inductance introduced by the SMA connector, and environmental scattering during measurement. Despite these influences, the measured bandwidths and resonance depths followed the simulated trends, confirming that the antenna design is reproducible under practical fabrication.

Radiation performance was measured in an anechoic chamber. Far-field results for realized gain patterns, efficiency, and peak gain were obtained at 2.45, 5.5, and 6.5 GHz and compared with simulations.

Figures 12–14 show the 2D realized gain patterns in the x–z and y–z planes. At 2.45 GHz, the radiation in the x–z plane was quasi-omnidirectional, with half-power beamwidth (HPBW) of 288° in simulation and 250° in measurement. The cross-polarization discrimination (XPD) was 2.5 dB. In the y–z plane, the HPBW narrowed to 74° (simulation) and 70° (measurement), with XPD of 2.1 dB. This indicates broad angular coverage at low frequency, with limited polarization purity.

At 5.5 GHz, the x–z plane beamwidth widened to 332° in simulation and 340° in measurement, while XPD improved to 11.2 dB. In the y–z plane, HPBW was 64° and 70°, with XPD reaching 20.3 dB. These results show stronger main polarization and reduced cross-polarization compared with the 2.4 GHz band.

At 6.5 GHz, the x–z plane beamwidth narrowed to 104° (simulation) and 120° (measurement), with XPD of 7.7 dB. In the y–z plane, beamwidth expanded to 244° and 310°, while XPD decreased to 5.4 dB. The wider spread reflects the higher-order modes, which generate more complex and less symmetric field distributions.

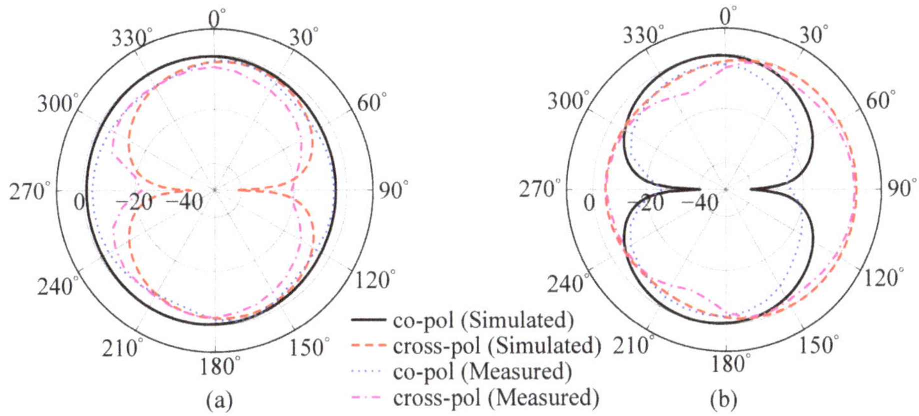

Figure 12. Simulated and measured realized gain patterns of the proposed antenna at 2.4 GHz on (**a**) x–z plane and (**b**) y–z plane. Unit: dBi.

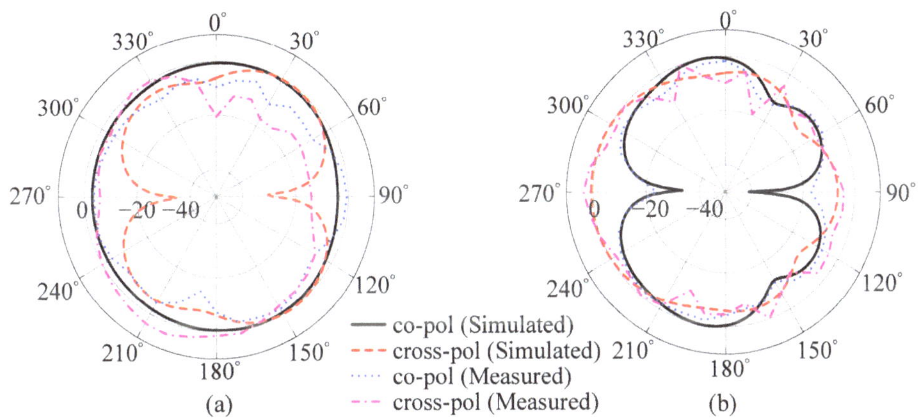

Figure 13. Simulated and measured realized gain patterns of the proposed antenna at 5.5 GHz on (**a**) x–z plane and (**b**) y–z plane. Unit: dBi.

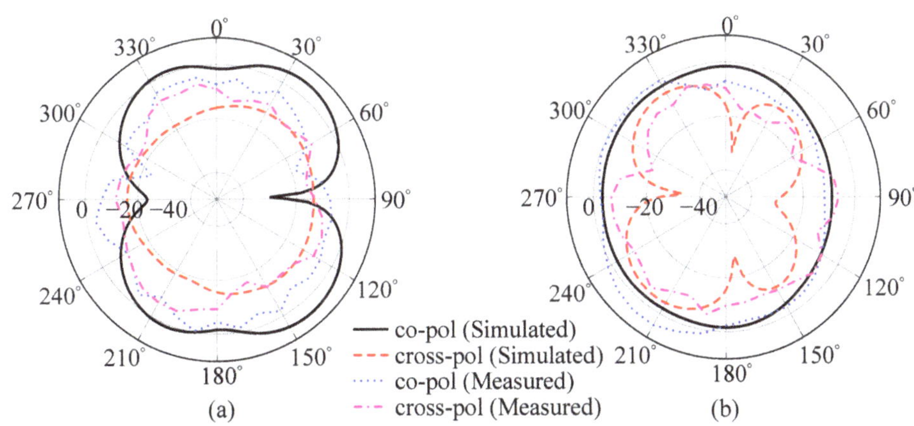

Figure 14. Simulated and measured realized gain patterns of the proposed antenna at 6.5 GHz on (**a**) x–z plane and (**b**) y–z plane. Unit: dBi.

Radiation efficiency is summarized in Figure 15. Simulated efficiencies were about 80% at 2.4 GHz and 75% in both higher bands. Measured values were 50–60% at 2.4 GHz and 50–80% across 5–7 GHz. The reduction is attributed to conductor and substrate losses, connector parasitics, and environmental factors. Measured results remain consistent with compact PCB-based antenna performance. While FR4 was adopted in this work to reflect cost-effective manufacturing, further efficiency improvements can be obtained by employing a lower-loss dielectric. Connector parasitics and temporary support structures also contribute to the measured reduction, which would be mitigated in final laptop integration through direct PCB feeding and optimized enclosure design. Nevertheless, the measured efficiencies of 50–60% at 2.4 GHz and 50–80% across 5–7 GHz already meet commercial WLAN antenna requirements, demonstrating that the proposed design offers a practical balance between low-cost implementation and reliable Wi-Fi 7 performance.

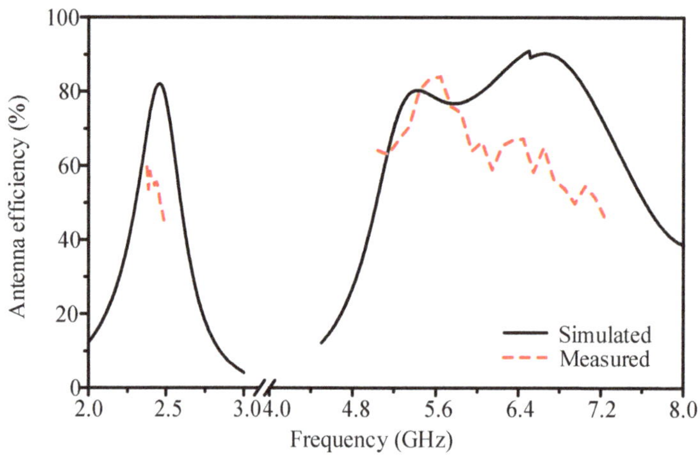

Figure 15. Simulated and measured radiation efficiency of the proposed antenna.

Figure 16 shows the measured realized peak gains. At 2.4 GHz, the gain ranged from 0.6 to 1.4 dBi. At 5 GHz and 6 GHz, it increased to 3.0–5.3 dBi. The simulated and measured realized peak gains (reported in dBi) differed by less than 1 dB in all cases, confirming the stability of the tri-band operation.

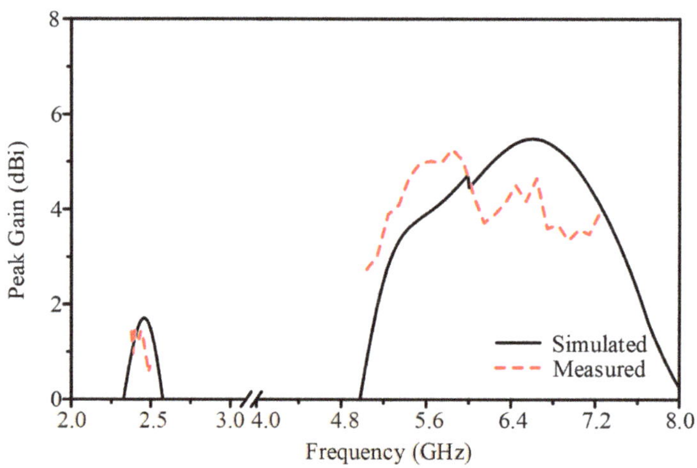

Figure 16. Simulated and measured realized peak gain of the proposed antenna.

Table 1 compares the proposed antenna with reported Wi-Fi 7 designs. Bandwidth definitions in the literature vary: some use −10 dB, others −6 dB, and some −7.4 dB. In terms of impedance bandwidth, the proposed antenna achieves a continuous bandwidth of 2.390–2.527 GHz in the lower band and 5.142–7.167 GHz in the upper band, fully covering the 802.11b/g/n and Wi-Fi 6E frequency allocations. Compared to prior works that either offer narrower bandwidths [14,16,18,19] or use more relaxed −6 dB criteria [15,16,19], the current design provides wider operating margins under a more stringent specification.

Table 1. Comparison of recent Wi-Fi 7 antenna designs.

No.	Impedance Bandwidth (GHz)	Antenna Efficiency (%)	Dimensions of the Antenna (mm^3)	Dimensions of the Ground Plane (mm^3)
[14]	2.400–2.484 [b] 5.150–7.125 [b]	802.11b/g: >55 802.11a/ac: 66–83 Wi-Fi 6E: 66–83	19.6 × 5.4 × 0.8	315.0 × 182.0 × 1.0
[15]	2.390–2.550 [c] 4.580–7.320 [c]	N/A	30.5 × 3.5 × 0.8	200.0 × 300.0 × 0.4
[16]	2.400–2.484 [c] 5.150–7.125 [c]	802.11b/g: 55–68 802.11a/ac: 66–78 Wi-Fi 6E: 66–78	15.0 × 4.6 × 0.8	315.0 × 182.0 × 1.0
[17]	2.400–2.485 [a] 4.470–7.260 [a]	802.11b/g: 40–45 802.11a/ac: 61–78 Wi-Fi 6E: 61–78	22.3 × 3.5 × 0.4	300.0 × 200.0 × 0.8
[18]	2.400–2.500 [b] 5.150–7.125 [b]	802.11b/g: 47–58 802.11a/ac: >70 Wi-Fi 6E: >70	16.0 × 4.0 × 0.4	315.0 × 182.0 × 1.0
[19]	2.400–2.500 [c] 5.150–7.125 [c]	802.11b/g: ≈60 802.11a/ac: >65 Wi-Fi 6E: >65	14.5 × 5.0 × 0.8	315.0 × 182.0 × 1.0
[20]	2.400–2.500 [b] 5.150–7.125 [b]	802.11b/g: >70 802.11a/ac: 77–86 Wi-Fi 6E: 77–86	30.0 × 6.0 × 0.8	315.0 × 182.0 × 1.0
[21]	2.400–2.484 [a] 5.150–7.125 [a]	802.11b/g: 63–69 802.11a/ac: 53–63 Wi-Fi 6E: 55–69	43.0 × 3.0 × 0.4	260.0 × 200.0 × 0.8
Proposed	2.390–2.527 [b] 5.142–7.167 [b]	802.11b/g: 74–81 802.11a/ac: 64–80 Wi-Fi 6E: 76–90	20.0 × 5.0 × 0.8	60 × 40 × 0.8

Note: The impedance bandwidths listed in the table are defined based on varying reflection coefficient thresholds reported in each study. Specifically, superscripts indicate the following criteria: [a] Bandwidth defined under return loss below −10 dB; [b] Bandwidth defined under return loss below −7.4 dB; [c] Bandwidth defined under return loss below −6 dB.

Regarding radiation efficiency, the proposed structure exhibits 74–81% efficiency at 2.4 GHz and 64–90% in the 5–7 GHz range. These results outperform or match the upper-band efficiency reported in [14,16,18], while also exceeding the lower-band values found in [17,18,21]. While [20] achieves higher upper-band performance (77–86%), its footprint is notably larger.

In terms of physical size, the proposed design occupies 20.0 × 5.0 × 0.8 mm^3, which balances compactness and performance. Compared to more compact structures such as [18,21], which adopt ultra-thin 0.4 mm substrates, the proposed antenna offers improved radiation performance while maintaining a thin-profile design suitable for laptop integration. Designs with similar or larger footprints, such as [14,16,20], do not consistently demonstrate simultaneous advantages in bandwidth and efficiency across all three Wi-Fi 7 bands.

In addition to radiator dimensions and bandwidth, Table 1 also includes the ground-plane sizes reported in earlier studies. It can be seen that most Wi-Fi 6E and Wi-Fi 7 antennas were validated on extended grounds such as 315×182 mm^2, 300×200 mm^2, or 260×200 mm^2, which are more than twenty times larger in area than the compact 60×40 mm^2 ground adopted in this work. The proposed antenna therefore demonstrates stable tri-band operation not only with a miniaturized radiator but also with a significantly reduced ground plane. This distinction highlights the novelty of the present design, where both the radiator and the ground are simultaneously constrained, directly addressing realistic laptop integration scenarios that earlier studies did not emphasize.

In summary, the proposed antenna design combines a compact footprint with tri-band coverage, competitive radiation efficiency, and a stricter impedance matching threshold. This performance combination is rarely achieved in the existing literature and supports practical integration in constrained device environments such as laptop enclosures.

4. Conclusions

This study presented a compact tri-band IFA for Wi-Fi 7 laptops that simultaneously addresses radiator miniaturization and reduced ground plane availability. Implemented within a $20 \times 5 \times 0.8$ mm^3 radiator and a $60 \times 40 \times 0.8$ mm^3 ground plane, the antenna achieves efficiencies of 74–81% at 2.4 GHz and 64–90% across 5–7 GHz, meeting a -7.4 dB matching criterion. Compared with existing compact designs, it demonstrates improved efficiency in the lower band and comparable upper-band coverage without relying on larger ground structures. The academic contribution lies in establishing and validating a design approach that ensures multiband performance under dual miniaturization constraints, thereby bridging the gap between compact radiator concepts and practical antenna integration in Wi-Fi 7 laptop platforms.

Author Contributions: Y.-K.H.: Data Curation, Formal Analysis, Investigation, Methodology, Validation. K.-H.T.: Funding Acquisition, Project Administration, Conceptualization. Y.-S.C.: Supervision, Writing—Original Draft Preparation, Writing—Review and Editing. All authors have read and agreed to the published version of the manuscript.

Funding: This work was supported in part by the National Science and Technology Council (NSTC), Taiwan, under Contract NSTC 113-2221-E-027-085-MY3, and in part by Micro-Star International Co., Ltd., under industry-academia collaboration projects.

Data Availability Statement: The data presented in this study are contained within the article.

Conflicts of Interest: Author Kuan-Hsueh Tseng is employed by Micro-Star International Co., Ltd. The authors declare that this study received funding from Micro-Star International Co., Ltd. under an industry-academia collaboration project. Author Kuan-Hsueh Tseng contributed to funding acquisition, project administration, and conceptualization. The funders did not influence the study design, data collection, data analysis, data interpretation, or the decision to submit the article for publication. The remaining authors declare that the research was conducted in the absence of any additional commercial or financial relationships that could be construed as a potential conflict of interest.

References

1. Hiertz, G.R.; Denteneer, D.; Stibor, L.; Zang, Y.; Costa, X.P.; Walke, B. The IEEE 802.11 universe. *IEEE Commun. Mag.* **2010**, *48*, 62–70. [CrossRef]
2. Noghanian, S. Dual-band wearable MIMO antenna for WiFi sensing applications. *Sensors* **2022**, *22*, 9257. [CrossRef] [PubMed]

3. Karmakar, R.; Chattopadhyay, S.; Chakraborty, S. Impact of IEEE 802.11n/ac PHY/MAC high throughput enhancements on transport and application protocols—A survey. *IEEE Commun. Surv. Tutor.* **2017**, *19*, 2050–2091. [CrossRef]
4. Khorov, E.; Kiryanov, A.; Lyakhov, A.; Bianchi, G. A tutorial on IEEE 802.11ax high efficiency WLANs. *IEEE Commun. Surv. Tutor.* **2019**, *21*, 197–216. [CrossRef]
5. Frascolla, V.; Cavalcanti, D.; Shah, R. Wi-Fi evolution: The path towards Wi-Fi 7 and its impact on IioT. *J. Mob. Multimed.* **2023**, *19*, 263–276. [CrossRef]
6. Lai, F.-P.; Mi, S.-Y.; Chen, Y.-S. Design and integration of millimeter-wave 5G and WLAN antennas in perfect full-screen display smartphones. *Electronics* **2022**, *11*, 957. [CrossRef]
7. Chen, J.; Tong, K.F.; Al-Armaghany, A.; Wang, J. A dual-band dual-polarization slot patch antenna for GPS and Wi-Fi applications. *IEEE Antennas Wirel. Propag. Lett.* **2016**, *15*, 406–409. [CrossRef]
8. Xiao, Y.; Qi, Y.; Li, F.; Fan, J.; Yu, W.; Lu, L. Dual-band directional slot antenna for Wi-Fi application. *IEEE Trans. Antennas Propag.* **2018**, *66*, 4277–4281. [CrossRef]
9. Wang, Z.; Wang, M.; Nie, W. Design of a dual-band WiFi antenna using the theory of characteristic modes and nested Chinese characters. *Electronics* **2023**, *12*, 3465. [CrossRef]
10. Jeong, G.T.; Kim, W.S.; Kwak, K.S. Dual-band Wi-Fi antenna with a ground stub for bandwidth enhancement. *IEEE Antennas Wirel. Propag. Lett.* **2012**, *11*, 1036–1039. [CrossRef]
11. Xu, Z.; Ding, C.; Zhou, Q.; Sun, Y.; Huang, S. A dual-band dual-antenna system with common-metal rim for smartphone applications. *Electronics* **2019**, *8*, 348. [CrossRef]
12. Paul, L.C.; Saha, H.K.; Rani, T.; Azim, R.; Islam, M.T.; Samsuzzaman, M. A dual-band semi-circular patch antenna for WiMAX and WiFi-5/6 applications. *Int. J. Commun. Syst.* **2023**, *36*, e5357. [CrossRef]
13. Sarkar, D.; Srivastava, K.V. Compact four-element SRR-loaded dual-band MIMO antenna for WLAN/WiMAX/WiFi/4G-LTE and 5G applications. *Electron. Lett.* **2017**, *53*, 1623–1624. [CrossRef]
14. Su, S.-W.; Magray, M.I.; Lu, J.-H. Metamaterial-inspired antenna for Wi-Fi 7 operation in wireless local area network bands. In Proceedings of the 2023 Asia-Pacific Microwave Conference (APMC), Taipei, Taiwan, 5–8 December 2023; pp. 76–79.
15. Sim, C.-Y.-D.; Ju, J.-W.; Chuang, P.-Y.; Chen, W.-C.; Chen, M.-G. A low profile laptop antenna design for Wi-Fi 7 operations. In Proceedings of the 2023 Asia-Pacific Microwave Conference (APMC), Taipei, Taiwan, 5–8 December 2023; pp. 539–541.
16. Magray, M.I.; Su, S.-W.; Lu, J.-H. Compact, hybrid slot-dipole-monopole Wi-Fi 7 antenna for slim-bezel notebook applications. In Proceedings of the 2023 Asia-Pacific Microwave Conference (APMC), Taipei, Taiwan, 5–8 December 2023; pp. 28–30.
17. Sim, C.-Y.-D.; Kulkarni, J.; Wang, S.-H.; Zheng, S.-Y.; Lin, Z.-H.; Chen, S.-C. Low-profile laptop antenna design for Wi-Fi 6E band. *IEEE Antennas Wirel. Propag. Lett.* **2023**, *22*, 79–83. [CrossRef]
18. Su, S.-W. Compact, small, chip-inductor-loaded Wi-Fi 6E monopole antenna. In Proceedings of the 2021 IEEE International Symposium on Antennas and Propagation and USNC-URSI Radio Science Meeting (APS/URSI), Singapore, 4–10 December 2021; pp. 937–938.
19. Su, S.-W.; Yusuf, D.P.; Chu, F.-H. Conjoined, Wi-Fi 6E MIMO antennas for laptops. In Proceedings of the 2021 International Symposium on Antennas and Propagation, Taipei, Taiwan, 19–22 October 2021; pp. 1–3.
20. Su, S.-W.; Wan, C.-C. Asymmetrical, self-isolated laptop antenna in the 2.4/5/6 GHz Wi-Fi 6E bands. In Proceedings of the 2021 International Symposium on Antennas and Propagation, Taipei, Taiwan, 19–22 October 2021; pp. 1–2.
21. Jhang, W.-C.; Sun, J.-S. Small antenna design of triple band for Wi-Fi 6E and WLAN applications in the narrow border laptop computer. *Int. J. Antennas Propag.* **2021**, *2021*, 7334206. [CrossRef]
22. Dong, Y.; You, J.; Keshmiri, F. Review on the new antenna evolution for next generation WiFi-7/8 systems. In Proceedings of the 2025 19th European Conference on Antennas and Propagation (EuCAP), Stockholm, Sweden, 30 March–4 April 2025; pp. 1–5.
23. Srinivas, G.; Reddy, B.R.S.; Sahoo, R. Broadband circularly polarized antenna for Wi-Fi 7 applications. *Int. J. Microw. Wirel. Technol.* **2025**, *17*, 447–454. [CrossRef]
24. Tu, D.T.T.; Van Hoa, N.; Nga, N.T.T. A high directional antenna at all three operating bands for wide-band IoT applications based on IEEE 802.11ax/be standard. *J. Commun.* **2025**, *20*, 483–490.
25. Le, H.D.; Ta, S.X. A compact dual-band dual-polarized antenna using non-uniform metasurface and mushroom structures for Wi-Fi 7 applications. *IEEE Antennas Wirel. Propag. Lett.* 2025, *early access*.
26. Diao, W.; Li, Y.; Chang, L.; Guan, Q.; Zhang, Z. A small-clearance high-efficiency dual-band Wi-Fi 6E antenna taking advantage of an all-metal laptop. *IEEE Trans. Antennas Propag.* **2025**, *73*, 6133–6138. [CrossRef]

electronics

MDPI

Article

Blind Source Separation for Joint Communication and Sensing in Time-Varying IBFD MIMO Systems

Siyao Li, Conrad Prisby and Thomas Yang *

Department of Electrical Engineering and Computer Science, Embry-Riddle Aeronautical University, Daytona Beach, FL 32114-3900, USA; lis14@erau.edu (S.L.); prisbyc@my.erau.edu (C.P.)
* Correspondence: yang482@erau.edu

Abstract

This paper presents a blind source separation (BSS)-based framework for joint communication and sensing (JCAS) in in-band full-duplex (IBFD) multiple-input multiple-output (MIMO) systems operating under time-varying channel conditions. Conventionally, self-interference (SI) in IBFD systems is a major obstacle to recovering the signal of interest (SOI). Under the JCAS paradigm, however, this high-power SI signal presents an opportunity for efficient sensing. Since each transceiver node has access to the original SI signal, its environmental reflections can be exploited to estimate channel conditions and detect changes, without requiring dedicated radar waveforms. We propose a blind source separation (BSS)-based framework to simultaneously perform self-interference cancellation (SIC) and extract sensing information in IBFD MIMO settings. The approach applies the Fast Independent Component Analysis (FastICA) algorithm in dynamic scenarios to separate the SI and SOI signals while enabling simultaneous signal recovery and channel estimation. Simulation results quantify the trade-off between estimation accuracy and channel dynamics, demonstrating that while FastICA is effective, its performance is fundamentally limited by a frame size optimized for the rate of channel variation. Specifically, in static channels, the signal-to-residual-error ratio (SRER) exceeds 22 dB with 500-symbol frames, whereas for moderately time-varying channels, performance degrades significantly for frames longer than 150 symbols, with SRER dropping below 4 dB.

Keywords: MIMO; IBFD; channel estimation; joint communication and sensing; blind source separation

1. Introduction

The relentless growth of mobile data traffic and the expansion into millimeter-wave (mmWave) frequencies demand transformative advances in wireless technology to overcome spectrum scarcity [1,2]. In-band full-duplex (IBFD) communication is a highly promising technique for enhancing spectral efficiency, as it allows a transceiver to transmit and receive simultaneously on the same frequency band [3,4]. However, the primary obstacle in realizing IBFD systems is the powerful self-interference (SI) signal, i.e., the leakage from the node's own transmitter into its receiver, which can be orders of magnitude stronger than the desired signal of interest (SOI) from a remote user.

Conventionally viewed as a detriment, SI finds a novel purpose within the emerging paradigm of joint communication and sensing (JCAS) [5]. The JCAS framework reinterprets the high-power SI signal as an invaluable "signal of opportunity" for environmental sensing. This dual-functionality is critical for next-generation applications, such as autonomous vehicle navigation, smart city infrastructure, and indoor localization, where real-time environmental awareness is as important as high-speed data links [6]. Since the waveform of the SI signal is perfectly known to the transceiver, its reflections off surrounding objects can be exploited to estimate the radio channel, detect environmental changes, and perform radar-like functions without needing dedicated sensing waveforms or hardware (e.g., see [6–10] and the references therein). This dual use of the radio signal promises to significantly improve resource utilization, efficiency, and system functionality.

To realize this dual functionality, the receiver must effectively separate the SOI from the SI and other signals. Blind source separation (BSS) techniques [11], particularly Independent Component Analysis (ICA) [12], are well suited for this task. BSS algorithms can distinguish and separate signals based on their statistical properties, such as statistical independence, making them ideal for untangling the SOI and the reflected SI components in a complex received signal mixture.

1.1. Related Work

The concept of using BSS for signal separation in communication systems is well established. Prior work has explored BSS for various JCAS and full-duplex applications, demonstrating its potential in managing overlapping signals and extracting valuable information [13]. In particular, a sparsity-enhanced source separation scheme is proposed in [14] to mitigate aliasing in joint communication and radar scenarios. Fouda et al. [15] further explored the design of BSS architectures tailored for full-duplex systems, emphasizing the challenges associated with maintaining computational efficiency while managing strong SI signals. Barneto [16] provides insights into waveform and hardware integration strategies, offering a foundational understanding of signal co-design in cellular JCAS systems.

1.2. Contributions

However, these studies primarily assume a stationary or quasi-stationary channel environment. Our work distinguishes itself by explicitly investigating the performance degradation of a BSS-based JCAS framework under continuously time-varying channel conditions. While some adaptive BSS techniques exist [17,18], the performance of the widely used and computationally efficient FastICA algorithm in such dynamic JCAS scenarios has not been systematically quantified. The primary objective of this paper is not to benchmark different BSS algorithms against one another but rather to examine the feasibility and performance implications of applying a BSS-based framework to the integrated communication and sensing scenario in time-varying IBFD MIMO systems. Our key contributions are the following:

1. We model a JCAS system where the known SI signal is used for sensing in a time-varying environment.
2. We systematically evaluate the system's sensing (ELMMSE) and communication (SRER) performance as a function of frame size and the rate of channel variation.
3. We identify and analyze the fundamental trade-off between statistical reliability (favoring longer frames) and channel stationarity (favoring shorter frames), revealing an optimal frame size that is dependent on channel dynamics.

1.3. Paper Organization

The remainder of this paper is organized as follows. Section 2 introduces the system model for IBFD MIMO with time-varying channels. Section 3 describes the BSS-based channel estimation framework using FastICA. Section 4 presents the simulation setup and analyzes the performance under various channel dynamics. Section 5 concludes this work and outlines future research directions.

1.4. Notations

We use $\mathbb{E}[X]$ to represent the expectation of the random variable X. Bold letters denote vectors or matrices. $\|\cdot\|^2$ represents the Frobenius norm. $\mathcal{N}(\mu, \sigma^2)$ denotes the Gaussian distribution with mean μ and variance σ^2. $\mathbf{M}^{n\times m}$ represents a generic matrix with n rows and m columns. $(\cdot)^T$ denotes transpose operation.

2. System Model

As illustrated in Figure 1, we consider an IBFD MIMO system comprising two transceiver nodes, each equipped with n transmit antennas and m receive antennas. Both nodes simultaneously transmit and receive signals over the same frequency band, enabling full-duplex communication. The received signal at each node is composed of multiple components: (1) the SOI, which includes both direct-path and reflected components from the remote node, and (2) the SI signal, which includes direct leakage between the local transmit and receive antennas and reflections from the environment. Unlike conventional channel models developed solely for communication systems (e.g., 3GPP TR 38.901 [19], WINNER II [20], COST 2100 [21]), our framework explicitly incorporates both direct and reflected SI components alongside the SOI. Non-JCAS models typically do not model the loopback SI path created by in-band full-duplex operation, nor do they account for environmental reflections of SI that are crucial for sensing functionality.

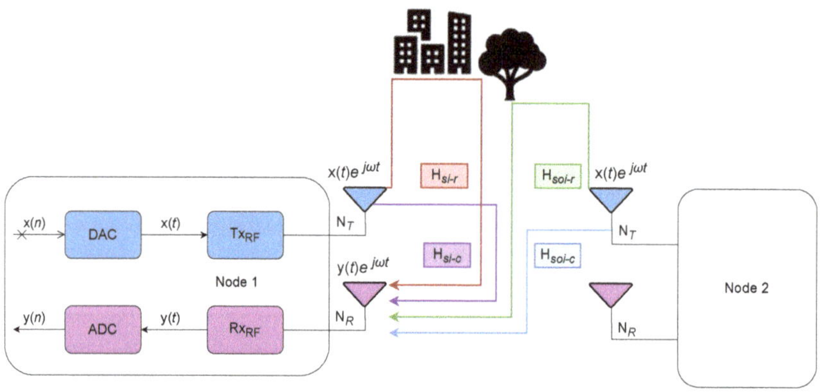

Figure 1. IBFD MIMO system model with self-interference and environmental reflections.

Crucially, in this work, we consider the impact of time-varying channel conditions, which are characteristic of many realistic wireless environments. The channel properties, including amplitude and phase responses, can change over time due to factors such as user mobility, changes in the environment, and multipath fading. Let $R_l(k)$ denote the time-domain received signal at the l-th receive antenna at time k. It is expressed as

$$\begin{aligned} R_l(k) = &\sum_{i=1}^{n}(H^i_{\text{si-c}}(k) + H^i_{\text{si-r}}(k))S^i_{\text{si}}(k) \\ &+ \sum_{i=1}^{n}(H^i_{\text{soi-c}}(k) + H^i_{\text{soi-r}}(k))S^i_{\text{soi}}(k) + N(k),\ l \le m. \end{aligned} \tag{1}$$

Here, $N(k) \sim \mathcal{N}(0, \sigma^2)$ is the additive Gaussian noise. For each node at time k, $H^i_{\text{si-c}}(k)$ and $H^i_{\text{soi-c}}(k)$ are the direct SI and SOI channels, respectively; $H^i_{\text{si-r}}(k)$ and $H^i_{\text{soi-r}}(k)$ denote the reflected channels at the i-th transmit antenna; and $S^i_{\text{si}}(k)$ and $S^i_{\text{soi}}(k)$ are the SI and SOI signals from the i-th transmit antenna, respectively. Each channel, now explicitly represented as a function of time k, is assumed to have independent, identically distributed (i.i.d.) entries for a given time instant. For convenience, we can rewrite the equation in matrix form at time k as follows:

$$\begin{aligned}\mathbf{R}(k) = (\mathbf{H}_{\text{si-c}}(k) + \mathbf{H}_{\text{si-r}}(k))\mathbf{S}_{\text{si}}(k) + (\mathbf{H}_{\text{soi-c}}(k) \\ + \mathbf{H}_{\text{soi-r}}(k))\mathbf{S}_{\text{soi}}(k) + \mathbf{N}(k)\end{aligned} \tag{2}$$

where $\mathbf{H}_{\text{si-c}}(k), \mathbf{H}_{\text{si-r}}(k), \mathbf{H}_{\text{soi-c}}(k), \mathbf{H}_{\text{soi-r}}(k) \in \mathbf{M}^{m\times n}$, $\mathbf{S}_{\text{si}}(k), \mathbf{S}_{\text{soi}}(k) \in \mathbf{M}^{n\times 1}$, and $\mathbf{R} \in \mathbf{M}^{m\times 1}$. In our model, the channels are explicitly time-dependent. We assume that the direct SI channel, $\mathbf{H}_{\text{si-c}}(k)$, is relatively stable or can be accurately calibrated, as it corresponds to the fixed geometry of the transceiver hardware. In contrast, the reflected channels, $\mathbf{H}_{\text{si-r}}(k)$ and $\mathbf{H}_{\text{soi-r}}(k)$, are stochastic and time-varying, capturing the dynamic nature of the propagation environment due to mobility and scattering. Note that $\mathbf{S}_{\text{si}}(k)$ is the transmitted signal from the same node at time k, implying that $\mathbf{S}_{\text{si}}(k)$ is known to the receiver. The direct self-interference channel $\mathbf{H}_{\text{si-c}}(k)$ is also generally known to the receiver and remains relatively stable compared to the reflected channels, although it can still exhibit some time dependency.

We can further rewrite (2) in a form that resembles a canonical BSS model at time k:

$$\begin{bmatrix}\mathbf{S}_{\text{si}}(k)\\ \mathbf{R}(k)\end{bmatrix} = \mathbf{H}(k)\begin{bmatrix}\mathbf{S}_{\text{si}}(k)\\ \mathbf{S}_{\text{soi}}(k)\end{bmatrix} + \begin{bmatrix}\mathbf{0}\\ \mathbf{N}(k)\end{bmatrix}, \tag{3}$$

where $\mathbf{H}(k)$ is a composite mixing matrix at time k, defined as

$$\mathbf{H}(k) = \begin{bmatrix}\mathbf{I} & \mathbf{0}\\ \mathbf{H}_{\text{si}}(k) & \mathbf{H}_{\text{soi}}(k)\end{bmatrix} \in \mathbf{M}^{2m\times 2n}, \tag{4}$$

$$\text{with } \mathbf{H}_{\text{si}}(k) = \mathbf{H}_{\text{si-c}}(k) + \mathbf{H}_{\text{si-r}}(k), \tag{5}$$

$$\text{and } \mathbf{H}_{\text{soi}}(k) = \mathbf{H}_{\text{soi-c}}(k) + \mathbf{H}_{\text{soi-r}}(k), \tag{6}$$

and $\mathbf{I}$ being the identity matrix of size $m \times n$. Equation (3) represents a key step in our framework. By stacking the known transmitted SI signal $\mathbf{S}_{\text{si}}(k)$ with the received signal $\mathbf{R}(k)$, we create an augmented observation vector. This transforms the problem into a canonical BSS form where the "sources" are the SI and SOI signals, and the "mixing matrix" $\mathbf{H}(k)$ contains the channel parameters we wish to estimate. Crucially, because $\mathbf{S}_{\text{si}}(k)$ is known, it can serve as a reference signal, anchoring the BSS algorithm and helping to resolve ambiguities in the separation progress, thereby enabling the estimation of both the unknown channel $\mathbf{H}_{\text{si-r}}(k)$ and the unknown signal $\mathbf{S}_{\text{soi}}(k)$.

3. BSS-Based Channel Estimation

We now describe the application of the FastICA algorithm [12] to separate the SI and SOI components from the observed signals. For simplicity, we assume $m = n$, meaning each node has an equal number of transmit and receive antennas. We treat the sum $\mathbf{H}_{\text{soi}}$ in (6) as the overall channel coefficient of the SOI. The goal is two-fold: (1) to estimate the unknown communication signal $\mathbf{S}_{\text{soi}}$ for data recovery and (2) to estimate the reflected SI channel $\mathbf{H}_{\text{si-r}}$ for environmental sensing.

3.1. Blind Source Separation for Sensing

The ICA algorithm is founded on the principle that many mixed signals are linear combinations of underlying source signals that are mutually statistically independent and non-Gaussian. In our model, the transmitted BPSK signals are non-Gaussian, satisfying this core requirement. The goal of ICA is to find a linear transformation (an "unmixing" matrix) that, when applied to the observed mixtures, maximizes the statistical independence of the outputs. This process allows us to recover the original source signals, or scaled versions of them. In particular, the FastICA algorithm exploits non-Gaussianity in the received signal mixture to estimate the independent source components iteratively. The algorithm follows the following key steps:

1. Preprocessing: The received signal matrix is first centered by subtracting its mean. Then, it is "whitened" using a technique like eigenvalue decomposition. Whitening is a linear transformation that removes any second-order correlations in the data, forcing the components to be uncorrelated and have unit variance. This simplifies the problem for the ICA algorithm, as the unknown mixing matrix $\mathbf{H}(k)$ is transformed into an orthogonal matrix, reducing the search space and improving convergence speed. It is important to note that for whitening to be effective, the number of samples (i.e., the frame size) must be sufficiently large relative to the number of signal dimensions to allow for the robust estimation of the covariance matrix. In highly dynamic environments, this requirement conflicts with the need for short frames to ensure channel stationarity, establishing a fundamental performance trade-off that we investigate in this paper.
2. Iterative Estimation: FastICA iteratively estimates the columns of the unmixing matrix by maximizing a measure of non-Gaussianity called negentropy, i.e., $J(y) = \mathbb{E}[G(y)] - \mathbb{E}[G(v)]$, where G is a non-quadratic function, y is the estimated source, and v is a Gaussian variable with the same variance as y. Negentropy is always non-negative and is zero only for a Gaussian distribution. Therefore, maximizing it drives the estimated source y away from Gaussianity and towards one of the independent source components. For computational simplicity, FastICA maximizes approximations of negentropy using non-quadratic functions $G(y)$ such as $G_1(y) = \log(cosh(y))$ and $G_2(y) = -\exp(-y^2/2)$.
3. Source Recovery: Estimate the unknown channel $\hat{\mathbf{H}}_{\text{si-r}}$ and signal $\hat{\mathbf{S}}_{\text{soi}}$ from the separated components using the known SI signal as a reference. The FastICA update rule for extracing one independent component is

$$\mathbf{w} = \mathbb{E}[\mathbf{z}g(\mathbf{w}^T\mathbf{z})] - \mathbb{E}[g'(\mathbf{w}^T\mathbf{z})]\mathbf{w},$$

where $\mathbf{z}$ is the whitened data, g is the derivative of G, and $\mathbf{w}$ is normalized after each iteration.

3.2. *Performance Metrics*

We evaluate the quality of channel estimation using the ergodic linear minimum mean squared error (ELMMSE), which quantifies the average estimation error of the sensed channel:

$$\text{ELMMSE} = \mathbb{E}[||\mathbf{H}_{\text{si-r}}(k) - \hat{\mathbf{H}}_{\text{si-r}}(k)||^2]. \tag{7}$$

The communication performance is quantified via the signal-to-residual-error ratio (SRER) of the extracted communication signal:

$$\text{SRER} = \frac{\mathbb{E}[||\mathbf{S}_{\text{soi}}(k)||^2]}{\mathbb{E}[||\mathbf{S}_{\text{soi}}(k) - \hat{\mathbf{S}}_{\text{soi}}(k)||^2]}. \tag{8}$$

A higher SRER implies more effective signal separation and cleaner communication signal extraction, directly reflecting the system's communication performance under the JCAS paradigm. Additionally, we can assess the ELMMSE of the overall SOI channel $\mathbf{H}_{\text{soi}}$ as

$$\text{ELMMSE} = \mathbb{E}[||\mathbf{H}_{\text{soi}}(k) - \hat{\mathbf{H}}_{\text{soi}}(k)||^2], \tag{9}$$

which accounts for both the direct transmission channel between the transmitter and receiver, as well as the reflected and scattered components from the surrounding environment. This channel state information can be leveraged to optimize the transmission schemes (e.g., precoder design in MIMO systems) and to maximize the signal-to-interference-plus-noise ratio (SINR), as follows:

$$\mathrm{SINR} = \frac{\mathbb{E}[||\mathbf{H}_{\mathrm{soi}}(k)\mathbf{S}_{\mathrm{soi}}(k)||^2]}{\mathbb{E}[||\mathbf{H}_{\mathrm{si}}(k)\mathbf{S}_{\mathrm{si}}(k)||^2] + \mathbb{E}[||\mathbf{N}(k)||^2]}.$$

In Section 4, we will also track the number of iterations required for the ICA algorithm to converge, which provides insights into the computational complexity and algorithmic efficiency of the proposed BSS approach under different signal block sizes and channel dynamics.

4. Simulation and Discussion

4.1. Simulation Setup

In this section, we evaluate the sensing and communication performance of the proposed system through numerical simulations based on a practical system model. Each node is equipped with two antennas. The reflected channel $\mathbf{H}_{\text{si-r}}$ is modeled as a Rician channel with $\mathbf{H}_{\text{si-r}} \sim \mathrm{Rice}(x, y)$, where x is the Rician factor and y is the average channel power. This is because the SI signal reflects off objects near the transceiver, creating a scenario with a dominant line-of-sight (LOS) path alongside multipath components, for which the Rician distribution is the standard model. In contrast, the overall communication channel $\mathbf{H}_{\mathrm{soi}}$ is modeled as Rayleigh-distributed. This represents a rich, scattering, non-LOS propagation environment, which is typical for communication between two separate nodes in urban settings where obstacles are common, which is consistent with 3GPP TR 38.901 urban microcell assumptions [19]. Our simplified model preserves these key statistical properties while reducing complexity, making it suitable for algorithm-level evaluations of BSS-based JCAS without introducing excessive simulation overhead.

The transmitted signals $\mathbf{S}_{\mathrm{si}}$ and $\mathbf{S}_{\mathrm{soi}}$ are BPSK-modulated, with each symbol randomly chosen from $\{-1, +1\}$. BPSK is chosen for its simplicity and non-Gaussian nature, which is a prerequisite for ICA, making it an ideal starting point for this analysis. The combined source matrix is defined as $\mathbf{S} = \left[\mathbf{S}_{\mathrm{si}}; \mathbf{S}_{\mathrm{soi}}\right] \in \mathbb{R}^{4 \times N}$, where N denotes the frame size (signal processing block length). The received signal is generated by linearly mixing the source signals through the corresponding channel matrices and adding a Gaussian noise of $\sigma^2 = 0.01$ (20 dB SNR).

To model time-varying channels, $\mathbf{H}_{\mathrm{si}}$ and $\mathbf{H}_{\mathrm{soi}}$, are initialized once and then linearly perturbed at each time frame. The chosen simulation values for channel variation (δ) and frame size were selected to systematically demonstrate the core trade-off between statistical reliability and channel stationarity in the proposed system. We test three time variation speeds corresponding to δ = 0, 0.0005, 0.001, where $\delta = 0$ represents an ideal, unchanging channel, and $\delta = 0.0005$ and 0.001 are selected to model realistic, dynamic mobile environments. The static channel serves as a crucial performance benchmark, showing how the FastICA algorithm performs when its core assumption of a stationary environment is perfectly met. This baseline allows for isolating performance limitations of the algorithm itself versus those caused by channel dynamics. In contrast, the non-zero small variation factors are chosen to be low enough to prevent immediate algorithm failure yet sufficiently high to introduce measurable degradation as the frame size increases. This design clearly demonstrates the existence of an optimal frame size that balances statistical reliability with the rate of channel variation. For each time frame $k \in [1, N_f]$ (where N_f is the number of frames), the time-varying channels are computed as

$$\mathbf{H}_{\text{si}}(k) = \mathbf{H}_{\text{si}}(0) + \delta \cdot k,$$
$$\mathbf{H}_{\text{soi}}(k) = \mathbf{H}_{\text{soi}}(0) + \delta \cdot k.$$

Since the mixing matrix is continuously changing with time, we consider the midpoint mixing matrix for evaluation, as follows:

$$\mathbf{H}_{\text{mid}} = \mathbf{H}(k = \lfloor N_f/2 \rfloor).$$

In cases where the mixing matrix varies sporadically over time, the average mixing matrix offers a more stable approximation than using a single midpoint. The average composite mixing matrix is defined as

$$\mathbf{H}_{\text{avg}} = \frac{1}{N_f} \sum_{k=1}^{N_f} \mathbf{H}(k).$$

The estimated mixing matrix is post-processed via column permutation to resolve inherent ICA ambiguities. The range of frame sizes (50 to 500 symbols) is selected to effectively capture and display the system's performance across different processing block lengths. The lower bound of 50 tests the system under conditions of limited statistical data, where performance is expected to be poor. The upper bound of 500 is large enough to achieve high performance in a static channel, demonstrating the benefit of having more data when the environment is stable. The step size of 50 provides enough data points to plot a smooth and clear performance curve, allowing for easy visualization of the performance trends and the identification of the optimal frame size regions.

4.2. Simulation Results

Figures 2–5 illustrate the system's performance as a function of frame size under three different time variation speeds. The time variation speeds $\delta = 0$, 0.0005, and 0.001 correspond to static, slowly time-varying, and moderately time-varying channel conditions, respectively. These plots evaluate how well the proposed FastICA-based joint communication and sensing framework performs as the rate of channel change increases.

Figure 2 shows the evolution of the ELMMSE for the SI reflection channel $\mathbf{H}_{\text{si-r}}$ across varying frame sizes. In the static case (i.e., $\delta = 0$), the estimation error decreases steadily with increasing frame length, reaching approximately 0.01 at 500 symbols. This improvement occurs because longer frames provide more statistical information for the ICA algorithm to exploit. However, for time-varying channels, the behavior is noticeably different. The slowly varying case ($\delta = 0.0005$) shows optimal performance around 200–250 frames, after which the error begins to increase. The moderately varying case ($\delta = 0.001$) exhibits even more pronounced degradation, with optimal performance occurring around 150 frames. This degradation occurs because the average composite mixing matrix becomes less representative of any individual time frame as the channel varies more significantly over the processing block.

A similar trend is observed in Figure 3, which shows the ELMMSE for the SOI channel $\mathbf{H}_{\text{soi}}$. The estimation error decreases rapidly for small-to-moderate frame sizes across all variation speeds but begins to degrade when the channel varies too quickly within a block. In particular, the curve for $\delta = 0.001$ demonstrates increasing error beyond 200–250 frames, highlighting the fundamental limitation of using batch ICA methods in dynamic environments. The SOI channel estimation appears more sensitive to time variations than the SI channel, likely due to the lack of prior knowledge about the SOI signal structure.

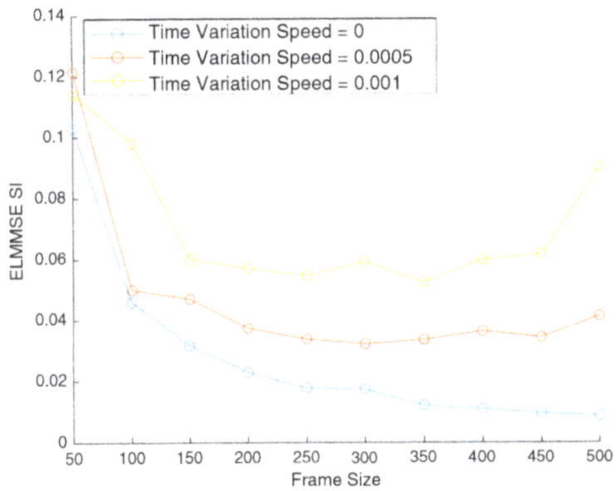

Figure 2. ELMMSE of channel sensing $\hat{\mathbf{H}}_{\text{si-r}}$ vs. frame size.

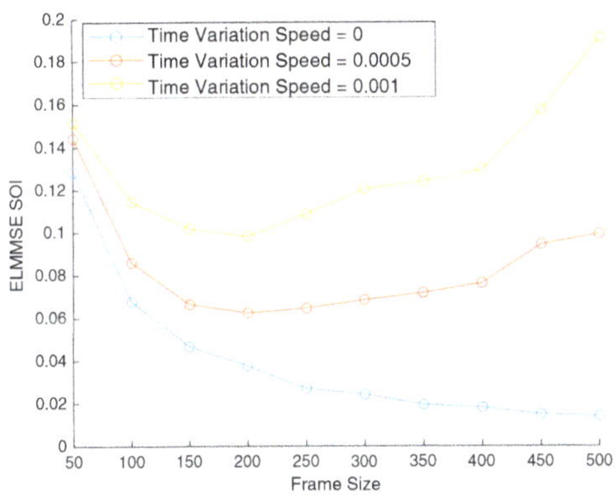

Figure 3. ELMMSE of channel sensing $\hat{\mathbf{H}}_{\text{soi}}$ vs. frame size.

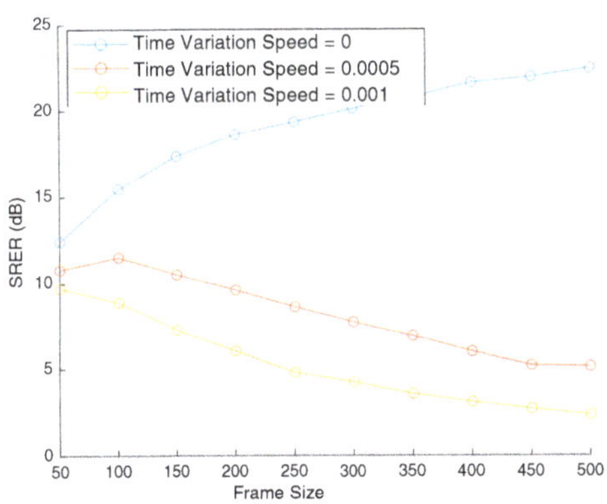

Figure 4. SRER for SOI in dB vs. frame size.

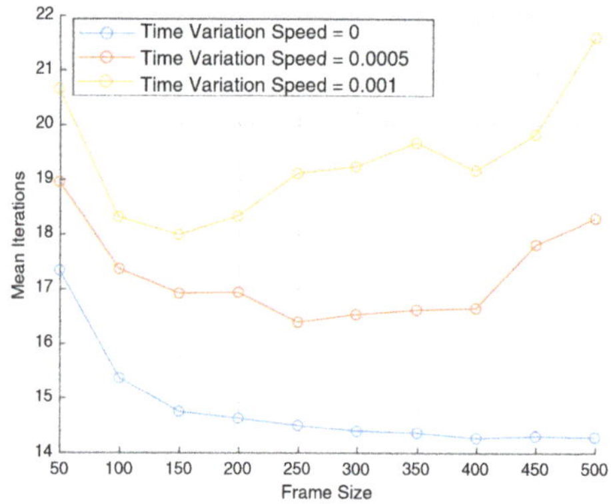

Figure 5. Number of FastICA iterations required for convergence vs. frame size.

Figure 4 displays the SRER in decibels, which quantifies the quality of the recovered communication signal. The static case ($\delta = 0$) shows a consistent increase in SRER with frame size, reaching over 22 dB by 500 symbols. However, in time-varying scenarios, performance peaks early and then degrades significantly. For $\delta = 0.0005$, the SRER begins to decline beyond 200 frames, while for $\delta = 0.001$, the degradation starts even earlier (around 150 frames) and is more severe, dropping below 4 dB. These results show that while longer frames improve statistical estimation in principle, they also introduce greater inconsistency between the assumed and actual mixing behavior, leading to poorer separation outcomes in dynamic channels.

Figure 5 depicts the average number of iterations required by the FastICA algorithm to converge. In the static case, iteration count decreases from about 15 at small frame sizes to approximately 14 at larger frame sizes, indicating improved convergence efficiency with more data. However, for time-varying cases, convergence behavior is different. The slowly varying case ($\delta = 0.0005$) maintains relatively stable iteration counts around 17–18. The moderately varying case ($\delta = 0.001$) shows the most challenging convergence behavior, requiring over 21 iterations at 500 symbols. The increased convergence time reflects the algorithm's difficulty in finding a consistent separation solution when the underlying mixing matrix is changing during the processing block.

4.3. Discussion and Insights

The simulation results reveal several important insights for the JCAS system's design in time-varying environments. A critical trade-off exists for the optimal frame size, which must balance statistical reliability with channel stationarity. For slowly varying channels ($\delta = 0.0005$), this optimum is around 200–250 symbols, whereas for faster variations ($\delta = 0.001$), it shifts to 150–200 symbols, meaning system designers must select frame sizes based on expected channel dynamics. This performance degradation stems from the mismatch between the block-based assumption of FastICA and the continuous evolution of the channel. As the frame size increases, this mismatch becomes more pronounced, leading to errors in both channel estimation and signal separation. Notably, the SOI channel estimation and signal recovery appear more sensitive to time variations than the SI channel estimation, which suggests that leveraging the known SI signal provides some robustness in dynamic environments. Furthermore, the increased iteration count in these scenarios indicates not only convergence difficulties but also higher computational requirements that must be considered for real-time implementation. Ultimately, these results demonstrate the limitations of traditional FastICA for time-varying channels and motivate the development of adaptive BSS algorithms capable of tracking such variations continuously.

5. Conclusions and Future Work

In this work, we proposed and evaluated a blind source separation (BSS)-enabled framework for joint communication and sensing (JCAS) in in-band full-duplex (IBFD) MIMO systems. By exploiting the known self-interference (SI) signal, we demonstrated how it can be repurposed for environmental sensing without requiring dedicated sensing waveforms. The system was formulated as a BSS problem and solved using the FastICA algorithm to achieve simultaneous SI suppression and channel estimation. Through extensive simulations under static and time-varying channel conditions, we evaluated the performance of the proposed framework using various metrics, including signal-to-residual-error ratio (SRER), ergodic linear minimum mean 124 squared error (ELMMSE) for both signal of interest (SOI) and SI channels, and convergence behavior of FastICA. Our results indicate that in static channels ($\delta = 0$), longer frame sizes consistently improve separation and estimation performance, with SRER exceeding 22 dB at 500-symbol frames and ELMMSE approaching 0.01. However, under time-varying conditions ($\delta = 0.0005$ and 0.001), performance degrades beyond certain optimal frame lengths. This degradation is due to the mismatch between the assumed stationary mixing matrix and the actual evolving one, which hampers both separation accuracy and algorithm convergence.

The average number of iterations required for convergence also increases significantly in highly dynamic channels, particularly for $\delta = 0.001$, highlighting the limitations of conventional FastICA in such settings.

To further enhance the practicality and robustness of the proposed JCAS framework, future research directions include the following:

- Hardware Implementation Considerations: As full-duplex JCAS systems move toward practical deployment, addressing implementation challenges such as I/Q imbalance, phase noise, and nonlinear distortions in the BSS framework becomes crucial.
- Dynamic Frame Size Adaptation: Developing reinforcement learning-based strategies that dynamically adjust processing block lengths based on real-time channel variation estimates could optimize the trade-off between statistical reliability and channel stationarity.
- Multi-User Scenarios: Extending the framework to multi-user MIMO scenarios where multiple communication pairs share the same spectrum while performing distributed sensing presents both theoretical and practical challenges worth exploring.
- Machine Learning Integration: Following recent trends in wireless communications, integrating deep learning approaches such as complex time-domain dilated convolutional recurrent networks could provide superior adaptation to time-varying channels while maintaining reasonable computational complexity.

Overall, the integration of adaptive intelligence and advanced BSS techniques holds significant promise for realizing spectrally efficient, environment-aware wireless networks. As 5G-Advanced and 6G standardization efforts progress, these technologies will play a crucial role in enabling applications ranging from autonomous vehicles to smart city infrastructure.

Author Contributions: Conceptualization, S.L. and T.Y.; methodology, S.L. and T.Y.; software, C.P.; validation, C.P., S.L. and T.Y.; formal analysis, C.P., S.L. and T.Y.; investigation, C.P.; data curation, C.P.; writing—original draft preparation, S.L. and C.P.; writing—review and editing, C.P., S.L. and T.Y.; visualization, C.P.; supervision, S.L. and T.Y.; project administration, S.L. and T.Y.; funding acquisition, S.L. and T.Y. All authors have read and agreed to the published version of the manuscript.

Funding: This research was supported by the Embry-Riddle Aeronautical University Internal Faculty Seed Fund (4.04) from The Boeing Center for Aviation and Aerospace Safety (BCAAS).

Data Availability Statement: The original contributions presented in this study are included in the article. Further inquiries can be directed to the corresponding author.

Conflicts of Interest: The authors declare no conflicts of interest.

Abbreviations

The following abbreviations are used in this manuscript:

MIMO	Multiple-Input Multiple-Output
IBFD	In-Band Full Duplex
BSS	Blind Source Separation
ICA	Independent Component Analysis
JCAS	Joint Communication and Sensing
SI	Self-Interference
SOI	Signal of Interest
ELMMSE	Ergodic Linear Minimum Mean Squared Error
SRER	Signal-to-Residual-Error Ratio
SINR	Signal-to-Interference-Plus-Noise Ratio

References

1. Heath, R.W.; Gonzalez-Prelcic, N.; Rangan, S.; Roh, W.; Sayeed, A.M. An overview of signal processing techniques for millimeter wave MIMO systems. *IEEE J. Sel. Top. Signal Process.* **2016**, *10*, 436–453. [CrossRef]
2. Belgium Completes 5G Spectrum Auction. Available online: https://www.rcrwireless.com/20220725/featured/belgium-completes-final-phase-spectrum-auction (accessed on 19 June 2025).
3. Kolodziej, K.E. *In-Band Full-Duplex Wireless Systems Handbook*; Artech House: Norwood, MA, USA, 2021.
4. Alves, H.; Riihonen, T.; Suraweera, H.A. *Full-Duplex Communications for Future Wireless Networks*; Springer: Cham, Switzerland, 2020.
5. Smida, B.; Alexandropoulos, G.C.; Riihonen, T.; Islam, M.A. In-band full-duplex MIMO systems for simultaneous communications and sensing: Challenges, methods, and future perspectives. *arXiv* **2024**, arXiv:2410.06512.
6. Liu, F.; Cui, Y.; Masouros, C.; Xu, J.; Han, T.X.; Eldar, Y.C.; Buzzi, S. Integrated sensing and communications: Toward dual-functional wireless networks for 6G and beyond. *IEEE J. Sel. Areas Commun.* **2022**, *40*, 1728–1767. [CrossRef]
7. Zhang, J.A.; Rahman, M.L.; Wu, K.; Huang, X.; Guo, Y.J.; Chen, S.; Yuan, J. Enabling joint communication and radar sensing in mobile networks—A survey. *IEEE Commun. Surv. Tutor.* **2022**, *24*, 306–345. [CrossRef]
8. Fang, X.; Feng, W.; Chen, Y.; Ge, N.; Zhang, Y. Joint communication and sensing toward 6G: Models and potential of using MIMO. *IEEE Internet Things J.* **2023**, *10*, 4093–4116. [CrossRef]
9. Li, S.; Caire, G. On the capacity and state estimation error of "beam-pointing" channels: The binary case. *IEEE Trans. Inf. Theory* **2023**, *69*, 5752–5770. [CrossRef]
10. Ahmadipour, M.; Kobayashi, M.; Wigger, M.; Caire, G. An information-theoretic approach to joint sensing and communication. *IEEE Trans. Inf. Theory* **2024**, *70*, 1124-1146. [CrossRef]
11. Cardoso, J.-F. Blind signal separation: Statistical principles. *Proc. IEEE* **1998**, *86*, 2009–2025. [CrossRef]
12. Bingham, E.; Hyvärinen, A. A fast fixed-point algorithm for independent component analysis of complex valued signals. *Int. J. Neural Syst.* **2000**, *10*, 1–8. [CrossRef] [PubMed]
13. Luo, Z.; Li, C.; Zhu, L. A comprehensive survey on blind source separation for wireless adaptive processing: Principles, perspectives, challenges and new research directions. *IEEE Access* **2018**, *6*, 66685–66708. [CrossRef]
14. Jin, B.; Sun, J.; Ye, P.; Zhou, F.; Lim, H.; Wu, Q.; Al-Dhahir, N. Data-driven sparsity-based source separation of the aliasing signal for joint communication and radar systems. *IEEE Trans. Veh. Technol.* **2023**, *72*, 2161–2174. [CrossRef]
15. Fouda, M.E.; Shen, C.-A.; Eltawil, A.E. Blind source separation for full-duplex systems: Potential and challenges. *IEEE Open J. Commun. Soc.* **2021**, *2*, 1379–1389. [CrossRef]
16. Baquero Barneto, C. Analysis and Design of Joint Communication and Sensing for Wireless Cellular Networks. Ph.D. Dissertation, Tampere University, Tampere, Finland, 2022.
17. Li, J.; Zhang, H.; Zhang, J. Fast adaptive BSS algorithm for independent/dependent sources. *IEEE Commun. Lett.* **2016**, *20*, 2221–2224. [CrossRef]

18. Thameri, M.; Abed-Meraim, K.; Belouchrani, A. New algorithms for adaptive BSS. In Proceedings of the 2012 11th International Conference on Information Science, Signal Processing and Their Applications (ISSPA), Montreal, QC, Canada, 2–5 July 2012; IEEE: Piscataway, NJ, USA, 2012; pp. 590–594.
19. Zhu, Q.; Wang, C.-X.; Hua, B.; Mao, K.; Jiang, S.; Yao, M. 3GPP TR 38.901 channel model. In *The Wiley 5G Reference: The Essential 5G Reference Online*; Wiley Press: Hoboken, NJ, USA, 2021; pp. 1–35.
20. Kyösti, P.; Meinilä, J.; Hentilä, L.; Zhao, X.; Jämsä, T.; Schneider, C.; Narandzić, M.; Milojević, M.; Hong, A.; Ylitalo, J.; et al. *IST-4-027756 WINNER II D1.1.2 V1.2 WINNER II Channel Models*; WINNER II Consortium: Brussels, Belgium, 2007.
21. Liu, L.; Oestges, C.; Poutanen, J.; Vainikainen, P.; Sarrazin, J.; Laitinen, M.; Costa, E.; Yin, X.; Wang, Y.; Kivinen, J.; et al. The COST 2100 MIMO channel model. *IEEE Wirel. Commun.* **2012**, *19*, 92–99. [CrossRef]

electronics

MDPI

Review

Radio Frequency Interference, Its Mitigation and Its Implications for the Civil Aviation Industry

Adnan Malik [1,2] and Muzaffar Rao [1,*]

[1] Department of Electronic and Computer Engineering, University of Limerick, V94 T9PX Limerick, Ireland; zee.adee@gmail.com
[2] AirNav Ireland, Ballycasey Cross, Co. Clare, V14 WN81 Shannon, Ireland
* Correspondence: muzaffar.rao@ul.ie

Abstract: Radio Frequency Interference has emerged as a growing challenge for aviation safety and system integrity due to the increasing spectral overlap between communication technologies and aviation systems. This paper investigates the sources, types, and consequences of RFI in Global Navigation Satellite Systems, Instrument Landing Systems, and altimeters used in civil aviation. A detailed examination of both intentional and unintentional interference is presented, highlighting real-world incidents and simulated impact models. The study analyzes technical mechanisms such as receiver desensitization, intermodulation, and cross-modulation, and further explores UAV-based interference detection frameworks. Mitigation strategies are reviewed, including regulatory practices, spectrum filters, shielding architectures, and dynamic UAV sensing systems. Comparative insights into simulation results, shielding techniques, and regulatory gaps are discussed. The paper concludes with recommendations for enhancing current aviation standards and suggests a hybrid validation model combining in-flight measurements with simulation-based assessments. This research contributes to the understanding of electromagnetic vulnerabilities in aviation and provides a basis for future mitigation protocols.

Keywords: radio frequency; UAV; VHF; UHF; navigation; aviation; flights; ILS

1. Introduction

In an era defined by the rapid evolution of wireless communication technologies, the aviation industry faces increasing challenges from RFI. As modern Aircraft heavily rely on electronic and wireless systems for navigation, communication, and surveillance, the integrity of the radio frequency spectrum becomes critical for operational safety and efficiency. However, RFI, whether intentional or unintentional, poses a significant risk to these systems, potentially disrupting flight operations and compromising passenger safety.

The Very High Frequency (VHF) band (118–137 MHz) and Ultra High Frequency (UHF) band (300 MHz–3 GHz) are the backbone of civil aviation communication and navigation systems. VHF is primarily utilized for air-to-ground communication, enabling seamless coordination between pilots and air traffic controllers. Meanwhile, UHF supports a variety of essential functions, including radar systems, transponders, and Instrument Landing Systems (ILSs). Despite strict management regulations, the growing proliferation of wireless technologies, such as 5G networks, Unmanned Aerial Vehicles (UAVs), and Internet of Things (IoT) devices, has increased the likelihood of interference within these critical bands.

RFI can arise from a multitude of sources. External sources include terrestrial communication towers, 5G base stations, improperly shielded electronic equipment, and atmospheric phenomena such as solar flares or sporadic E-layer activity. Internally, onboard electronic devices, including malfunctioning avionics or improperly used passenger devices, can also contribute to interference. Natural sources, such as ionospheric activity or solar flares, can further degrade or distort signals in the VHF and UHF bands. Regardless of the source, the impacts of RFI on aviation systems are profound, leading to communication failures, navigation errors, and surveillance system malfunctions. For instance, interference with radar altimeters or GPS signals can disrupt flight paths, particularly during critical phases such as landing or take-off, where precision is paramount.

Additionally, intentional electromagnetic interference (IEMI) carried out by malicious actors poses a growing threat to civil aviation. Such threats often involve directed high-power RF signals or electromagnetic pulse devices that can disrupt avionics. Interference may enter systems via front-door coupling (e.g., antennas) or back-door coupling (e.g., power lines, cabling, or structural seams), each requiring different mitigation strategies. According to Jie et al. (2024) [1], the immunity of Aircraft systems must span a wide spectral range—from MHz to several GHz—to ensure resilience across all potential interference scenarios. These requirements are consistent with aviation standards such as RTCA DO-160 [2] and MIL-STD-461 [3], which provide test protocols for radiated and conducted susceptibility in airborne equipment.

Mitigating RFI has therefore become a key area of focus within the aviation industry. Effective mitigation strategies range from advanced technological solutions to stringent regulatory measures. Filters and shielding techniques are integral to protecting onboard systems from interference, while spectrum management ensures that frequency bands allocated to aviation are protected from unauthorized or overlapping transmissions. Additionally, real-time monitoring and enforcement by aviation authorities play a crucial role in identifying and addressing interference sources promptly.

Despite these efforts, significant challenges persist. The global adoption of 5G technology, for example, has brought renewed concerns about its potential to interfere with radar altimeters operating in adjacent frequency bands. Moreover, the increasing use of UAVs and IoT devices introduces complexities in managing the radio frequency spectrum, particularly in urban environments where RF activity is dense. The challenge is further compounded by the limited availability of alternative navigation aids, as many aviation systems rely predominantly on satellite-based technologies like GPS.

Considering these developments, the need for comprehensive research into RFI, its mitigation, and its implications for the aviation industry is more pressing than ever. Understanding the mechanisms of interference, exploring innovative mitigation techniques, and developing robust regulatory frameworks are essential steps in safeguarding the future of aviation. This research aims to contribute to these efforts by investigating the sources, impacts, and mitigation strategies for RFI, with the ultimate goal of enhancing safety and efficiency in the aviation sector.

By addressing these challenges, the aviation industry can not only maintain operational integrity but also adapt to the rapidly evolving technological landscape. This paper will explore these aspects in detail, shedding light on the critical role of RFI management in ensuring the continued safety and reliability of global air travel.

The reliance on advanced electronic and wireless systems in aviation has made RFI a critical challenge. RFI disrupts essential communication, navigation, and surveillance systems, posing risks to safety and efficiency. With emerging technologies like 5G, IoT, and UAVs increasing the likelihood of interference in VHF and UHF bands, addressing RFI has become vital. This research focuses on identifying RFI sources, analyzing its impacts, and proposing mitigation strategies using real-time monitoring and simulations to ensure aviation system safety and reliability. The primary research objectives of this study are as follows:

- Review multiple research papers covering RFI interference to figure out the methodologies and technologies used in identifying the RFI.
- Identify the sources of RFI in aviation, including external factors like terrestrial transmitters, on-board electronic equipment, portable electronic devices, and weather.

This paper is organized as follows: Section 2 provides the theoretical background, Section 3 provides a literature review, Section 4 provides a discussion of the results, and Section 5 concludes the paper.

2. Theoretical Background

RFI is a phenomenon where unwanted electromagnetic signals disrupt the operation of electronic systems. In aviation, the consequences of such interference are particularly critical, as industry relies heavily on precise and reliable communication, navigation, and surveillance systems. Understanding the theoretical underpinnings of RFI is essential for developing effective mitigation strategies to ensure aviation safety and efficiency.

RF interference in aviation systems may result from several technical mechanisms. One such mechanism is receiver desensitization, where strong out-of-band signals suppress the sensitivity of adjacent weak signals, such as those from navigation beacons. Intermodulation occurs when two or more signals mix in a non-linear component (e.g., an amplifier), generating spurious frequencies that fall within the receiver band. Additionally, cross-modulation may arise when an unwanted signal modulates the amplitude of a desired carrier, leading to distorted reception. These effects are particularly relevant in densely packed RF environments around airports, where multiple emitters coexist in proximity.

2.1. The Role of the RF Spectrum in Aviation

The radio frequency spectrum is the foundation for aviation communication and navigation. The VHF band supports air-to-ground voice communication, enabling pilots and air traffic controllers to coordinate operations effectively. The Ultra High Frequency (UHF) band is used for functions such as radar, transponder communication, and (ILS). These systems ensure safe take-off, cruising, and landing, particularly in poor visibility or adverse weather conditions. RFI can couple into Aircraft systems through two primary mechanisms: front-door and back-door coupling. Front-door coupling refers to interference that enters through designed input ports, such as antennas or receiver terminals. This type is often filtered but still susceptible to strong nearby emissions. In contrast, back-door coupling occurs through unintended paths, including power cables, structural seams, or signal lines, where electromagnetic waves bypass front-end protection. Both coupling types can disrupt onboard avionics, especially if shielding or grounding measures are inadequate. RFI impacts aviation systems through several pathways:

- Adjacent Channel Interference: Signals from neighboring frequency bands bleed into aviation channels, reducing signal clarity.
- Harmonic Interference: Harmonics from non-aviation frequencies can coincide with aviation bands, creating noise and disrupting system performance.
- Receiver Desensitization: Strong external signals can overwhelm receivers, making them less sensitive to desired signals.
- Multipath Propagation: Reflected signals from terrain, buildings, or other surfaces can create signal distortion and loss.

2.2. Implications for Aviation Systems

The impacts of RFI on aviation systems are broad and significant:

- Communication Systems: Interference can result in noise, signal loss, or communication delays between pilots and air traffic controllers.
- Navigation Systems: Disruptions to GPS or radar altimeters can compromise flight path accuracy, particularly during critical operations such as landing or approach.
- Surveillance Systems: Interference with radar systems or transponders can obscure Aircraft positioning, increasing collision risks.
- Recent incidents of radio RFI have significantly impacted aviation operations, highlighting the critical need for robust navigation systems and international cooperation. Two notable examples are given below.

1. GPS Interference Affecting Finnish Airports (November 2024) [4]: In November 2024, three airports in Eastern Finland—Joensuu, Savonlinna, and Lappeenranta—experienced disruptions in GPS signals, suspected to be caused by Russian interference. This interference compromised the accuracy of navigation systems, prompting Finavia, the Finnish airport operator, to reinstate traditional radio navigation equipment to assist with Aircraft landings. This situation underscores the vulnerability of satellite-based navigation systems to external disruptions and the necessity for alternative navigation aids to ensure aviation safety.

2. Unauthorized Radio Transmissions in Melbourne, Australia (September 2024): Between July and September 2024, a 45-year-old man in Melbourne allegedly made unauthorized radio transmissions, including fake mayday calls, to commercial Aircraft operating at Melbourne Airport. These transmissions had the potential to endanger Aircraft operations by causing confusion and diverting attention from genuine emergencies. The Australian Federal Police arrested the individual, charging him with offenses that carry significant penalties, emphasizing the serious nature of such interference.

The above-mentioned incidents illustrate the diverse sources of RFI—from suspected state-sponsored GPS jamming to individual malicious acts—and their profound impact on aviation safety. They highlight the importance of maintaining and updating traditional navigation aids, enforcing stringent regulations against unauthorized transmissions, and fostering international collaboration to mitigate and respond to RFI threats effectively.

To better understand the safety implications of RFI, fault modes can be classified based on severity and likelihood. For instance, interference with radar altimeters during landing presents a high-severity, moderate-likelihood event due to its impact on altitude accuracy. GPS jamming during enroute flight is moderate in severity but higher in likelihood, especially near conflict zones. Unauthorized radio transmissions fall under moderate severity and low likelihood but can still create confusion. A structured risk matrix can help prioritize mitigation, with high-risk scenarios requiring immediate attention through shielding, redundancy, or alternative aids.

Aviation communication and navigation systems are vulnerable to both unintentional noise and intentional interference. Common noise sources include atmospheric discharge, cosmic background radiation, and electronic emissions from nearby equipment. These noise elements can degrade the signal-to-noise ratio (SNR), leading to reduced positioning accuracy or missed alerts. To mitigate these issues, systems often employ filtering, low-noise amplifiers (LNAs), and error correction algorithms. For safety-critical operations like ILS or GNSS-based approaches, spectrum management and guard band allocation are also essential in reducing noise overlap with active aviation frequencies.

2.3. Mitigation Strategies

Addressing RFI requires a combination of technological, regulatory, and operational approaches: Technological Solutions: Filters, shielding, and adaptive antenna designs help mitigate RFI at the system level. Advanced algorithms are also being developed to detect and counteract interference dynamically. Spectrum Management: Regulatory bodies such as EUROCONTROL in Europe enforce strict frequency allocation and emission standards to protect aviation bands from unauthorized use or overlap. Monitoring and Detection: Continuous monitoring of the RF spectrum using tools like spectrum analyzers and UAVs aids in identifying and addressing interference sources.

Recent work also demonstrates the effectiveness of structured safety approaches. For example, Zhang et al. (2021) [5] proposed a three-stage enhancement framework for UAV systems involving electromagnetic shielding, digital filtering, and real-time diagnostics. Similar architectures may be adapted for aviation platforms where mission-critical electronics must withstand burst noise and high-power emissions. Additionally, frequency-hopping spread spectrum (FHSS) has emerged as a resilient technique for reducing susceptibility to narrowband interference, especially in airborne communication links exposed to hostile RF conditions.

2.4. Challenges in Mitigating RFI

Despite advancements in mitigation techniques, several challenges persist [6]. The rapid expansion of wireless technologies, including 5G networks and Internet of Things (IoT) devices, has increased RF spectrum congestion, raising the likelihood of interference. Additionally, the reliance on satellite-based navigation systems like GPS makes aviation systems vulnerable to jamming and spoofing. Developing robust alternatives and enhancing existing systems is critical to overcoming these challenges.

Test Equipment Used to Investigate RFI in Aviation. The detection and mitigation of RFI in aviation require the use of specialized test equipment. These tools are integral for identifying sources of interference, analyzing their impact on aviation systems, and validating mitigation strategies to ensure the safe and efficient operation of Aircraft.

The detection and mitigation of RFI in aviation rely on specialized tools that ensure system integrity and operational safety. These tools help identify interference sources, analyze signal disruptions, and validate mitigation strategies. Their role is crucial in maintaining reliable communication, navigation, and surveillance systems. Details of these tools are discussed below.

2.4.1. Test Equipment Used to Investigate RFI

A spectrum analyzer [7] is a critical tool for monitoring radio frequencies. It measures the power of electromagnetic signals over a range of frequencies, making it possible to detect unauthorized or unintended transmissions that could disrupt aviation communication or navigation systems. Spectrum analyzers are widely used to identify interference in both VHF and UHF bands. RF signal generators [8] are essential for producing controlled electromagnetic signals to test and calibrate aviation systems. These devices help simulate interference scenarios, allowing engineers to evaluate how avionics, such as radar altimeters and communication systems, respond to specific types of interference.

2.4.2. Network Analyzer

A network analyzer [9] is employed to measure the performance of RF components, including antennas and filters. It evaluates parameters such as signal loss and reflection, ensuring that these components are effectively reducing or rejecting unwanted frequencies that could cause interference.

2.4.3. Directional Antennas

Directional antennas [10] are used to locate sources of interference. By focusing on the directionality of signals, these antennas allow the precise tracking of unauthorized transmissions, which is critical for resolving interference issues in airport environments or urban settings.

2.4.4. RF Site Master

An RF Site Master [11] is a portable device used to test the integrity of cable and antenna systems. It identifies issues such as signal degradation or physical damage that could compromise the system's ability to handle interference effectively. These tools are particularly useful for maintaining ground-based navigation aids like ILS.

2.4.5. Radio Communication Test Sets

Radio test sets, such as the CMA 180, are comprehensive testing tools [12] designed for aviation communication systems. These devices assess the performance of air-to-ground radios and their resilience to interference, ensuring the clarity and reliability of voice and data transmissions.

2.4.6. EMC Chambers

EMC chambers are shielded environments where aviation equipment is tested for compliance with electromagnetic interference (EMI) standards. These chambers simulate controlled interference scenarios to determine how well systems perform in real-world conditions.

To ensure that Aircraft systems meet acceptable electromagnetic performance thresholds, industry standards have been developed. The RTCA DO-160 standard is widely used in commercial aviation to define test procedures for environmental conditions, including electromagnetic emissions and susceptibility. In defense applications, MIL-STD-461 outlines detailed EMI/EMC requirements for airborne equipment. These standards guide design, testing, and certification processes to improve interference resilience across mission profiles.

2.4.7. GPS Jamming and Spoofing Detectors

Specialized tools for detecting GPS interference are used to ensure the accuracy of satellite-based navigation systems. These detectors monitor signal quality and identify instances of jamming or spoofing that could disrupt critical navigation functions.

2.4.8. Time Domain Reflectometer (TDR)

TFRs [13] are used to diagnose faults in RF transmission lines. By analyzing the time delay of reflected signals, these tools help locate disruptions in signal pathways that could be contributing to interference.

2.4.9. UAV-Based RF Monitoring Platforms

UAVs equipped with RF sensors provide a mobile platform for interference detection [14]. These systems are especially useful for monitoring interference in areas where traditional ground-based equipment may not be effective, such as airport perimeters or remote locations. The use of these tools ensures that aviation systems remain resilient to interference, enhancing safety and operational reliability. By enabling the precise detection, comprehensive analysis, and validation of mitigation strategies, this equipment plays a pivotal role in protecting critical aviation infrastructure from the risks posed by RFI.

3. Literature Review

RFI is a very broad yet interesting topic; however, when it comes to its implications for the aviation industry, the sourcing of recent research work becomes quite challenging. A lot of effort was put into finding relevant and recent research papers for consideration in this review of the literature (see the citation references at the end of this document).

Bukhari et al. [15] examine the significant concerns about interference of fifth generation (5G) telecommunications technology with aviation altimeters, particularly due to the overlap of frequency bands. This interference poses serious risks to flight operations and navigational accuracy, potentially compromising aviation safety. In response to these challenges, regulatory authorities around the world have enacted various measures, including the issuance of Airworthiness Directives (AD) and restrictions on the effective isotropic radiated power (EIRP) of 5G base stations near airports. In addition, exclusion zones have been established to mitigate the potential risks associated with 5G operations in proximity to aviation activities. The paper advocates for a collaborative approach among regulators, industry stakeholders, and researchers to develop compatible technologies and regulatory frameworks. It emphasizes the need for ongoing research to better understand interference dynamics and to create comprehensive guidelines that balance the advancement of 5G technology with the critical requirements of aviation safety. Figure 1 shows the proposed Aircraft's altimeter in boresight of a 5G antenna. The orange ellipse in the diagram represents the zone of maximum interference, where the main beam of the 5G interfering signal overlaps with the aircraft's approach path and altimeter signal. This region signifies the critical spatial area in which radio altimeter receivers are most vulnerable to potential signal corruption or degradation due to high-power emissions from 5G base stations. The intersection of the interfering signal's main beam with the aircraft's glide slope and altitude measurement path highlights the likelihood of interference during final approach, especially in low-visibility or instrument landing conditions.

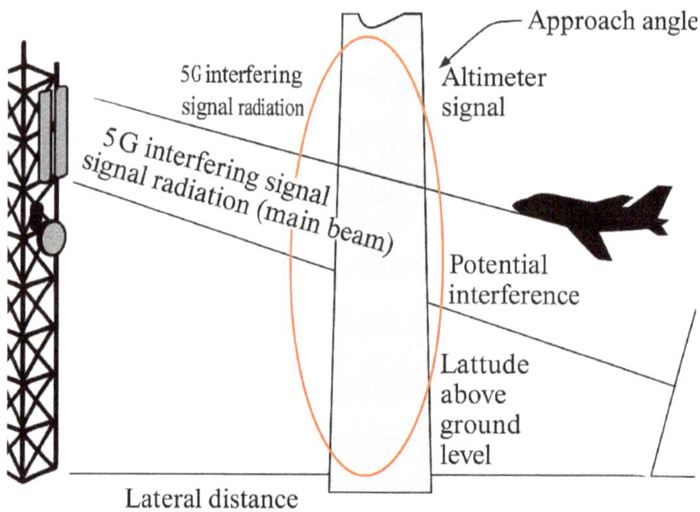

Figure 1. Aircraft's altimeter in boresight of 5G antenna used by Bukhari et al. [14].

Li et al. [16] highlight the substantial interference that 5G base stations can cause to radar altimeters operating within the 4.2–4.4 GHz frequency band, raising significant safety concerns for aviation. A detailed mathematical model is created to systematically assess this interference in various environments, including rural, suburban, and urban areas. The model takes into account different altitudes and operational scenarios to accurately evaluate the interference-to-noise ratio experienced by radar altimeters.

In addition, the study examines the impact of two types of antennas—omnidirectional and multiple input–multiple output (MIMO) antennas—on interference dynamics. This analysis is crucial to understanding how antenna design influences interference. The results emphasize the need for effective mitigation strategies to ensure the safe and reliable coexistence of 5G communication systems and radar altimeters.

Son [17] investigates the potential interference between 5G systems operating within the 3.7–4.0 GHz frequency band and aeronautical radio altimeters operating in the adjacent 4.2–4.4 GHz band in South Korea. This analysis is crucial because South Korea plans to extend the 5G spectrum into the 3.7–4.0 GHz range. The primary objective of the research is to establish technical conditions that ensure compatibility between 5G systems and radio altimeters. This involves evaluating the interference impact of 5G base stations on altimeter operations. The researchers propose an interference analysis scenario and methodology grounded in Monte Carlo simulations. These simulations are designed to replicate real-world conditions, including Aircraft approaches and 5G network deployments. The study identifies the main challenge to determining the necessary restrictions on 5G base station emissions to protect radio altimeter functionality. The findings indicate that unwanted 5G emissions must be 35 dB more stringent than the current 3GPP standards to prevent interference with altimeters. Figure 2 shows the interference scenario in the above-discussed paper, while Figure 3 shows the analysis method.

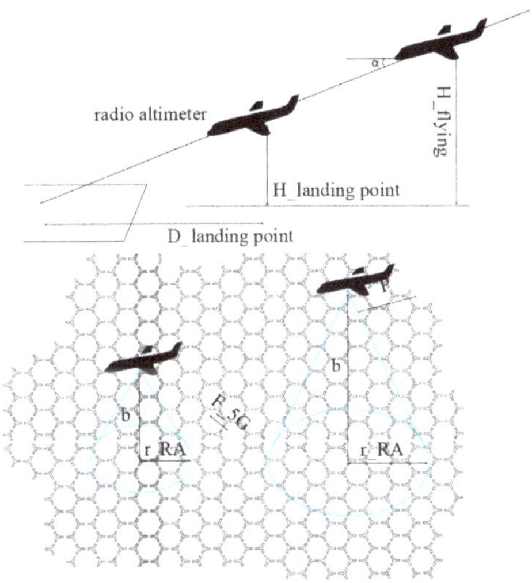

Figure 2. An interference scenario by Son [17]. Schematic representation of the radio altimeter beam spread over a hexagonally divided ground surface, illustrating signal coverage at two different aircraft altitudes. The blue cones indicate the area covered by the altimeter signal, which expands as the altitude increases. In this figure, H represents the altitude of the aircraft above ground level, while D denotes the diameter of the corresponding radio altimeter footprint on the ground. The terms $H_{landing}$ point and $D_{landing}$ point refer to the aircraft's altitude and signal footprint, respectively, at the moment of landing.

Zhou et al. [18] conducted a study on identifying sources of civil aviation radio interference using four UAVs. It highlights the limitations of traditional ground-based monitoring methods, which are often slow and ineffective in tracking aerial interference. The study aims to develop an innovative time–frequency difference positioning method to improve monitoring accuracy and efficiency. The research successfully establishes a synchronized model for UAVs and demonstrates its effectiveness through simulations, achieving balanced positioning performance and minimal errors across various deployment configurations. The findings underscore the method's potential for practical applications in aviation safety.

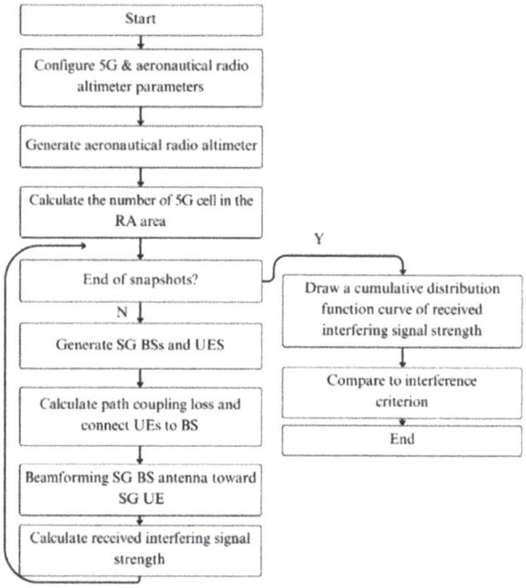

Figure 3. The analysis method used by Son [6].

Kircher, Daniel, and Deutschmann [19] examine how RFI influences the electromagnetic emissions of ICs used in automotive applications. The primary objective is to elucidate how RFI can compromise the functionality and adherence of ICs to EMC standards. The intended readership comprises IC manufacturers and automotive suppliers, particularly those involved in safety-critical applications. A significant challenge addressed is the sequential nature of EMC testing, which frequently overlooks the interplay between electromagnetic immunity and emissions. The results indicate substantial increases in electromagnetic emissions due to injected disturbance signals, resulting in the functional degradation of ICs. Consequently, the findings emphasize the necessity of integrated testing approaches to ensure both safety and regulatory compliance in automotive electronics. Figure 4 provides an example of EMI emissions, while Figure 5 shows the architecture of EMI emissions used by Kircher, Daniel, and Deutschmann [19].

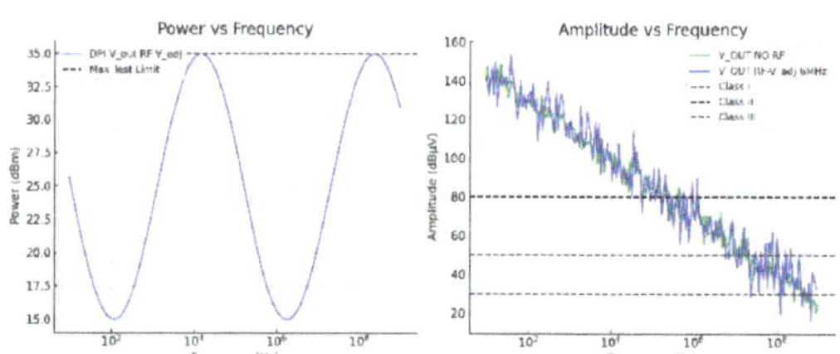

Figure 4. Example of EMI emissions in study by Kircher, Daniel, and Deutschmann [19].

K. Hussain [20] presents a detailed view and explores the critical role of electrical and electronic technology in the twenty-first century, alongside the escalating issue of EMI resulting from the widespread use of electronic devices. It clarifies the often-confused terms EMI and EMC, providing their precise definitions and detailing their interrelationship. The paper underscores that EMI is a disturbance that causes malfunctions or undesired responses in electronic systems, whereas EMC denotes the ability of equipment to operate as intended within its electromagnetic environment. Additionally, it examines the assessment of EMI/EMC and strategies for mitigating EMI issues, with a particular focus on the significance of PCB design in addressing these challenges. For interference mitigation this paper emphasizes implementing bonding, grounding, shielding, and filtering procedures regularly to address EMI and achieve EMC. It also highlights the significance of best PCB design, the proper construction of structural systems, and addressing communication and radar systems for EMC in naval ships and marine platforms. Furthermore, the paper stresses achieving EMC through design, manufacturing, maintenance, training, and awareness initiatives for improved technological device usage and maintenance.

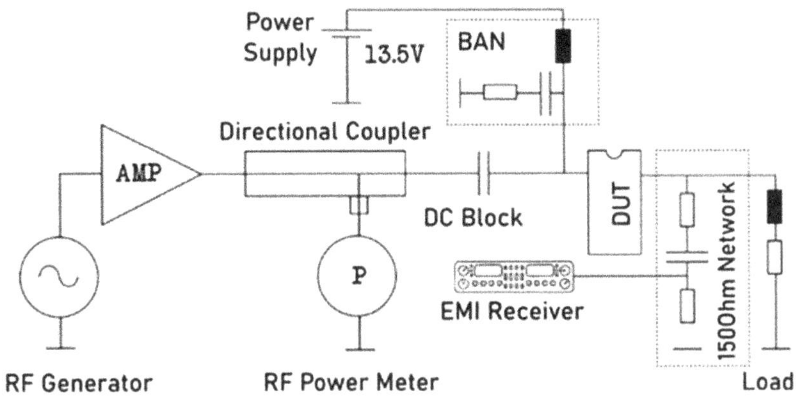

Figure 5. Architecture used by Kircher, Daniel, and Deutschmann [19] to show emissions.

Hegarty et al. [21] examine various mechanisms that cause out-of-band interference, which can disrupt GNSS receivers. It identifies key disruptive factors such as saturation, desensitization, reciprocal mixing effects, intermodulation products, aliasing, and in-band emissions. The paper provides detailed explanations of these mechanisms and discusses strategies to mitigate each one. The study is set against the backdrop of the growing dependence on GNSS technology and the susceptibility of GNSS receivers to out-of-band interference. The findings offer a comprehensive understanding of these disruptive mechanisms and effective mitigation strategies, which are essential for enhancing the reliability and accuracy of GNSSs.

Monitoring interference is an important topic, and Zhou et al. [22] outline a method for monitoring RFI in civil aviation using UAVs, emphasizing four key contributions. Firstly, it introduces an angle-of-arrival (AOA) algorithm that improves the accuracy of RFI location by using directional antennas to determine the azimuth of radio signals. Secondly, it describes the development of a lightweight UAV monitoring platform that meets airworthiness standards, ensuring safe data collection. Thirdly, it details a ground analysis system designed for real-time data processing and the intuitive visualization of RFI locations. Lastly, the method is validated through various test scenarios, showing superior performance in terms of radius monitoring and positioning accuracy compared to traditional ground-based methods. Figure 6 shows the process of Zhou et al. [22] for RFI monitoring using a UAV.

Hosokawa et al. [23] introduce a monitoring network established in Japan to observe the anomalous propagation (EsAP) of aeronautical VHF radio waves, which are influenced by sporadic layers of E (Es). Operational since May 2019, the network comprises six stations equipped with software-defined radios that capture signals within the 98 to 118 MHz frequency range. This system facilitates near-real-time monitoring, providing essential data for aviation navigation and communication. The network has successfully identified multiple instances of EsAP, with notable variations in signal intensity across different stations. This capability enables the distinction between equipment malfunctions and interference caused by Es, thereby enhancing operational safety. Ongoing efforts to develop algorithms for automatic data analysis aim to further improve the visualization and understanding of Es impacts on VHF communications.

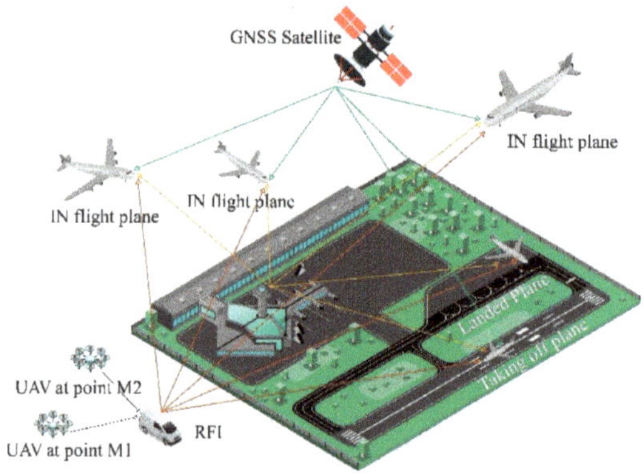

Figure 6. RFI monitoring using UAV in study by Zhou et al. [22].

Joseph et al. [24] explore the growing dangers that RFI poses to Global Navigation Satellite System (GNSS) receivers in commercial Aircraft. They highlight problems such as jamming and spoofing, which can result in navigation failures and safety hazards. To counter these issues, the authors suggest a three-part approach: detecting RFI, accurately measuring the extent of the errors caused, and ensuring that receivers can still provide genuine position information even during interference. Furthermore, the paper stresses the importance of combining GNSS data with other navigation sensors like Inertial Reference Systems (IRSs) and Distance Measuring Equipment (DME) to improve overall navigation integrity and reliability in the face of RFI threats. Figure 7 shows the process of RFI on the GNSS receivers used by the authors.

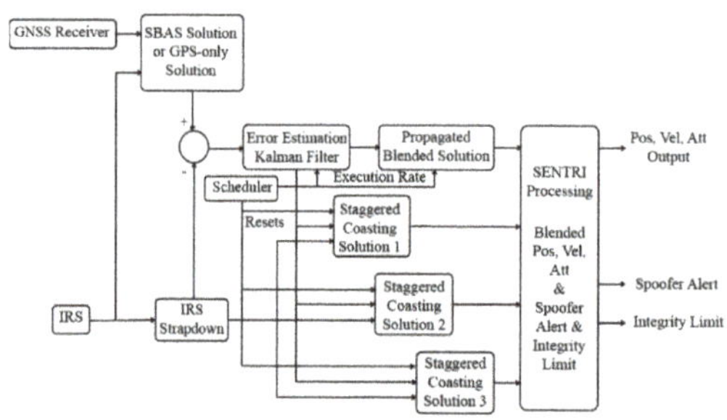

Figure 7. RFI on GNSS receivers used by Joseph et al. [24].

Researchers at MIT [14] discussed three novel techniques to mitigate interference in aeronautical communication: Pulse Blanking, Space Time Block Coding (STBC), and Decision-Directed Noise Estimation (DDNE). Using cognitive radio technology, Aircraft spectrum access was enhanced through dynamic identification and the utilization of available spectral bands, addressing the problem of spectrum scarcity. However, interference from DME presents significant challenges to the LDACS sXignal, as the high power of DME signals can substantially degrade communication quality. The proposed methods aim to effectively reduce this interference, thereby ensuring reliable air-to-ground communication.

Reddy [25] describes an RF front-end combinational filter for airborne V/UHF communication systems which integrates low-pass, high-pass, and band-stop filters, covering a frequency range from 30 to 600 MHz. A specific band-stop filter between 88 and 118 MHz mitigates interference from commercial FM broadcasts, enhancing communication reliability in software-defined radio (SDR) applications. This design ensures the selection of desired signals while rejecting unwanted interference, preserving signal quality, and improving the overall sensitivity and rejection of interference of the system. Implemented on a Polytetrafluoroethylene (PTFE) Printed Circuit Board (PCB), the filter's compact form factor is suitable for military and defense applications. Comprehensive simulations optimized the filter's performance, and real-world tests closely matched these results, demonstrating effective FM broadcast interference mitigation and high signal integrity. The robust design of the filter improves communication reliability, safety, and operational efficiency in aviation. Figure 8 shows the RF front-end combinational filter used by Reddy [25].

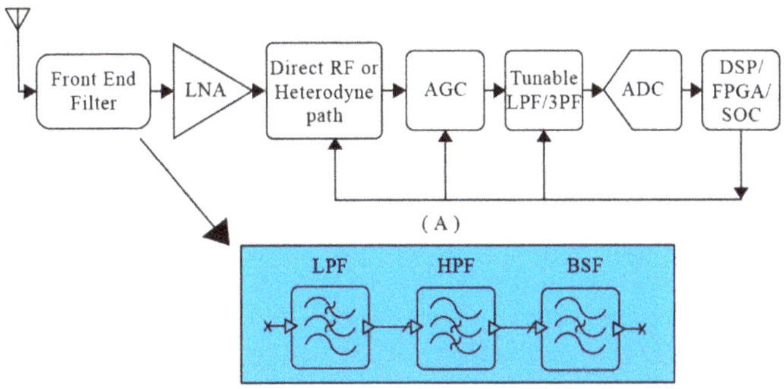

Figure 8. RF front-end combinational filter used by Reddy [25].

Vogel [26] highlights that co-site interference in Aircraft systems arises from several mechanisms, including adjacent channel interference, harmonic frequencies, receiver blocking, intermodulation products, and out-of-band transmission. These issues are particularly prevalent when systems operate across a broad frequency range. To address these challenges, electromagnetic simulation plays a vital role by analyzing the electromagnetic environment and quantifying antenna coupling through S parameters. This study highlights that even modest coordination between systems can significantly improve receiver sensitivity. It underscores the importance of enhancing transmitter electronics and implementing practical mitigation strategies such as narrow bandpass filters and meticulous spectrum management. Ultimately, a thorough understanding of the interference mechanisms, coupled with advanced simulation techniques, leads to effective solutions to optimize communication and sensing systems in Aircraft. Figure 9 shows the co-site interference in an Aircraft system described by Vogel [26]. Table 1 provides the signal strength measurements (db).

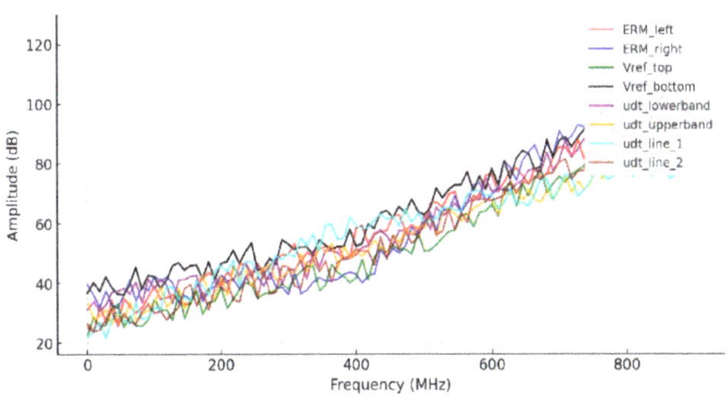

Figure 9. Co-site interference in Aircraft system used by Vogel [26].

Table 1. Signal strength measurements (DB).

Rx Tx	VHF Top	HF	VHF Bottom	UHF Top 1	UHF Top 2	UHF Bottom 1	UHF Lowerband	UHF Bottom 2
VHF top	25.3	−0.8	26.9	28.9	21.8	53.5	45.8	50.1
HF	−0.8	9.8	22.4	22.4	21.8	28.5	28.6	28.6
ESM left	13.4	49.9	28.8	25.8	29.7	38.5	39.5	39.5
ESM right	14.3	43.0	22.0	25.4	31.6	29.3	46.7	46.7
VHF bottom	29.4	27.9	44.3	44.3	52.8	39.5	37.0	27.1
UHF top 1	30.0	39.5	39.8	19.3	19.3	39.8	40.0	40.0
UHF top 2	20.8	39.3	12.3	63.4	−47.4	38.9	39.8	−61.4
UHF bottom 1	39.7	39.9	30.4	63.4	−47.4	40.0	25.6	30.4
UHF lowerband	36.8	39.9	25.7	40.0	40.0	25.6	28.2	30.4
UHF bottom 2	38.9	39.9	25.0	40.0	−30.5	−30.5	27.8	30.4

Ilive [27] describes Intermodulation products (IMPs) arise when multiple signals interact within a non-linear medium, such as electronic components or antennas. This interaction can create new frequencies that are combinations of the original signals, potentially disrupting intended communication or navigation signals. The research highlights that third-order and fifth-order IMPs are particularly troublesome for navigation systems like the ILS and VHF Omnidirectional Range (VOR). These IMPs can produce frequencies that are very close to the primary frequencies used by these systems, thereby increasing the risk of interference. To mitigate these issues, the study emphasizes the need for robust measurement and assessment techniques to identify and evaluate potential conflicts between different communication systems in aviation. This approach aims to ensure that navigational signals remain reliable and safe for Aircraft operations, especially during critical phases such as landing and approach.

Mahama et al. [28] stipulate findings that UAVs utilize specific radio frequency bands, such as 72–73 MHz, 902–928 MHz, and 2400–2483.5 MHz, which are also employed by technologies like Wi-Fi and ZigBee. This overlap can cause radio frequency (RF) interference, particularly in urban settings with numerous devices, increasing the risk of signal disruption. During bridge inspections, this interference can weaken signal strength, degrade video quality, and impair GPS functionality, thus compromising UAV control and safety. To address these challenges, it is essential to choose UAVs with robust RF immunity capable of effectively managing interference. Furthermore, UAVs should have minimal switching delays between redundant communication channels to ensure seamless operation. Conducting a comprehensive assessment of the inspection environment to identify potential RF noise sources is also crucial. By adopting these measures, the reliability and safety of UAV operations during bridge inspections can be improved, leading to successful outcomes even in demanding environments. Figure 10 shows the utilization of a specific frequency band by a UAV.

Kim, Jaesin, and Inkyu Lee [29] examine path loss in long-range air-to-ground (AG) communication systems using the UHF band. They find that the measured path loss aligns well with the predictions of the established models. The UHF band has been shown to be more reliable than higher frequencies, particularly in reducing multipath fading and atmospheric effects, which are crucial for long-distance communication. The analysis reveals that path loss behavior varies with different ground station environments, such as sea and ground reflections, and that multipath components can occur even at long ranges. Flight tests were conducted to observe path loss and multipath characteristics in realistic scenarios. The setup included UHF antennas, a high-power amplifier, and a channel sounding configuration, allowing data collection over distances of hundreds of kilometers. These findings suggest that UHF communications can be effectively used in UAV systems for military and civilian applications, especially in challenging environments. This research highlights the importance of empirical measurements in developing accurate models for long-range AG communications, aiding in the design and optimization of AG communication systems for reliable UAV connectivity. Figure 11 shows the path loss in long-range A/G communication system.

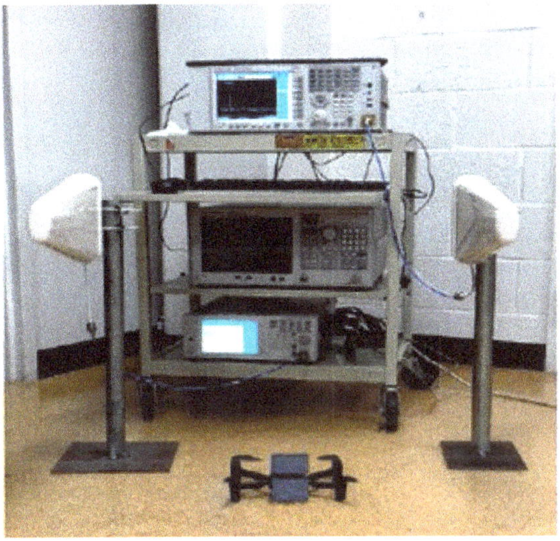

Figure 10. Utilization of specific frequency band by UAV in study by Mahama et al. [28].

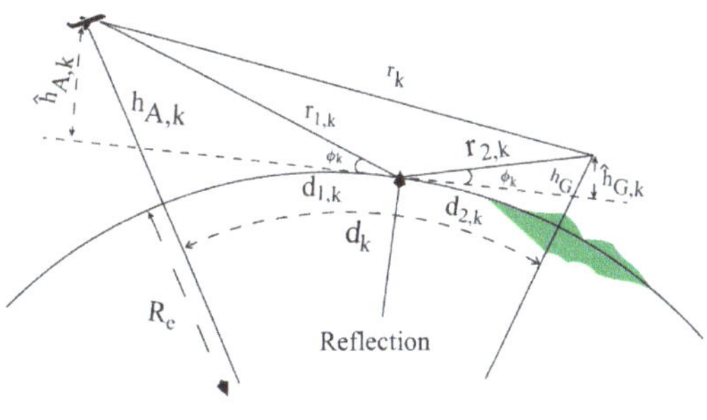

Figure 11. Path loss in long-range A/G communication system in study by Kim, Jaesin, and Inkyu Lee [29].

Xiaoyi [30] describes an ILS, crucial for civil aviation, helping pilots land safely in poor visibility and adverse weather. It uses wireless signals and instruments for precise guidance. Traditionally, maintenance relied on time domain analysis, which can be inaccurate, especially for weak signals. The study highlights the advantages of frequency domain analysis, which accurately measures signal strength across frequencies. Using a spectrum analyzer, maintenance personnel can evaluate the spectral distribution of signals, identifying frequency, power, harmonic components, and noise. This method enhances ILS debugging and calibration, particularly for single-antenna performance and transmitter output adjustments. It also stresses the importance of training to avoid electromagnetic interference and adhering to regulations to reduce external disruptions.

Chunqing [31] highlights the distribution of near-field (NF) signals in localizer antenna systems within a radius of 150 m. It contrasts the NF behavior with far-field (FF) characteristics, underscoring the complexity and intricacy of NF measurements. The study constructs a three-dimensional model, refined through phase analysis and on-site data, to evaluate signal distributions against theoretical predictions. The results indicate that the NF distributions vary significantly with elevation angles and reflection coefficients, with unequal amplitude signals producing higher relative amplitudes at lower angles. In addition, the research examines the effects of Fresnel loss and directional weakening on signal transmission.

According to Elsayem et al. [32] the primary concern with 5G technology is its potential to interfere with radar altimeters due to overlapping frequency ranges (3.7–3.98 GHz for 5G and 4.2–4.4 GHz for radar altimeters). This overlap could pose safety risks during critical flight phases, such as landing and take-off. The study evaluated compatibility by simulating realistic interference scenarios, taking into account various factors, such as the positioning of 5G base stations and the characteristics of radar altimeters. The findings indicated a higher likelihood of interference at lower altitudes, where precise altitude measurements are crucial. To mitigate these risks, the article recommends reducing the transmission power of 5G base stations in specific areas, especially near airports. It also suggests additional strategies such as antenna down-tilting and creating exclusion zones.

Leslie [33] explores the conflict between US regulatory bodies regarding the deployment of 5G technology and its potential effects on Aircraft safety, specifically focusing on radio altimeters. It begins by outlining the planned expansion into C-band frequencies, which has sparked concerns about possible interference with aviation instruments. The methodology includes an analysis of data from various stakeholders, such as the FAA and FCC, to evaluate the risks posed by 5G transmissions. The discussion highlights the importance of cooperation between regulators and industry players to address safety issues while progressing with 5G implementation. It also points out that other countries have managed to roll out 5G with necessary precautions, in contrast to the regulatory hurdles faced in the US. Ultimately, the article advocates for a more data-driven strategy to resolve future disputes over spectrum allocation.

Ivanov et al. [34] developed a deep learning method for detecting GNSS interference using Aircraft data from GPS and AHRS. The model operates in near real-time, analyzing sensor data to identify RFI. It combines temporal sequence modeling with anomaly detection to detect subtle interference patterns missed by rule-based systems. Achieving 83.5% accuracy, it outperforms traditional techniques. Amairah et al. [35] conducted similar work. However, it heavily depended on large labeled training data, which is often unavailable, limiting scalability and deployment.

Studies such as Cevik at al. [36], Joseph et al. [37], and Kennedy et al. [38] used machine learning algorithms with radio spectrum monitoring to classify and mitigate interference in aviation channels. They applied CNNs and RNNs to distinguish legitimate signals from interference, reducing false positives. A multi-tier framework enhanced spectral efficiency and clarity by adjusting detection thresholds based on conditions. However, its high computational cost limits its suitability for low-latency, resource-constrained aviation systems.

Novella et al. [39] introduced a GNSS-jamming protection method that refines interference mask models to improve the detection of intentional jamming. By using probabilistic modeling and adaptive filtering, the method creates accurate interference masks, enhancing jamming protection around critical navigation zones. Simulated tests showed a major decrease in false jamming alerts while maintaining high sensitivity to real threats. However, optimization of the model's computational complexity is needed for real-time use.

Liu et al. [40] studied UAV interference in urban settings using deep learning detection methods, noting the vulnerability of command links to electromagnetic environments. They relied on predefined models, lacking adaptive real-world testing. Wang et al. [41] investigated cognitive radio for mitigating aviation communication interference dynamically with a real-time adaptive prototype. The system is effective but has a high demand for computational resources, limiting its onboard application. Rhoads [42] introduced AIMDS for FAA flight inspection, integrating real-time spectral monitoring with machine learning to enhance RFI tracking and impact assessment on communication and navigation systems. AIMDS used onboard sensors and ground stations for accurate interference localization and classification. However, it did not fully address the challenge of detecting low-power interference below conventional equipment noise floors, which can go unnoticed until avionics performance degrades.

Table 2 presents a comprehensive analysis of the research papers discussed above, addressing various aspects of RFI in aviation. It outlines the technologies utilized, research focus areas, identified gaps, and the equipment or simulations used in each study. The entries cover a wide range of RFI sources, including 5G's impacts on radar altimeters, GNSS system vulnerabilities, and interference from co-located transmitters and receivers. Key themes include the importance of developing effective mitigation techniques, implementing real-time monitoring systems, and improving EMC standards. Identified gaps, such as a lack of field validation, limited automation for anomaly detection, and inconsistencies in regulatory standards, are emphasized. The testing methods highlighted in the table include spectrum analysis, EMI/EMC simulations, UAV-based monitoring, and advanced filtering techniques.

Table 2. Comparison table of various research papers and gap identification.

Ref.	Key Technology Used	Area of Research	Gaps Identified	Test Equipment/Simulation Used
[14]	5G Interference Analysis	Impact of 5G on aviation altimeters	Insufficient real-world implementation data	Airworthiness Directives, EIRP simulation
[15]	Mathematical Modeling	Compatibility of 5G systems with radar altimeters	Need for in-depth field validation	Monte Carlo Simulations
[16]	Interference Scenario Analysis	Compatibility between 5G and aeronautical systems	Challenges in establishing global standards	Simulation tools for operational scenarios
[17]	Time Difference of Arrival (TDOA)	Locating civil aviation RFI sources	Limited deployment in diverse environments	UAV-based synchronization models
[18]	EMC Standards Evaluation	RFI's impact on integrated circuits	Integrated immunity-emission testing methods	EMC spectrum analyzers
[19]	PCB Design for EMI Mitigation	EMC in electronic systems	Integration of PCB-based EMI mitigation techniques	Simulation tools for EMI shielding analysis
[20]	Out-of-Band Interference Mitigation	Impact of interference on GNSS receivers	Broader scenarios for GNSS testing	GNSS signal simulators
[21]	UAV Platforms for RFI Monitoring	Monitoring RFI in aviation using UAVs	Challenges in UAV airworthiness compliance	Angle of Arrival (AOA) algorithms

Table 2. *Cont.*

Ref.	Key Technology Used	Area of Research	Gaps Identified	Test Equipment/Simulation Used
[22]	Anomalous VHF Wave Monitoring	Sporadic E propagation's impact on VHF signals	Automation in anomaly detection algorithms	Software-defined radios (SDRs)
[23]	GNSSs and Inertial Navigation	Mitigation of RFI in GNSS receivers	Integration of GNSSs with alternative systems	GNSS-DME hybrid models
[24]	Interference Cancelation Techniques	Cognitive radio systems in aeronautical communication	Effectiveness of new algorithms in real-world scenarios	Pulse Blanking, DDNE Simulations
[25]	RF Combinational Filters	Mitigation of FM interference in V/UHF communication	Limited studies on compact, multi-band filter designs	Polytetrafluoroethylene (PTFE) PCB testing
[26]	Electromagnetic Simulations	Co-site interference in Aircraft communication systems	Lack of robust transmitter improvements	S-parameter-based simulations
[27]	Intermodulation Product Analysis	Impact of intermodulation on navigation signals	Better assessment methods for third/fifth-order IMPs	Frequency domain analysers
[28]	RF Immunity Evaluation	Testing UAVs for RF interference during inspections	Limited multi-channel redundancy assessments	RF Noise Source Analysis Tools
[29]	Channel Measurement Tools	Air-to-ground communication using UHF band	Need for extensive testing in varied environments	Channel Sounders, Path Loss Simulations
[30]	Frequency Domain Testing	Instrument Landing System (ILS) maintenance	Broader training for ILS maintenance personnel	Spectrum Analyzers
[31]	Near-Field Signal Modeling	Localizer antenna system signal behavior	Limited understanding of NF vs. FF signal interactions	3D Phase Analysis Tools
[32]	C-Band Interference Mitigation	5G interference with radar altimeters	Implementation challenges for exclusion zones	Simulations of Base Station Placement
[33]	Regulatory Impact Analysis	5G rollout and aviation safety concerns	Data-driven strategies for spectrum allocation	Policy Analysis Tools
[43]	Structured Shielding and Filtering	Multi-layer EMI defense for UAV gimbal systems	Needs adaptation for larger-scale airborne platforms	Real-world prototype, diagnostic filters

4. Discussion and Future Work

The growing reliance on advanced communication technologies, such as 5G, UAVs, and GNSSs, has introduced significant challenges for aviation safety and system performance. RFI poses risks that require a systematic and collaborative approach to mitigate. One of the key challenges is the lack of real-world data on 5G interference mitigation near airports. While theoretical models exist, comprehensive field trials and real-time monitoring are essential to understand the actual impact of 5G signals on radar altimeters and aviation navigation systems. Establishing exclusion zones and validating their effectiveness through data-driven methodologies can help minimize risks.

4.1. Current Research Challenges

Another significant gap is the limited validation of theoretical models assessing interference-to-noise ratios. Many existing studies rely on simulations rather than real-world implementation, making it necessary to conduct field experiments in diverse geographical and environmental conditions to ensure accurate validation. Additionally, disparate regulatory standards across regions create inconsistencies in managing spectrum usage and interference control. A unified framework through international collaborations and shared data can help in establishing standardized mitigation practices, ensuring global aviation safety.

The effectiveness of UAV-based RFI monitoring systems is also an area that needs further research. These systems have been proposed for interference detection, but their testing in varied terrains, such as urban, rural, and high-altitude environments, remains inadequate. Developing modular UAV platforms equipped with adaptive RF monitoring algorithms can optimize interference detection. Additionally, fragmented immunity and emission testing in integrated circuits (ICs) is a concern, as existing methods fail to evaluate immunity and emissions simultaneously. AI-driven predictive models and dual-channel analyzers could enhance these testing processes to ensure compliance with EMC standards.

Another major challenge is the insufficient automation of VHF anomaly detection. It is also important to note that while many mitigation strategies emphasize front-door coupling—such as filtering and antenna shielding—back-door coupling presents more persistent challenges. These pathways often include unintentional entry points through power lines, cable harnesses, or structural seams. Effective protection requires comprehensive grounding, multi-layer shielding, and EMI-aware layout design, which are more complex to implement and validate. The literature still lacks field-tested frameworks specifically focused on this harder-to-control mechanism.

While Mahama et al. [28] propose a multi-layer shielding model for UAVs, their simulations assume ideal ground reflection models and ignore Aircraft cabin resonance. In contrast, Li et al. demonstrate the real-world validation of interference thresholds but only for static scenarios. Our analysis suggests a hybrid validation approach that combines field testing with simulation to achieve scalable, reliable EMI certification. Future standards should mandate real-time in-flight data logging for evaluating signal integrity under operational loads.

Traditional methods rely on manual monitoring, which can be inefficient and error-prone. Deploying machine learning models trained on historical data can help classify and predict anomalies in VHF aviation communications. Similarly, GNSS interference mitigation remains a challenge, as GNSS signals are highly vulnerable to jamming, spoofing, and out-of-band interference. Integrating inertial navigation systems (INS) and advanced signal processing algorithms can enhance resilience against such disruptions.

Managing co-site interference in Aircraft communication systems is another ongoing issue. Existing transmitter and antenna designs are not optimized to minimize interference. Implementing electromagnetic simulations to refine antenna placement and spectrum usage, along with employing adaptive filtering systems, can help mitigate overlapping-frequency problems. Likewise, understanding near-field interactions in localizer antenna systems requires further research, as current studies do not sufficiently differentiate between near-field and far-field behaviors. Conducting real-world experiments alongside simulations would improve knowledge in this domain.

Another gap exists in multi-band RF filtering, as current designs lack versatility. Advanced materials such as Polytetrafluoroethylene (PTFE) could improve filter efficiency. Testing these filters in both laboratory and operational settings will ensure their effectiveness. Furthermore, creating 5G exclusion zones near airports presents logistical challenges.

Using predictive tools to define interference zones dynamically and collaborating with telecom providers to optimize base station configurations could provide a viable solution.

The limited testing of GNSSs under various conditions also poses risks to aviation reliability. Conducting controlled experiments in high-interference environments, such as those caused by multipath effects, would allow for better GNSS validation. Similarly, UAV-based RFI monitoring systems lack certification for airworthiness in aviation. Developing lightweight UAVs with certified airframes for regulatory compliance is essential before they can be fully deployed in aviation applications.

Further research is required in understanding UHF path loss in long-range air-to-ground communication systems, as current models lack sufficient empirical data. Field experiments that assess signal behavior over various environmental conditions can improve aviation communication reliability. Moreover, cognitive radio systems, which dynamically adjust frequency usage, require operational testing to ensure their effectiveness in managing real-time interference.

Another challenge is insufficient training for ILS maintenance personnel in handling advanced signal analysis techniques. Implementing standardized training programs focused on spectrum analyzer usage and interference mitigation strategies could improve maintenance efficiency. Additionally, higher-order intermodulation effects on navigation systems need more in-depth analysis, as interference from intermodulation products could disrupt critical signals like ILS and VOR. Employing frequency domain analyzers for in-depth assessment would help in mitigating such risks.

Testing UAV RF immunity in high-interference environments is another underdeveloped area, as UAVs must be designed to withstand radio frequency noise while performing sensitive aviation-related tasks. Introducing multi-channel redundancy can improve UAV resilience in these environments. Additionally, real-time 5G interference mitigation during flight requires further exploration. Onboard interference filtering mechanisms could help dynamically adjust altimeter readings in response to signal disruptions.

The spectrum allocation strategies in aviation often lack data-driven insights. Establishing a centralized repository for RFI incidents and applying predictive analytics to spectrum allocation policies can ensure more efficient frequency management. By addressing these gaps with advanced research, improved regulations, and technological innovations, aviation systems can become more resilient to RFI threats, ensuring safe and reliable global air travel.

4.2. Future Work

- The field validation of 5G interference exclusion zones using real-time altimeter feedback.
- The certification and deployment of airworthy UAV-based RFI detection platforms.
- Standardized VHF anomaly detection using AI-based classifiers.
- The development of adaptive back-door coupling suppression systems.
- Broader testing of GNSSs in multipath-heavy conditions.
- Lightweight RF immunity designs for UAV and urban inspection tasks.

5. Conclusions

The rising prevalence of RFI in aviation underscores the pressing need for comprehensive strategies to address its impact on safety and operational efficiency. The integration of modern technologies such as 5G, GNSS, and UAV systems has introduced new challenges that demand innovative solutions, rigorous testing, and international collaboration. This study has identified some critical gaps in cited research papers and outlined practical approaches to bridge them.

The issue of 5G interference with radar altimeters highlights the importance of real-world testing and extensive field trials. By leveraging simulation models alongside empirical data, the aviation sector can develop more effective mitigation techniques. Similarly, GNSSs and UAV systems require robust evaluation under diverse environmental and operational conditions to improve their reliability and resilience to interference.

The lack of unified international regulations creates inconsistencies in addressing RFI across different regions. Establishing collaborative platforms for data sharing and policy development can harmonize global standards, and foster consistency and interoperability. Furthermore, integrating artificial intelligence and machine learning into areas such as anomaly detection, spectrum management, and testing protocols offers significant potential to improve efficiency and accuracy.

Technological advancements in materials and simulation tools can help optimize co-site interference management, intermodulation analysis, and multi-band filtering. Compact, high-performance designs and innovative testing methods can enhance system compatibility without compromising efficiency. Additionally, targeted training programs for maintenance personnel and the development of real-time in-flight mitigation tools can strengthen operational capabilities.

Collaboration among industry stakeholders—including regulators, manufacturers, and researchers—is essential for implementing these strategies. By pooling resources and expertise, the aviation industry can establish a unified approach to managing RFI, ensuring technological progress aligns with the need for safety and reliability. Addressing the challenges of RFI requires a proactive, multidisciplinary strategy that incorporates advanced research, innovative technologies, and collaborative policymaking. By following the recommended approaches, the aviation industry can effectively mitigate RFI risks, enhance system robustness, and ensure a safer and more efficient future for global air travel.

Author Contributions: Conceptualization, A.M. and M.R.; methodology, A.M.; software, A.M.; validation, A.M. and M.R.; formal analysis, A.M.; investigation, A.M.; resources, A.M.; data curation, A.M.; writing—original draft preparation, A.M.; writing—review and editing, A.M. and M.R.; visualization, A.M.; supervision, M.R.; project administration, M.R.; funding acquisition, M.R. All authors have read and agreed to the published version of the manuscript.

Funding: This research received no external funding.

Acknowledgments: To the authors thank Airnav Ireland and the Department of Electronic and Computer Engineering at the University of Limerick, Ireland, for supporting this work.

Conflicts of Interest: Author Adnan Malik was employed by the company AirNav Ireland, Ballycasey Cross, Co. Clare. The remaining authors declare that the research was conducted in the absence of any commercial or financial relationships that could be construed as a potential conflict of interest.

References

1. Jie, H.; Zhao, Z.; Zeng, Y.; Chang, Y.; Fan, F.; Wang, C.; See, K.Y. A review of intentional electromagnetic interference in power electronics: Conducted and radiated susceptibility. *IET Power Electron.* **2024**, *17*, 1487–1506. [CrossRef]
2. *RTCA DO-160G*; Environmental Conditions and Test Procedures for Airborne Equipment. RTCA, Inc.: Washington, DC, USA, 2010.
3. *MIL-STD-461G*; Requirements for the Control of Electromagnetic Interference Characteristics of Subsystems and Equipment. U.S. Department of Defense: Washington, DC, USA, 2015.
4. Mänttäri, J.; Vihavainen, T. Efficient Deployment of Satellite Navigation Systems in Finland: Action Plan 2021–2025. 2021. Available online: https://julkaisut.valtioneuvosto.fi/handle/10024/163699 (accessed on 11 March 2025).
5. Zhang, M.; Li, X.; Wang, R. Incipient fault diagnosis of batch process based on deep time series feature extraction. *Arab. J. Sci. Eng.* **2021**, *46*, 10125–10136. [CrossRef]
6. Śliwa, R.E.; Dymora, P.; Mazurek, M.; Kowal, B.; Jurek, M.; Kordos, D.; Rogalski, T.; Flaszynski, P.; Doerffer, P.; Doerffer, K. The latest advances in wireless communication in aviation, wind turbines and bridges. *Inventions* **2022**, *7*, 18. [CrossRef]

7. Griffiths, H.; Cohen, L.; Watts, S.; Mokole, E.; Baker, C.; Wicks, M.; Blunt, S. Radar spectrum engineering and management: Technical and regulatory issues. *Proc. IEEE* **2014**, *103*, 85–102. [CrossRef]
8. Zatuchny, D.A.; Negreskul, G.G.; Sauta, O.I.; Shatrakov, A.Y.; Shatrakov, Y.G. *Aerospace Radionavigation Systems: Electromagnetic Compatibility;* Springer Aerospace Technology; Springer Nature: Singapore, 2022; ISBN 978-981-19-6340-7.
9. Bilen, T.; Ahmadi, H.; Canberk, B.; Duong, T.Q. Aeronautical networks for in-flight connectivity: A tutorial of the state-of-the-art and survey of research challenges. *IEEE Access* **2022**, *10*, 20053–20079. [CrossRef]
10. Nasser, H.; Berz, G.; Gómez, M.; de la Fuente, A.; Fidalgo, J.; Li, W.; Pattinson, M.; Truffer, P.; Troller, M. GNSS interference detection and geolocalization for aviation applications. In Proceedings of the 35th International Technical Meeting of the Satellite Division of The Institute of Navigation (ION GNSS+ 2022), Denver, CO, USA, 19–23 September 2022; pp. 192–216. Available online: https://www.ion.org/publications/abstract.cfm?articleID=18358 (accessed on 11 March 2025).
11. Belous, A. Methods and Means of Ensuring Reliability of Radar and Communication Systems. In *Handbook of Microwave and Radar Engineering;* Springer International Publishing: Cham, Switzerland, 2021; pp. 661–775, ISBN 978-3-030-58698-0.
12. Beziuk, G.; Krajewski, A.; Baum, T.C.; Nicholson, K.J.; Ghorbani, K. Electromagnetic and electronic aerospace conformal load-bearing smart skins: A review. *IEEE J. Microw.* **2023**, *4*, 13–42. [CrossRef]
13. Möhring, B.N.; Siart, U.; Eibert, T.F. Fast chirp frequency-modulated continuous-wave reflectometer for monitoring fast varying discontinuities on transmission lines. *IEEE Trans. Instrum. Meas.* **2021**, *70*, 1–11. [CrossRef]
14. Wang, J.; Liu, Y.; Song, H. Counter-unmanned aircraft system (s)(C-UAS): State of the art, challenges, and future trends. *IEEE Aerosp. Electron. Syst. Mag.* **2021**, *36*, 4–29. [CrossRef]
15. Bukhari, J.; Mérida, W. 5G interference with aviation altimeters: Technology and policy recommendations for coexistence. *Transp. Policy* **2024**, *153*, 1–11. [CrossRef]
16. Li, J.; Liu, S.; Lu, H.; Hwang, S.-H. Coexistence of 5G communication systems with radar altimeters. *IEEE Access* **2024**, *12*, 32554–32568. Available online: https://ieeexplore.ieee.org/abstract/document/10445175/ (accessed on 11 March 2025). [CrossRef]
17. Son, H.; Chong, Y. Interference analysis for compatibility between 5G system and aeronautical radio altimeter. In Proceedings of the 2020 International Conference on Information and Communication Technology Convergence (ICTC), Jeju, Republic of Korea, 21–23 October 2020; IEEE: Piscataway, NJ, USA, 2020; pp. 1553–1556. Available online: https://ieeexplore.ieee.org/abstract/document/9289432/ (accessed on 11 March 2025).
18. Zhou, C.; Zhu, X.; Xiong, R.; Hu, K.; Ouyang, F.; Huang, C.; Huang, T. Research on a Method of Locating Civil Aviation Radio Interference Sources Based on Time Difference of Arrival and Frequency Difference of Arrival for Four Unmanned Aerial Vehicles. *Sensors* **2023**, *23*, 7939. [CrossRef] [PubMed]
19. Kircher, D.; Deutschmann, B.; Czepl, N. Influence of radio frequency interference on the electromagnetic emission of integrated circuits. In Proceedings of the 2022 International Symposium on Electromagnetic Compatibility–EMC Europe, Gothenburg, Sweden, 5–8 September 2022; IEEE: Piscataway, NJ, USA, 2022; pp. 257–261. Available online: https://ieeexplore.ieee.org/abstract/document/9901202/ (accessed on 11 March 2025).
20. Hossain, K.A. Study on electromagnetic interference (EMI) and electromagnetic compatibility (EMC): Sources and design concept for mitigation of EMI/EMC. *J. Lib. Arts Humanit. (JLAH)* **2023**, *4*, 68–96.
21. Hegarty, C.J.; Bobyn, D.; Grabowski, J.; Van Dierendonck, A.J. An overview of the effects of out-of-band interference on GNSS receivers. *Navig. J. Inst. Navig.* **2020**, *67*, 143–161. [CrossRef]
22. Zhou, C.; Xiong, R.; Zeng, H.; Xiao, J.; Wang, Y.; Jia, P.; Ye, J.; Zhao, T.; Hu, K. Aerial Locating Method Design for Civil Aviation RFI: UAV Monitoring Platform and Ground Terminal System. *J. Intell. Robot. Syst.* **2021**, *103*, 29. [CrossRef]
23. Hosokawa, K.; Sakai, J.; Tomizawa, I.; Saito, S.; Tsugawa, T.; Nishioka, M.; Ishii, M. A monitoring network for anomalous propagation of aeronautical VHF radio waves due to sporadic E in Japan. *Earth Planets Space* **2020**, *72*, 88. [CrossRef]
24. Joseph, A.; Bartolone, P.; Griggs, J.; Schnaufer, B.; Phan, H.; Malhotra, V. GNSS radio frequency interference mitigation in Collins commercial airborne receivers. *Eng. Proc.* **2023**, *54*, 18.
25. Reddy, R.J. RF Front End Combinational Filter for Airborne V/UHF Communication System. In Proceedings of the 2024 IEEE Wireless Antenna and Microwave Symposium (WAMS), Visakhapatnam, India, 29 February–3 March 2024; IEEE: Piscataway, NJ, USA, 2024; pp. 1–4. Available online: https://ieeexplore.ieee.org/abstract/document/10528138/ (accessed on 11 March 2025).
26. Vogel, M. Analysis and Mitigation of Co-site Interference Problems on Aircraft. In Proceedings of the 2024 IEEE Aerospace Conference, Big Sky, MT, USA, 2–9 March 2024; IEEE: Piscataway, NJ, USA, 2024; pp. 1–8. Available online: https://ieeexplore.ieee.org/abstract/document/10521430/ (accessed on 11 March 2025).
27. Iliev, T.B.; Stoyanov, I.S.; Mihaylov, G.Y.; Ivanova, E.P. Study the influence of intermodulation products on navigation signals. In Proceedings of the IOP Conference Series: Materials Science and Engineering, Borovets, Bulgaria, 26–29 November 2020; IOP Publishing: Bristol, UK, 2021; Volume 1032, p. 012013. Available online: https://iopscience.iop.org/article/10.1088/1757-899X/1032/1/012013/meta (accessed on 11 March 2025).

28. Mahama, E.; Walpita, T.; Karimoddini, A.; Eroglu, A.; Goudarzi, N.; Cavalline, T.; Khan, M. Testing and evaluation of radio frequency immunity of unmanned aerial vehicles for bridge inspection. In Proceedings of the 2021 IEEE aerospace conference (50100), Big Sky, MT, USA, 6–13 March 2021; IEEE: Piscataway, NJ, USA, 2021; pp. 1–8. Available online: https://ieeexplore.ieee.org/abstract/document/9438457/ (accessed on 11 March 2025).
29. Kim, J.; Lee, I. Channel measurements and characterizations for long range air-to-ground communication systems in the UHF band. *IEEE Access* **2022**, *10*, 101880–101888. [CrossRef]
30. Ge, X. Frequency domain testing and of Instrument landing System signals. In Proceedings of the 2024 IEEE 6th Advanced Information Management, Communicates, Electronic and Automation Control Conference (IMCEC), Chongqing, China, 24–26 May 2024; IEEE: Piscataway, NJ, USA, 2024; Volume 6, pp. 1724–1727. Available online: https://ieeexplore.ieee.org/abstract/document/10575130/ (accessed on 11 March 2025).
31. Qu, C. Research on Near Field Signal Behaviour of Localizer Antenna System. In Proceedings of the Journal of Physics: Conference Series, Online, 26–27 November 2022; IOP Publishing: Bristol, UK, 2023; Volume 2449, p. 012046. Available online: https://iopscience.iop.org/article/10.1088/1742-6596/2449/1/012046/meta (accessed on 11 March 2025).
32. Elsayem, A.; Elghamrawy, H.; Massoud, A.; Noureldin, A. Airborne safety in the age of 5G: Assessing the potential interference between C-band and aeronautical radar altimeter. *Int. Arch. Photogramm. Remote Sens. Spat. Inf. Sci.* **2023**, *48*, 889–894. [CrossRef]
33. Leslie, M. US regulators clash over 5G rollout and aircraft safety. *Engineering* **2022**, *15*, 6–8. [CrossRef]
34. Ivanov, V.; Scaramuzza, M.; Wilson, R.C. Deep temporal semi-supervised one-class classification for GNSS radio frequency interference detection. *J. Navig.* **2024**, *77*, 59–81. [CrossRef]
35. Amaireh, A.; Zhang, Y. Machine Learning-Based Identification and Mitigation of 5G Interference for Radar Altimeters. *IEEE Access* **2024**, *12*, 102425–102439. Available online: https://ieeexplore.ieee.org/abstract/document/10606464/ (accessed on 11 March 2025). [CrossRef]
36. Çevik, N.; Akleylek, S. Sok of machine learning and deep learning based anomaly detection methods for automatic dependent surveillance-broadcast. *IEEE Access* **2024**, *12*, 35643–35662. Available online: https://ieeexplore.ieee.org/abstract/document/10443932/ (accessed on 11 March 2025). [CrossRef]
37. Shukla, S.; Mishra, S.; Anandan, V.; Mishra, D. Deep Learning-Based Radio Frequency Interference Classification and Mitigation in Radar Images: A CNN and DnCNN Approach. In Proceedings of the 2024 IEEE Space, Aerospace and Defence Conference (SPACE), Bangalore, India, 22–23 July 2024; IEEE: Piscataway, NJ, USA, 2024; pp. 419–423. Available online: https://ieeexplore.ieee.org/abstract/document/10667825/ (accessed on 11 March 2025).
38. Okokpujie, K.; Nwokolo, I.O.; Adenugba, A.V.; Awomoyi, M.E. Development of a Machine Learning Based Fault Detection Model for Received Signal Level in Telecommunication Enterprise Infrastructure. *Int. J. Saf. Secur. Eng.* **2024**, *14*, 679–690. Available online: https://www.researchgate.net/profile/Kennedy-Okokpujie/publication/381694354_Development_of_a_Machine_Learning_Based_Fault_Detection_Model_for_Received_Signal_Level_in_Telecommunication_Enterprise_Infrastructure/links/667d11990a25e27fbc0e7154/Development-of-a-Machine-Learning-Based-Fault-Detection-Model-for-Received-Signal-Level-in-Telecommunication-Enterprise-Infrastructure.pdf (accessed on 11 March 2025). [CrossRef]
39. Novella, G.; Macabiau, C.; Garcia-Pena, A.; Martineau, A.; Ladoux, P.; Estival, P.; Troubet-Lacoste, O.; Fleury, C.; Ronfle-Nadaud, C. From ICAO GNSS Interference Mask to Jamming Protection Area For Safe Civil Aviation Operation. In Proceedings of the 34th International Technical Meeting of the Satellite Division of The Institute of Navigation (ION GNSS+ 2021), St. Louis, MO, USA, 20–24 September 2021; pp. 834–854. Available online: https://www.ion.org/publications/abstract.cfm?articleID=17936 (accessed on 11 March 2025).
40. Li, J.; Alvestegui, D.G.-G.; Rodriguez-Morales, F.; Leuschen, C. Testing and Assessment of Small Unmanned Aerial Vehicle Susceptibility to Environmental Radio Frequency Interference. *IEEE Trans. Electromagn. Compat.* **2024**, *66*, 728–748. Available online: https://ieeexplore.ieee.org/abstract/document/10507177/ (accessed on 11 March 2025). [CrossRef]
41. Zhou, M.; Kong, M.; Ye, Y.; Deng, B.; Tang, Y. Identifying sources of interference in civil aviation radio communication. *EURASIP J. Adv. Signal Process.* **2024**, *2024*, 88. [CrossRef]
42. Rhoads, D.M. United States Aeronautical Radio Spectrum Interference Investigations, Sustaining Air Travel Safety, Present and Future. Available online: https://www.icasc.co/wp-content/uploads/2023/02/United_States_Aeronautical_Radio_Spectrum_Interference_Investigations_Substaining_Air_Travel_Safety_Present_and_Future.pdf (accessed on 11 March 2025).
43. Jie, H.; Zhao, Z.; Li, H.; Gan, T.H.; See, K.Y. A systematic three-stage safety enhancement approach for motor drive and gimbal systems in unmanned aerial vehicles. *IEEE Trans. Power Electron.* **2025**, *40*, 9329–9342. Available online: https://ieeexplore.ieee.org/abstract/document/10925478/ (accessed on 11 March 2025). [CrossRef]

electronics

MDPI

Article

Generalized Maximum Delay Estimation for Enhanced Channel Estimation in IEEE 802.11p/OFDM Systems

Kyunbyoung Ko and Sungmook Lim *

Department of Electronics Engineering, Korea National University of Transportation, 50 Daehak-ro, Chungju-si 27469, Republic of Korea; kbko@ut.ac.kr
* Correspondence: smlim@ut.ac.kr

Abstract: This paper proposes a generalized maximum access delay time (MADT) estimation method for orthogonal frequency division multiplexing (OFDM) systems operating over multipath fading channels. The proposed approach derives a novel log-likelihood ratio (LLR) formulation by exploiting the correlation characteristics introduced by the cyclic prefix (CP) in received OFDM symbols, thereby enabling the efficient approximation of the maximum likelihood (ML) MADT estimation. A key contribution of this study is represented by the unification and generalization of existing MADT estimation methods by explicitly formulating the bias term associated with the geometric mean. Within this framework, a previously reported scheme is shown to be a special case of the proposed method. The effectiveness of the proposed MADT estimator is evaluated in terms of correct and good detection probabilities, illustrating not only improved detection accuracy but also robustness across varying channel conditions, in comparison with existing methods. Furthermore, the estimator is applied to both noise-canceling channel estimation (NCCE) and time-domain least squares (TDLS) methods, and its practical effectiveness is verified in IEEE 802.11p/OFDM system scenarios relevant to vehicle-to-everything (V2X) communications. Simulation results confirm that when integrated with NCCE and TDLS, the proposed estimator closely approaches the performance bound of ideal MADT estimation.

Keywords: MADT; OFDM; LLR; CP; ML; NCCE; TDLS

1. Introduction

Cooperative intelligent transportation systems (C-ITSs) have emerged as pivotal components in the development of smart transportation infrastructures, enabling real-time communication between vehicles and roadside infrastructure to enhance road safety and traffic efficiency [1]. As part of this evolution, vehicle-to-everything (V2X) communication has gained significant attention, necessitating reliable wireless communication protocols capable of operating in highly dynamic vehicular environments. To address these requirements, the IEEE 802.11p standard was introduced, defining the medium access control (MAC) and physical layer specifications for wireless access in vehicular environments (WAVE) [2]. IEEE 802.11p is derived from the IEEE 802.11a standard, with a key adaptation that reduces the channel bandwidth from 20 MHz to 10 MHz in order to improve robustness against multipath fading and Doppler effects prevalent in high-speed vehicular scenarios [2]. However, this reduction in bandwidth results in a corresponding decrease in the number of pilot subcarriers (i.e., only four pilot subcarriers) available per orthogonal frequency division multiplexing (OFDM) symbol. This limited number of pilots poses a significant challenge in accurately estimating channel variations in the frequency domain, especially under rapidly changing wireless conditions. As a result, ensuring the reliable and timely dissemination of traffic information requires enhanced channel estimation methods tailored to the constraints of IEEE 802.11p.

Various channel estimation (CE) techniques have been investigated for IEEE 802.11p/OFDM systems [3–12]. To enhance the performance of the construct data pilot (CDP)-based scheme proposed in [1], the authors in [9] and refs. [4,5] have introduced the time-domain reliability test and frequency-domain interpolation (TRFI) scheme, as well as modified TRFI schemes utilizing virtual subcarriers for channel estimation. While these methods demonstrate performance enhancements, they are accompanied by relatively high computational complexity and inherent performance limitations. In particular, the studies in [4,5] assume that the channel impulse response (CIR) length is either perfectly known or set to $N_g - 1$, where N_g denotes the length of the cyclic prefix (CP), when estimating the channel frequency response (CFR) of virtual subcarriers.

In OFDM systems, low-complexity CE techniques are typically categorized into frequency-domain and time-domain approaches. In time-domain methods, the least square (LS) channel estimates, obtained by using pilot symbols or training sequences, are first transformed into the time domain via the inverse discrete Fourier transform (IDFT). Linear estimation is then applied in the time domain to reduce noise or perform interpolation, and the result is transformed back into the frequency domain to yield the CFR. Accordingly, the method proposed in [5] can be classified as a time-domain approach. Moreover, most practical OFDM systems employ null subcarriers at the spectrum edges to form guard bands, which can cause channel energy leakage even in sample-spaced channels. Several methods have been proposed to mitigate this effect [10–12]. Unlike these leakage mitigation approaches, the methods in [4,5] estimate the CFRs of virtual subcarriers directly.

Recently, the authors in [13] introduced a noise-canceling channel estimation (NCCE) technique that leverages the estimated channel length obtained from the algorithms presented in [6,7]. The NCCE scheme is based on a structure that performs inverse fast Fourier transform (IFFT), nulling, and fast Fourier transform (FFT) without relying on any interpolation techniques. However, its performance is highly sensitive to the accuracy of the channel length estimation, particularly in highly frequency-selective fading environments with long delay spreads.

The main contributions of this paper are summarized as follows:

1. A novel log-likelihood ratio (LLR) expression is derived by exploiting the inherent correlation characteristics introduced by the cyclic prefix (CP) in the received OFDM symbols. This formulation enables the robust detection of the channel delay profile.
2. An approximated maximum likelihood (ML) estimator for the maximum access delay is proposed. The proposed estimator adopts a generalized form incorporating a weighting function applied to the observation random variables, thus enhancing detection accuracy under various channel conditions.
3. The unification and generalization of existing methods are achieved by explicitly formulating the bias term associated with the geometric mean. It is analytically demonstrated that the estimator proposed in [6] represents a special case within the proposed framework, thereby extending and generalizing the prior approach.
4. Performance advantages are demonstrated through extensive simulations, where the proposed scheme is applied to both the NCCE and the time-domain least square (TDLS) channel estimation methods. Simulation results under IEEE 802.11p/OFDM system conditions confirm the superiority of the proposed estimator in terms of correct and good detection probabilities. Notably, with respect to error rates, the proposed method closely approaches the performance bound of the ideal MADT estimation across all three considered channel environments.

Through these contributions, the proposed method not only improves the robustness of channel delay estimation but also enhances the overall performance of CE techniques in practical vehicular communication scenarios.

The remainder of this paper is organized as follows: Section 2 describes OFDM systems over multipath fading channels. The proposed channel delay estimation scheme is described in Section 3. Section 4 shows the simulation results, and concluding remarks are given in Section 5.

2. OFDM System and Discrete Signal Model

In OFDM systems, the source data are grouped and mapped onto modulated symbols $X_m(k)$, where $k \in \{0, 1, \cdots, N-1\}$, N is the IDFT size, and the symbols satisfy $E\{X_m(k)\} = 0$ and $E\left\{|X_m(k)|^2\right\} = 1$, with $E\{\cdot\}$ denoting the expectation operator. The time-domain transmitted signal corresponding to the nth sample of the mth OFDM symbol is then obtained by applying the IDFT over N subcarriers and is given by

$$x_m(n) = \frac{1}{\sqrt{N}} \sum_{k=0}^{N-1} X_m(k) e^{j2\pi kn/N}, \quad n = 0, 1, 2, \cdots, N-1 \tag{1}$$

where $E\{x_m(n)\} = 0$ and $E\left\{|x_m(n)|^2\right\} = 1$ [14–16].

To mitigate inter-symbol interference (ISI), a guard interval is appended to each OFDM symbol, comprising a CP that duplicates the last N_g samples of the IDFT output. The length of the guard interval, N_g, is chosen to exceed the maximum delay spread of the channel. The transmitted signal propagates through a multipath fading channel, whose low-pass equivalent impulse response is given by

$$h(t;\tau) = \sum_{l=0}^{L-1} h_l(t)\delta(\tau - \tau_l) \tag{2}$$

where t, τ, $\delta(\cdot)$, L, and τ_l are the time, the delay, a Dirac delta function, the number of multipaths, and the propagation delay of the lth path, respectively [14]. The statistical characteristics of the multipath fading channel can be described by the wide-sense stationary uncorrelated scattering (WSSUS) model [15]. This model assumes that individual propagation paths are mutually uncorrelated and that the channel exhibits wide-sense stationarity in its correlation properties.

After removing the cyclic prefix, the discrete-time received signal can be expressed, from (1) and (2), as

$$y_m(n) = \sum_{l=0}^{L-1} h_{l,m}(n)\sqrt{E_s}x_m((n - d_l)_N) + w_m(n) \tag{3}$$

where E_s is the transmitting signal power, $(\cdot)_N$ represents a cyclic shift in the base of N, $w_m(n) \sim \mathcal{N}(0, \sigma^2)$ is complex additive white Gaussian noise (AWGN), $h_{l,m}(n) = h_l(t)|_{t=[m(N_g+N)+n]T_s}$ is the lth path channel gain of the nth sample for the mth OFDM symbol, and $d_l = \lfloor \tau_l / T_s \rfloor$ is the delay normalized by the sampling time T_s [16]. For simplicity, we round d_l to an integer without considering leakage. However, the correlation approach in this paper may also be extended to fractional d_l [17]. When we assume the perfect synchronization as $d_l = l$ (i.e., $\tau_l = lT_s$) and the time-invariant channel within two consecutive OFDM symbols, the indexes m and (n) in $h_{l,m}(n)$ of (3) can be omitted; then, we define h_l as $h_l = \sqrt{E_s}h_{l,m}(n)$, where $E\{h_l\} = 0$, $E\left\{|h_l|^2\right\} = \sigma_l^2$, and $\sigma_h^2 = \sum_{l=0}^{L-1} \sigma_l^2 = E_s$.

The signal-to-noise ratio (SNR) can be written as $\mathrm{SNR} = \sigma_h^2/\sigma^2 = E_s/\sigma^2$. In addition, let us define the maximum number of paths including the zero channel gain path, from (3), as

$$L_{\max} = \max\left\{ d_l \big|_{l=0}^{L-1} \right\} + 1 \tag{4}$$

where the maximum delay time, normalized by T_s, is $d_{\max} = \max\left\{ d_l \big|_{l=0}^{L-1} \right\} = L_{\max} - 1$.

2.1. Joint PDFs

For $-N_g \le -k \le -1$ (i.e., $N_g \ge k \ge 1$), the received signal samples at the border between two OFDM symbols can be written, from (3), as

$$\begin{aligned} y_m(-k) &= \sum_{l=0}^{L-1} h_l x_m(N-k-d_l) U\big((N_g-k)-d_l\big) \\ &\quad + \sum_{l=0}^{L-1} h_l x_{m-1}(N_s-k-d_l) U\big((d_l-1)-(N_g-k)\big) + w_m(-k) \end{aligned} \tag{5}$$

where $N_s = N + N_g$ and $U(\cdot)$ is the unit step function [17].

From (5), it is shown that the received signal has the ISI terms, related with $x_{m-1}(\cdot)$, when $N_g \ge k \ge (N_g - d_{\max} - 1)$. Therefore, the correlation between each received signal over CP duration and its corresponding sample at the end of the OFDM symbol can be expressed as

$$\begin{aligned} &E\{y_m(-k)y_m^*(N-k)\} \\ &= \begin{cases} \sigma_h^2 & \text{if} \quad N_g - d_{\max} \ge k \ge 1 \\ \sum_{l=0}^{L-1} \sigma_l^2 U(N_g - k - d_l) & \text{else} \quad (N_g \ge k \ge N_g - d_{\max} + 1) \end{cases} \end{aligned} \tag{6}$$

where $k \in \{1, 2, \cdots, N_g\}$ [7,8,17,18]. Note that the expectation in (6) is taken with regard to the random variables of $\{h_l\}$, $\{x_m(n), x_{m-1}(n)\}$, and $\{w(n)\}$, which are assumed to be mutually independent for different indices l, m, and n.

When L is large, the signal $y_m(n)\big|_{n=0}^{N-1-N_g}$ can be approximately modeled as a complex Gaussian random variable based on the central limit theorem [18,19]. The corresponding probability density function (PDF) is given from (6) as

$$f(y_m(n)) = \frac{\exp\left(-\frac{|y_m(n)|^2}{\sigma_h^2+\sigma^2}\right)}{\pi(\sigma_h^2+\sigma^2)}. \tag{7}$$

For $k \in \{1, 2, \cdots, N_g\}$, the samples $y_m(-k)$ and $y_m(N-k)$ are jointly complex Gaussian. The joint PDF is given from (6) by

$$f(y_m(-k), y_m(N-k)) = \frac{\exp\left(-\frac{|y_m(-k)|^2 + |y_m(N-k)|^2 - 2\rho_k \Re\{y_m(-k)y_m^*(N-k)\}}{(\sigma_h^2+\sigma^2)(1-\rho_k^2)}\right)}{\pi^2(\sigma_h^2+\sigma^2)(1-\rho_k^2)} \tag{8}$$

where $\Re\{\cdot\}$ denotes the real part of a complex-valued input and ρ_k represents the correlation coefficient between $y_m(-k)$ and $y_m(N-k)$. It yields

$$\rho_k = \left| \frac{E\{y_m(-k)y_m^*(N-k)\}}{\sqrt{E\left\{|y_m(-k)|^2\right\}E\left\{|y_m(N-k)|^2\right\}}} \right| = \frac{\sum_{l=0}^{L-1} \sigma_l^2 U(N_g - k - d_l)}{\sigma_h^2 + \sigma^2}. \tag{9}$$

Note that $0 < \rho_k < 1$ and $\rho_{k+1} \leq \rho_k$; i.e., ρ_k is a non-increasing function with respect to k. Furthermore, we can find from (6) and (9) that the set $\{\rho_k\}_{k=1}^{N_g-d_{\max}}$ consists of identical values, given by

$$\rho_{N_g-d_{\max}} = \cdots = \rho_2 = \rho_1 = \frac{\sigma_h^2}{\sigma_h^2 + \sigma^2}.$$

2.2. N_g Noise Power Estimators

Assuming perfect synchronization at the receiver and a time-invariant channel over the duration of two consecutive OFDM symbols, N_g noise variance estimators can be formulated from (5) and (6) as

$$\begin{aligned} \hat{\sigma}^2 &\leftarrow J(u),\ 1 \leq u \leq N_g \\ J(u) &= \frac{1}{N_g - u + 1} \sum_{l=N_g}^{u} z(l) \end{aligned} \tag{10}$$

where

$$\begin{aligned} z(l) &= \frac{1}{2M} \sum_{m=1}^{M} |y_m(-k) - y_m(N-k)|^2 \Bigg|_{k=N_g-l+1} \\ &= \frac{1}{2M} \sum_{m=1}^{M} |y_m(-N_g+l-1) - y_m(N-N_g+l-1)|^2 \end{aligned} \tag{11}$$

and M denotes the number of OFDM symbols within the observation window (e.g., a packet or frame). It can be readily shown that the noise variance estimator achieves its minimum variance when $u = L_{\max}(= d_{\max}+1)$ [6–8].

For $L_{\max} \leq l \leq N_g$, from (5), (6), and (9), $z(l)$ of (11) can be represented as

$$\begin{aligned} z(l)|_{l=L_{\max}}^{N_g} &= \frac{1}{2M} \sum_{m=1}^{M} |y_m(-N_g+l-1) - y_m(N-N_g+l-1)|^2 \\ &= \frac{1}{2M} \sum_{m=1}^{M} |w_m(-N_g+l-1) - w_m(N-N_g+l-1)|^2 \end{aligned} \tag{12}$$

where notice that $z(l)$ represents a random variable corresponding to the sum of exponentially distributed random variables. According to the central limit theorem (CLT) [19], for sufficiently large M, the distribution of $z(l)$ can be approximated by a Gaussian distribution as

$$z(l)|_{l=L_{\max}}^{N_g} \sim \mathcal{N}\left(\sigma^2, \frac{\sigma^4}{M}\right) \tag{13}$$

where $E\left\{z(l)|_{l=L_{\max}}^{N_g}\right\} = \sigma^2$ and $Var\left\{z(l)|_{l=L_{\max}}^{N_g}\right\} = \sigma^4/M$.

It is important to note that for $1 \leq l < L_{\max}$, $z(l)$ of (11) must be expressed in a form that includes ISI terms related with (6) and (9), rather than solely in terms of noise components, as in (12).

3. Proposed $L_{\max}$ Estimation Method in a Generalized Framework

From here, we propose the general method for $L_{\max}$ estimation based on the observation variables of $\left\{z(l)_{l=u}^{N_g}\right\}$. When it is assumed that the given u is $L_{\max}$ (i.e., $u \leftarrow L_{\max}$), we can approximate $\sigma^2 \simeq J(u)$; then, $f(z(l)|u \leftarrow L_{\max})$ is computed, from (10) and (13), as

$$z(l)|_{l=u, u\leftarrow L_{\max}}^{N_g} \sim \mathcal{N}\left(J(u), \frac{J(u)^2}{M}\right) \tag{14}$$

and

$$f_z(z(l)|u \leftarrow L_{\max})\Big|_{l=u}^{N_g} \approx \frac{1}{\sqrt{2\pi J(u)^2/M}} \exp\left(-\frac{|z(l)-J(u)|^2}{2J(u)^2/M}\right). \tag{15}$$

The LLR for (15) is written as

$$\begin{aligned}\lambda^z_{l|u} &= \log\left(f_z(z(l)|u \leftarrow L_{\max})\right) \\ &= -\frac{|z(l)-J(u)|^2}{2J(u)^2/M} - \frac{\log(2\pi J(u)^2/M)}{2}.\end{aligned} \tag{16}$$

Figure 1 illustrates examples of the PDF $f_z(z(u)|u \leftarrow L_{\max})$, as defined in (15) with $l = u$, and the corresponding LLR $\lambda^z_{u|u}$ defined in (16) under the same condition, $l = u$. For $L_{\max} \leq u \leq N_g$, the PDF value can exceed 1, as illustrated in Figure 1a, and the corresponding LLR is greater than 0, as depicted in Figure 1b. On the other hand, for $1 \leq u < L_{\max}$, the PDF values are close to zero, and the corresponding LLRs are negative. Therefore, our goal is to identify the point, specifically $u = L_{\max}$, at which the PDF value exhibits a sharp decline as u decreases from N_g to 1. In order to present this concept in a generalized form, the random variable $\epsilon(l)$ is defined as

$$\epsilon(l) = \frac{1}{g(l)} z(l) \tag{17}$$

where $g(l) > 0$ denotes a weighting function that transforms the random variable $z(l)$ into $\epsilon(l)$. Furthermore, $g(l)$ is assumed to be a monotonically decreasing function with respect to the index l, such that the smallest value of u is estimated as $L_{\max}$.

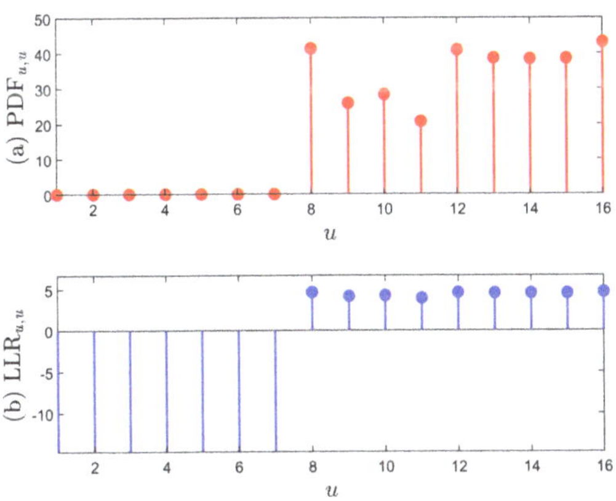

Figure 1. Example for PDF in (15) and LLR in (16) (SNR = 10 dB, $N_g = 16$, $L_{\max} = 8$, Highway NLOS, 252 km/h, QPSK, CR = 1/2, $M = 100$).

3.1. PDF Multiplication and Geometric Mean

Notice that $\epsilon(l)$ can be approximated, for $L_{\max} \leq l \leq N_g$ and large M, as the random variable of $\epsilon(l) \sim \mathcal{N}\left(\frac{\sigma^2}{g(l)}, \frac{\sigma^4}{Mg(l)^2}\right)$ [6,8,19]. Since the different $z(l)$ are independent, the observation variables of $\left\{\epsilon(l)_{l=u}^{N_g}\right\}$ are also mutually independent. Thus, $\hat{L}_{\max}$ can be obtained as the likelihood function of

$$\hat{L}_{\max} = \arg\max_{1\leq u\leq N_g} \left[\prod_{l=N_g}^{u} f_\epsilon(\epsilon(l)|u \leftarrow L_{\max})\right]^{1/(N_g-u+1)} \tag{18}$$

where $f_\epsilon(\epsilon(l)|u \leftarrow L_{\max})$ is computed from (15) and (17) with the approximation that $\sigma^2 \simeq J(u)$. By assuming $L_{\max} = u$ and considering the interval $u \leq l \leq N_g$, we can express

$$\epsilon(l) \sim \mathcal{N}\left(\frac{J(u)}{g(l)}, \frac{J(u)^2}{Mg(l)^2}\right) \tag{19}$$

and

$$f_\epsilon\left(\epsilon(l)\middle|u \leftarrow L_{\max}\right) = g(l) f_z\left(z(l)\middle|u \leftarrow L_{\max}\right). \tag{20}$$

As discussed in [6], the expression in (18) represents an average likelihood function, where the likelihood is computed over a set of observation variables $\left\{\epsilon(l)\big|_{l=u}^{N_g}\right\}$ of (17). These variables are generally modeled as independent random variables, and their corresponding PDFs can vary in number depending on the delay spread or the estimated maximum channel length. Due to this variable-length observation set, the joint likelihood of the observations cannot be expressed by using a fixed-length product of PDFs. Instead, to ensure statistical consistency and comparability across different lengths, the geometric mean of the individual PDFs is used as in (18), which maintains the overall scale of the likelihood irrespective of the number of observations. This formulation preserves the multiplicative structure of independent observations while normalizing the outcome to allow for a meaningful comparison across different hypotheses or models with different numbers of taps. This approach reflects a fundamental principle in statistical signal processing, where the likelihood function must account for both the independence and variable dimensionality of the underlying random observations to remain statistically robust.

3.2. LLR Summation and Average

From (17) and (19), we can rewrite (18) as the LLR form of

$$\begin{aligned}\Lambda_u^\epsilon = \Lambda_u^\epsilon\left(\{\epsilon(l)\}_{l=u}^{N_g}\middle|u \leftarrow L_{\max}\right) &= \log\left[\prod_{l=N_g}^{u} f_\epsilon(\epsilon(l)|u \leftarrow L_{\max})\right]^{1/(N_g-u+1)} \\ &= \frac{1}{N_g-u+1}\sum_{l=N_g}^{u} \log\left(f_\epsilon(\epsilon(l)|u \leftarrow L_{\max})\right)\end{aligned} \tag{21}$$

where $1 \leq u \leq N_g$. From $\lambda_{l|u}^z$ of (16), it yields

$$\Lambda_u^\epsilon = \Lambda_u^\epsilon\left(\{\epsilon(l)\}_{l=u}^{N_g}\middle|u \leftarrow L_{\max}\right) = \frac{\Lambda_u^z}{N_g-u+1} + B_u \tag{22}$$

where

$$\begin{aligned}\Lambda_u^z = \Lambda_u^z\left(\{z(l)\}_{l=u}^{N_g}\middle|u \leftarrow L_{\max}\right) &= \sum_{l=N_g}^{u} \lambda_{l|u}^z \\ &= -\sum_{l=N_g}^{u} \frac{|z(l)-J(u)|^2}{2J(u)^2/M} - (N_g-u+1)\frac{\log(2\pi J(u)^2/M)}{2}.\end{aligned} \tag{23}$$

and

$$B_u = \frac{1}{N_g-u+1}\sum_{l=N_g}^{u} \log\left(g(l)\right). \tag{24}$$

Then, from (18) and (22), the maximum access delay time can be alternatively estimated as

$$\hat{L}_{\max} = \arg \max_{1 \leq u \leq N_g} \left[\Lambda_u^{\epsilon} \left(\{\epsilon(l)\}_{l=u}^{N_g} \middle| u \leftarrow L_{\max} \right) \right]. \tag{25}$$

Fundamentally, the CIR length estimator, expressed by (18) and (25), is based on the N_g noise estimator $J(u)$ of (10). In (18) and (25), the given u (i.e., $u \leftarrow L_{\max}$) indicates that $z(l)|_{l=u}^{N_g}$ of (12) can be each noise estimator. Accordingly, $J(u)$ among $\left\{ J(l)|_{l=u}^{N_g} \right\}$ can be regarded as the estimator with the smallest variance. Therefore, as confirmed in (19), not $J(l)$ but $J(u)$ can be utilized.

3.3. *Special Cases for $g(l)$*

To illustrate the impact of the function $g(l)$ in the proposed scheme, the $L_{\max}$ estimation method presented in [6] is first considered.

3.3.1. Method in [6] with $g(l) = (N_g - l + 1)$

From (10) and (11), the uth noise variance estimator of $J(u)$ can be represented with regard to $\{\epsilon(l)\}$ as

$$J(u) = \frac{1}{N_g - u + 1} \frac{N_g - u}{1} J(u+1) + \epsilon(u) \tag{26}$$

and

$$\epsilon(l) = \frac{1}{N_g - l + 1} z(l). \tag{27}$$

Note that the estimator described in [6] uses $g(l) = (N_g - l + 1)$ obtained from (26). From (22) and (24), the LLR summation is expressed as a function of $z(l)|_{l=u}^{N_g}$ and has the bias term of

$$B_u = \frac{\log[(N_g - u + 1)!]}{(N_g - u + 1)}. \tag{28}$$

As a result, the bias term involving $\log[(N_g - u + 1)!]$ is attributed to the structure of $g(l) = (N_g - l + 1)$, while the corresponding geometric mean produces a scaling factor of $1/(N_g - u + 1)$.

The operation of the proposed method, as formulated in (18) and (25), is described as follows: Figure 1 shows the examples for $f_z(z(l)|u \leftarrow L_{\max})_{l=u}$ of (15) and $\left(\lambda_{l|u}^z\right)_{l=u}$ of (16). Examples of Λ_u^z, $\frac{1}{N_g - u + 1}\Lambda_u^z$, and Λ_u^{ϵ}, as defined in (22), (23), and (22) with (28), are depicted in Figure 2 for cases (a), (b), and (c), respectively.

For $L_{\max} \leq u \leq N_g$, the values of $\left\{ f_z(z(l))|_{l=N_g}^{u} \right\}$ are observed to be similar, as shown in Figure 1a. Accordingly, the values of $\left\{ f_{\epsilon}(\epsilon(l))|_{l=N_g}^{u} \right\}$ in (20) increase as l decreases from N_g to u, due to the scaling effect of $g(l) = (N_g - l + 1)$. As a result, the method proposed in [6], utilizing $\{\epsilon(l)\}$ derived from $\{z(l)\}$, can be interpreted as a specific instance of the proposed approach, wherein the weight function $g(l) = (N_g - l + 1)$ is employed. In (27), the weighting factors $\{1/(N_g - l + 1)\}$ are chosen based on the relationship among $\{J(u)\}$. This mechanism leads to the bias term described in (22), which is further detailed in (28), resulting in an increase in the LLR summation as u decreases from N_g to u, as illustrated in Figure 2c.

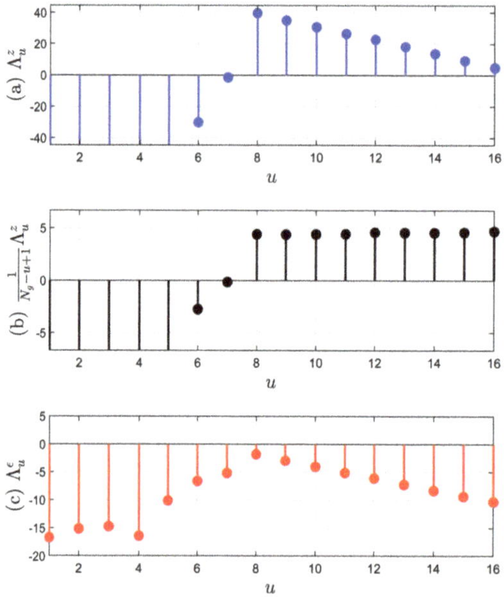

Figure 2. Example for cost functions (SNR = 10 dB, Λ_u^ϵ by $g(l) = (N_g - l + 1)$, $N_g = 16$, $L_{\max} = 8$, Highway NLOS, 252 km/h, QPSK, CR = 1/2, $M = 100$).

Figure 2a shows that Λ_u^z of (23) increases as u decreases from N_g to $L_{\max}$. In this range, each decrement in u results in the multiplication of a PDF value greater than 1, as shown in Figure 1a, which corresponds to the addition of a positive LLR value in (23), as seen in Figure 1b. Conversely, when u is reduced from $d_{\max}(= L_{\max} - 1)$ to 1, a PDF value less than 1, as shown in Figure 1a, is successively multiplied. This is equivalent to adding a negative LLR value in (23), as illustrated in Figure 1b, thereby causing a sharp decline in the LLR summation, as depicted in Figure 2a. For $\frac{1}{N_g-u+1}\Lambda_u^z$, Figure 2b shows a plateau starting at $u = L_{\max}$. The width of this plateau is equal to $N_g - L_{\max} + 1$. Nevertheless, the method in [6], utilizing $\{\epsilon(l)\}$ of (27), does not exhibit a plateau in the region $L_{\max} \leq u \leq N_g$, as shown in Figure 2c. This behavior is attributed to the presence of the bias term described in (28). From Figure 2c, the estimated value $\hat{L}_{\max} = 8$ is obtained by using the criterion in (25).

3.3.2. Proposed $g(l)$

In this paper, we propose the estimators utilizing $g(l)$ as follows:

- (Prop.1) $g(l) = \sqrt{N_g - l + 1} \rightarrow B_u = \frac{\log[(N_g-u+1)!]}{2(N_g-u+1)}$.
- (Prop.2) $g(l) = (N_g - l + 1)^2 \rightarrow B_u = \frac{2\log[(N_g-u+1)!]}{(N_g-u+1)}$.
- (Prop.3) $g(l) = e^{N_g - l} \rightarrow B_u = \frac{N_g-u+1}{2}$.
- (Prop.4) $g(l) = [f_z(z(l)|u \leftarrow L_{\max})]^{N_g-u} \rightarrow B_u = \frac{N_g-u}{N_g-u+1}\Lambda_u^z$.

From the derived results, it can be concluded that the function $g(l)$, in (17) and (20), serves as a weighting factor for $z(l)$, enabling the identification of the endpoint of the plateau observed in Figure 2b. This mechanism can be interpreted such that as the estimation scheme evolves from 'Prop.1' to 'Prop.3' (including the case where $g(l) = (N_g - l + 1)$ in [6]), the proposed estimator employs a progressively increasing weight for smaller values of u. This strategy facilitates the selection of the smallest possible u as the estimate $\hat{L}_{\max}$. A detailed performance comparison among these schemes is provided in Section 4.

For a special case, let us consider 'Prop.4', under which (21) can be reformulated as follows:

$$\Lambda_u^{\epsilon} = \sum_{l=N_g}^{u} \log\left(f_z(z(l)|u \leftarrow L_{\max})\right) = \sum_{l=N_g}^{u} \lambda_{l|u}^{z} = \Lambda_u^{z} \tag{29}$$

Figure 2a presents the result corresponding to 'Prop.4'. Although (29) does not appear to represent an average likelihood in the form of a geometric mean of the individual PDF elements, the proposed approach demonstrates that the weight function $g(u)$ is incorporated prior to averaging the LLR summation.

3.4. Computational Complexity

Table 1 compares the computational complexity of the proposed schemes with those of the existing methods presented in [6,7]. As shown in Table 1, for both the proposed schemes and the method in [6], the dominant term determining the complexity of the multiplication operations is a function of $2MN_g$. In contrast, the dominant term for the method in [7] is a function of $2MN$. Therefore, even when excluding the operations required to compute the real roots of N_g cubic equations, the method in [7] exhibits the highest overall complexity. Meanwhile, the complexity of the proposed schemes is comparable to that of the method in [6].

Table 1. Computational complexity comparison.

Term	[1] Operation	Number
Prop.1 and 2, $\hat{L}_{\max}$ [6]	$+/-$	$(8M+3N_g-1)N_g/2$
	$\times/\div$	$(2M+N_g+6)N_g$
	log	$2N_g$
[2] Prop.3 and 4	$+/-$	$(8M+3N_g-1)N_g/2$
	$\times/\div$	$(4M+N_g+15)N_g/2$
	log	N_g
[3] $\hat{L}_{\max}$ [7]	$+/-$	$(3N_g+N)M+$ $(3N_g+13)N_g/2-3$
	$\times/\div$	$(4M+4N_g+7)N_g+$ $(2M+1)N$
	log	N_g

[1] Not complex but real operation. [2] In Prop.3 and Prop.4 of (29), no numerical operations related to B_u are required. [3] Additionally, it is necessary to determine the real roots of N_g cubic equations [8].

4. Simulation Results

In this paper, a generalized $L_{\max}$ estimation method is proposed and applied to OFDM channel estimation techniques, specifically to TDLS in [3] and NCCE in [13]. The performance of the proposed approach is evaluated through simulation. It is worth noting that NCCE can be implemented by using a sequence of IFFT/nulling/FFT operations, as shown in Table 2, resulting in lower computational complexity compared with TDLS.

In this section, simulation results are presented to evaluate the detection probabilities, error rate, and MSE performance of the proposed scheme. The simulations are conducted based on the IEEE 802.11p standard with system parameters set to $N = 64$, $N_g = 16$, and $T_s = 0.1$ μs [2,8]. Both the transmitter and receiver employ a convolutional encoder and a Viterbi decoder with a constraint length of 7 [2,8]. Each packet consists of $M(= 100)$ OFDM symbols, with quadrature phase shift keying (QPSK) and 16-quadrature amplitude modulation (QAM) modulation schemes used at a coding rate of 1/2. All performance metrics are averaged over 5×10^5 packet transmissions to ensure statistical reliability.

Additional simulation parameters are provided in Table 3 of [8]. Among five scenarios of 'CohdaWireless V2V channel model' in [20], we consider the 'Highway NLOS with 252 km/h', 'Street Crossing NLOS with 126 km/h', and 'Highway LOS with 252 km/h' channel environments, the channel profiles of which are presented in Table 3. The other parameters, such as the Doppler spectrum for each channel tap, are listed in [20].

Table 2. Legend description for simulation results.

Comments	Legend	$\hat{L}_{\max}$	IFFT/Nulling/FFT
Bounds	`Ideal`		
	`TDLS` + `Ideal`$(L_{\max})$	$L_{\max}$	Not applicable
	`NCCE` + `Ideal`$(L_{\max})$	$L_{\max}$	[2] O
[3] + [6]	`TDLS` + $\hat{L}_{\max}$ "Houcke+2008"	By (5) in [6]	Not applicable
[3] + [7]	`TDLS` + $\hat{L}_{\max}$ "Ko+2022"	By (16) in [7]	Not applicable
[3] + Prop.X [1]	`TDLS` + $\hat{L}_{\max}$ Prop.X [1]	By (18) and (25)	Not applicable
[13] + [6]	`NCCE` + $\hat{L}_{\max}$ "Houcke+2008"	By (5) in [6]	[2] O
[13] + [7]	`NCCE` + $\hat{L}_{\max}$ "Ko+2022"	By (16) in [7]	[2] O
[13] + Prop.X [1]	`NCCE` + $\hat{L}_{\max}$ Prop.X [1]	By (18) and (25)	[2] O

[1] $X \in \{1,2,3,4\}$ in Section 3.3.2. [2] NCCE can be implemented by using a sequence of IFFT/nulling/FFT operations [13].

Table 3. Channel profile due to scenarios in [20] (see Table 4 in [8]).

Ch. Type	Item	Tap 0	Tap 1	Tap 2	Tap 3	Unit
Highway	Power	0	−2	−5	−7	dB
NLOS	Delay(τ_l)	0	200	433	700	ns
(252 km/h,	d_l	$d_0 = 0$	$d_2 = 2$	$d_4 = 4, d_5 = 5$	$d_7 = 7$	$\times T_s$
$L_{\max} = 8$)	Doppler	0	689	−492	886	Hz
Street Crossing	Power	0	−3	−5	−10	dB
NLOS	Delay(τ_l)	0	267	400	533	ns
(126 km/h,	d_l	$d_0 = 0$	$d_2 = 2, d_3 = 3$	$d_4 = 4$	$d_5 = 5, d_6 = 6$	$\times T_s$
$L_{\max} = 7$)	Doppler	0	295	−98	591	Hz
Highway	Power	0	−10	−15	−20	dB
LOS	Delay(τ_l)	0	100	167	500	ns
(252 km/h,	d_l	$d_0 = 0$	$d_1 = 1$	$d_1 = 1, d_2 = 2$	$d_5 = 5$	$\times T_s$
$L_{\max} = 6$)	Doppler	0	689	−492	886	Hz

As presented in Table 2, three performance bounds, namely, 'Ideal', 'TDLS+Ideal$(L_{\max})$', and 'NCCE+Ideal$(L_{\max})$', are introduced for comparison purposes. 'Ideal' denotes the case in which the channel coefficients are obtained by applying the FFT to the actual time-varying channel response at the midpoint of each OFDM symbol. 'TDLS+Ideal$(L_{\max})$' and 'NCCE+Ideal$(L_{\max})$' denote the TDLS CE method in [3] and the NCCE method in [13], respectively, both operating under the assumption of a perfectly estimated $L_{\max}$. The notations 'TDLS+$\hat{L}_{\max}$ [6]' and 'TDLS+$\hat{L}_{\max}$ [7]' refer to TDLS channel estimation methods employing the channel length estimate $\hat{L}_{\max}$ from [6,7], respectively. Similarly, 'NCCE+$\hat{L}_{\max}$ [6]' and 'NCCE+$\hat{L}$max [7]' denote NCCE-based methods utilizing the corresponding $\hat{L}_{\max}$ estimators. In addition, '$\hat{L}_{\max}$ Prop.X', where $X \in \{1,2,3,4\}$, indicates the proposed estimators presented in Section 3.3.2.

4.1. Simulation Results for Detection Probabilities

From this point, we can define the probabilities of correct detection (CD), good detection (GD), erroneous detection (ED), and bad detection (BD) for $\hat{L}_{\max}$, respectively, as follows [7,8]:

$$
\begin{aligned}
P_{CD} &= \Pr\{\hat{L}_{\max} = L_{\max}\} \\
P_{GD} &= \Pr\{L_{\max} \leq \hat{L}_{\max} < (L_{\max} + N_g)/2\} \\
P_{ED} &= \Pr\{\hat{L}_{\max} < L_{\max}\} \\
P_{BD} &= \Pr\{(L_{\max} + N_g)/2 \leq \hat{L}_{\max}\}
\end{aligned} \tag{30}
$$

Furthermore, to effectively highlight the advantages of the proposed method, the error-free probability associated with (30) is defined as follows:

$$
\begin{aligned}
P_{CD}^{ef} &= \Pr\{\text{error-free } \& \ \hat{L}_{\max} = L_{\max}\} \\
P_{GD}^{ef} &= \Pr\{\text{error-free } \& \ L_{\max} \leq \hat{L}_{\max} < (L_{\max} + N_g)/2\} \\
P_{ED}^{ef} &= \Pr\{\text{error-free } \& \ \hat{L}_{\max} < L_{\max}\} \\
P_{BD}^{ef} &= \Pr\{\text{error-free } \& \ (L_{\max} + N_g)/2 \leq \hat{L}_{\max}\}
\end{aligned} \tag{31}
$$

Figures 3–5 illustrate the detection probability performance of various CE schemes in the 'Highway NLOS with 252 km/h' scenario. Figures 6–8 present the detection probability comparison in the 'Street Crossing NLOS' scenario at 126 km/h. Figures 9–11 show the detection probability results for the 'Highway LOS' scenario at 252 km/h. Figures 3a, 6a, and 9a depict the detection probability performance corresponding to (30) for QPSK modulation. Figures 3b, 6b, and 9b show the corresponding error-free detection probability of (31) under the same modulation scheme. Similarly, Figures 4a, 7a, and 10a illustrate the detection probability performance according to (30) for 16QAM modulation. Figures 4b, 7b, and 10b present the error-free detection probability corresponding to (31) for 16QAM. Furthermore, Figures 5a, 8a, and 11a demonstrate the error-free GD+ED probability performance for QPSK modulation. Figures 5b, 8b, and 11b provide a comparative analysis of the error-free GD+ED probability for 16QAM modulation.

In the high-SNR region of the Highway NLOS environment, as shown in Figures 3a and 4a, the proposed methods (from 2 to 4) exhibit higher CD and GD probabilities compared with other methods, while the BD probability remains close to zero. It is noted that a non-zero BD probability can lead to reduced CD and GD probabilities. In the order of '$\hat{L}_{\max}$ Prop.1', '$\hat{L}_{\max}$ by (27)', and '$\hat{L}_{\max}$ Prop.2', the BD probability gradually decreases from a high value and approaches zero. In the low-SNR region of Figures 3a and 4a, the methods in [6,7] appear to exhibit higher CD, GD, and ED performance. It is important to note that the purpose of estimating $L_{\max}$ in this paper is to enhance channel estimation performance by enabling interference cancellation based on the estimated value of $\hat{L}_{\max}$.

Furthermore, based on the error-free ED probability P_{ED}^{ef} shown in Figures 3b and 4b for the Highway NLOS environment, it can be observed that even when ED events occur in the proposed methods, channel estimation is still performed successfully without error. This situation may occur when the channel power of the last tap in the instantaneous channel is very small, causing $L_{\max}$ to be estimated within the ED region. Nevertheless, these cases may still result in error-free symbol detection. Consequently, it is confirmed that the proposed methods utilizing $g(l)$ in (17) are capable of handling the ED case while minimizing the BD probability to zero as much as possible. A comparison of Figures 3b, 6b, and 9b (as well as Figures 4b, 7b, and 10b) reveals that the error-free ED probability P_{ED}^{ef} increases inversely with the channel power of the last tap, given the specific channel environment.

A similar detection performance trend is observed in both the Street Crossing NLOS (see Figures 6–8) and Highway LOS (see Figures 9–11) environments. Prop.3 and Prop.4 exhibit nearly identical performance, while Prop.2 outperforms the existing schemes proposed in [6,7]. In contrast, Prop.1 demonstrates inferior detection performance compared with the conventional techniques. In addition, as illustrated in Figures 5, 8, and 11, Prop.1, Prop.2, and Prop.3 demonstrate superior performance compared with the conventional schemes in terms of the error-free GD+ED probability.

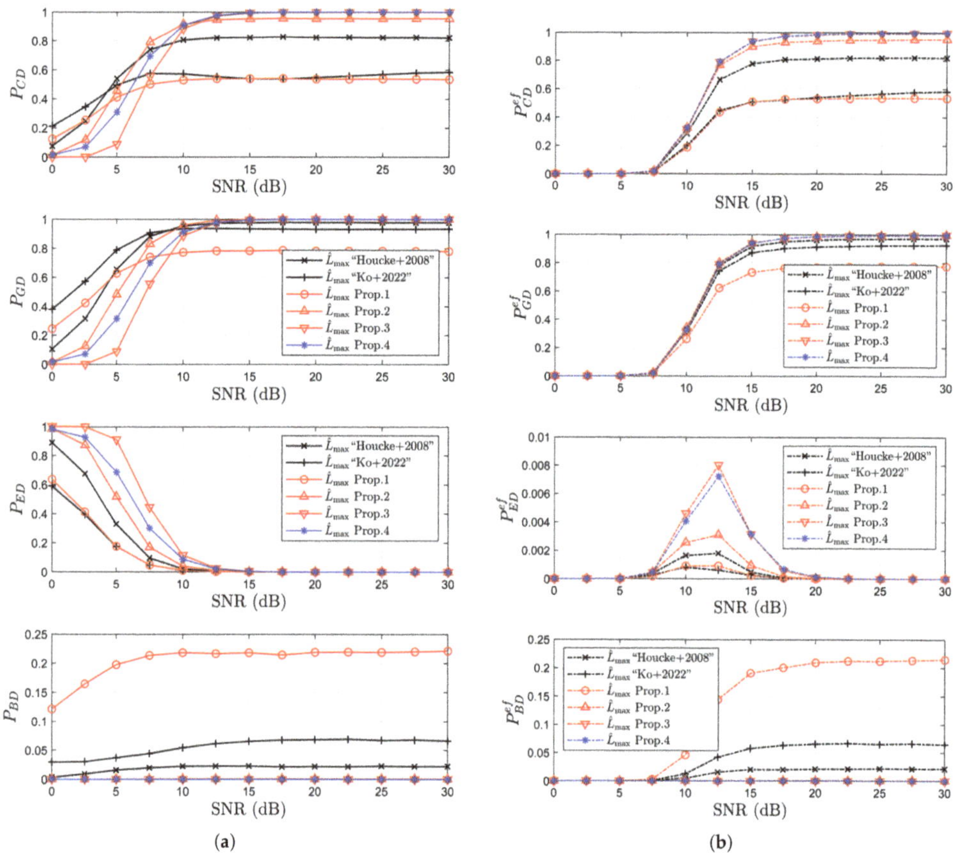

Figure 3. Detection probability and error-free detection probability for $L_{\max}$ with QPSK for Highway NLOS (252 km/h, $L_{\max} = 8$, CR = 1/2, $M = 100$): (**a**) detection probability vs. SNR (dB) and (**b**) error-free detection probability vs. SNR (dB).

4.2. Simulation Results for MSE

From here, let us define the mean square error (MSE) of the kth CFR and the average MSE (AMSE) of CFRs as

$$\mathrm{MSE}_k = E\left[\left|\hat{H}_i(k) - H_i(k)\right|^2\right]$$
$$\mathrm{AMSE} = \frac{1}{N_u}\sum_{k\in\mathbb{S}_u} E\left[\left|\hat{H}_i(k) - H_i(k)\right|^2\right],$$

where $\hat{H}_i(k)$ is the estimated kth CFR for the given CE scheme, $H_i(k)$ is the ideal kth CFR obtained by the FFT on the actual time-varying channel coefficient at the middle position of each OFDM symbol, and $\mathbb{S}_u$ is the index set of useful subcarriers for pilot and data subcarriers [13].

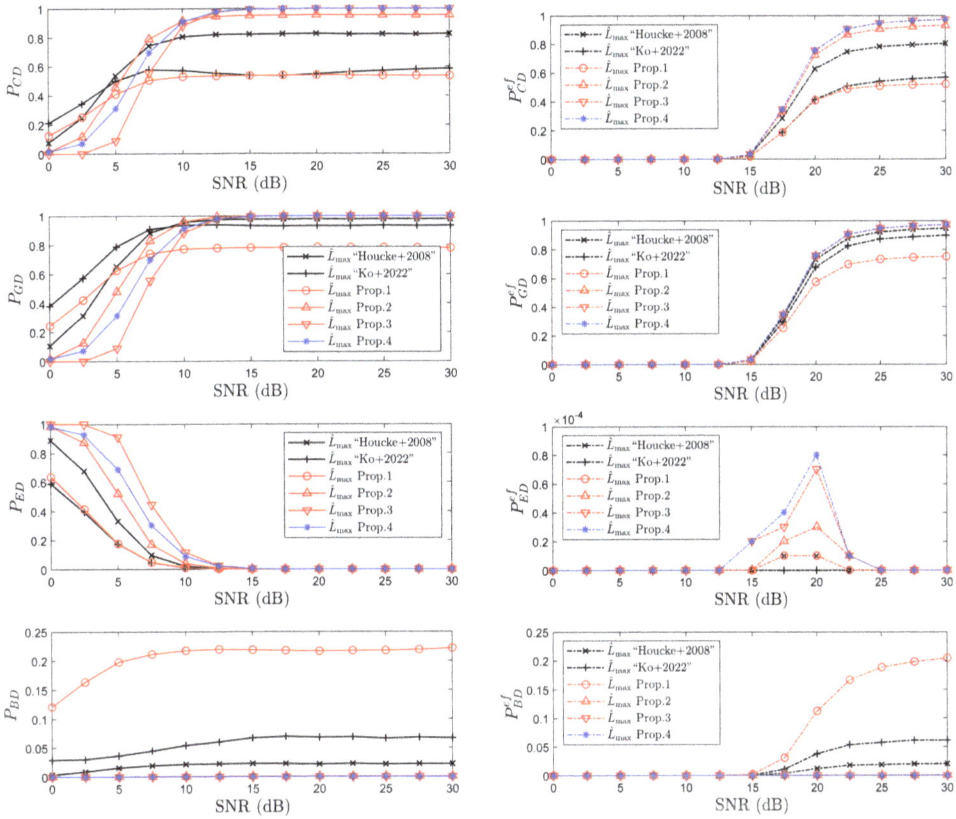

(a) (b)

Figure 4. Detection probability and error-free detection probability for $L_{\max}$ with 16QAM for Highway NLOS (252 km/h, $L_{\max}$ = 8, CR = 1/2, M = 100): (**a**) detection probability vs. SNR (dB) and (**b**) error-free detection probability vs. SNR (dB).

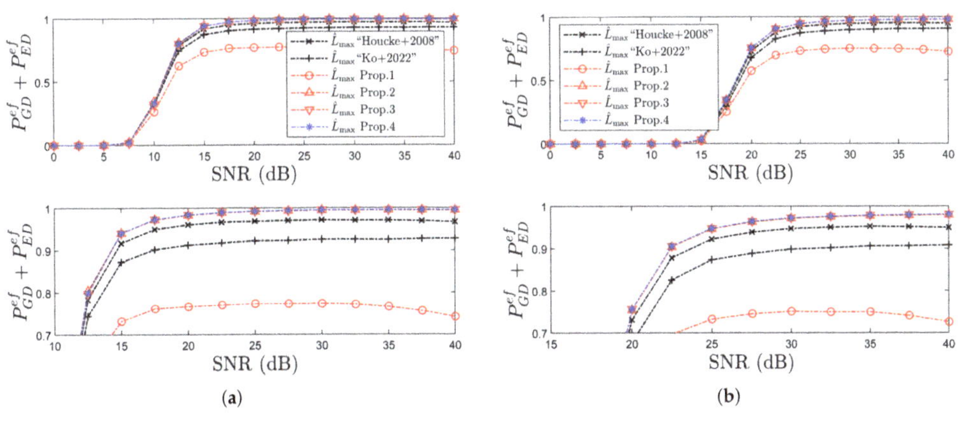

(a) (b)

Figure 5. Error-free GD+ED probability for $L_{\max}$ for Highway NLOS (252 km/h, $L_{\max}$ = 8, CR = 1/2, M = 100): (**a**) QPSK and (**b**) 16QAM.

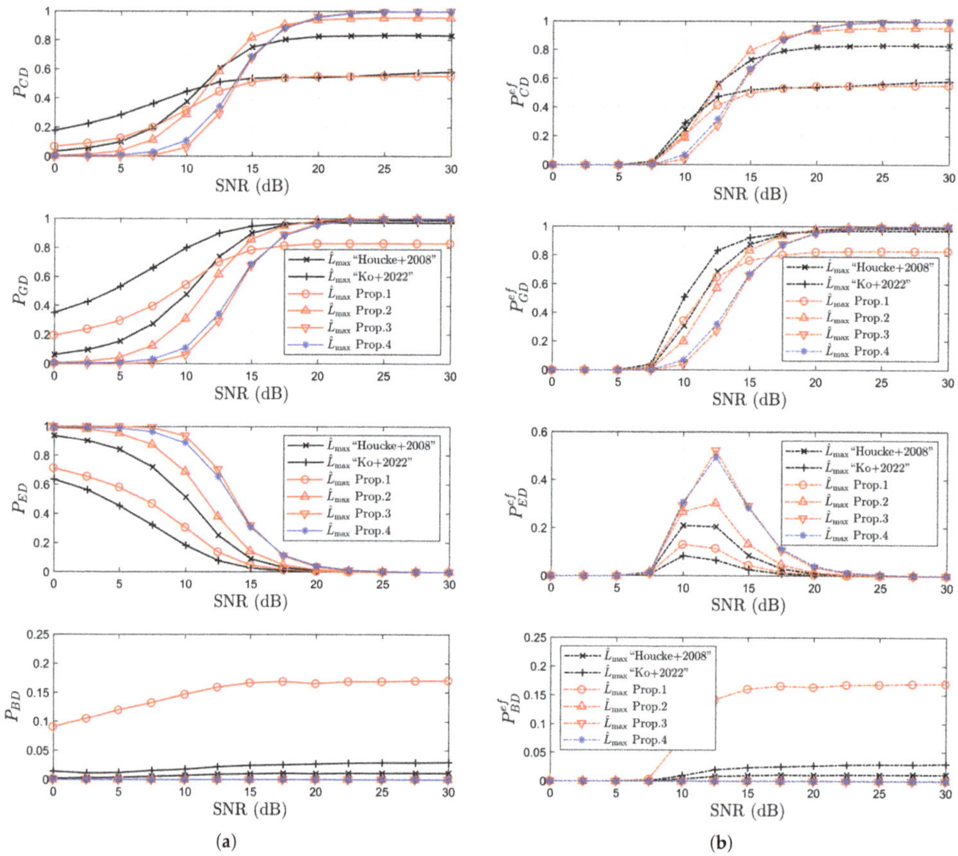

Figure 6. Detection probability and error-free detection probability for L_{max} with QPSK for Street Crossing NLOS (126 km/h, $L_{max} = 7$, CR = 1/2, $M = 100$): (**a**) detection probability vs. SNR (dB) and (**b**) error-free detection probability vs. SNR (dB).

Figures 12–17 show the AMSE and MSE performance comparison for 'Highway NLOS with 252 km/h'. Figures 18–23 illustrate the AMSE and MSE performance comparison for 'Street Crossing NLOS with 126 km/h'. Figures 24–29 present the AMSE and MSE performance comparison for 'Highway LOS with 252 km/h'. Figures 12 and 13 compare the AMSE performance of various CE schemes for QPSK and 16QAM for 'Highway NLOS with 252 km/h'. Figures 18 and 19 show the AMSE comparison for 'Street Crossing NLOS with 126 km/h'. Figures 24 and 25 illustrate the AMSE comparison for 'Highway LOS with 252 km/h'. Figures 12a, 13a, 18a, 19a, 24a, and 25a illustrate the AMSE performance of the NCCE schemes. Figures 12b, 13b, 18b, 19b, 24b, and 25b present the AMSE comparison of the TDLS schemes. For 'Highway NLOS with 252 km/h', the kth CFR MSE results for QPSK and 16QAM are shown in Figures 14–17, where the CE is conducted by using the NCCE and TDLS methods in conjunction with different L_{max} estimators. Figures 20–23 display the corresponding kth CFR MSE results for the Street Crossing NLOS environment with a velocity of 126 km/h. Similarly, Figures 26–29 illustrate the kth CFR MSE results in the Highway LOS scenario at 252 km/h.

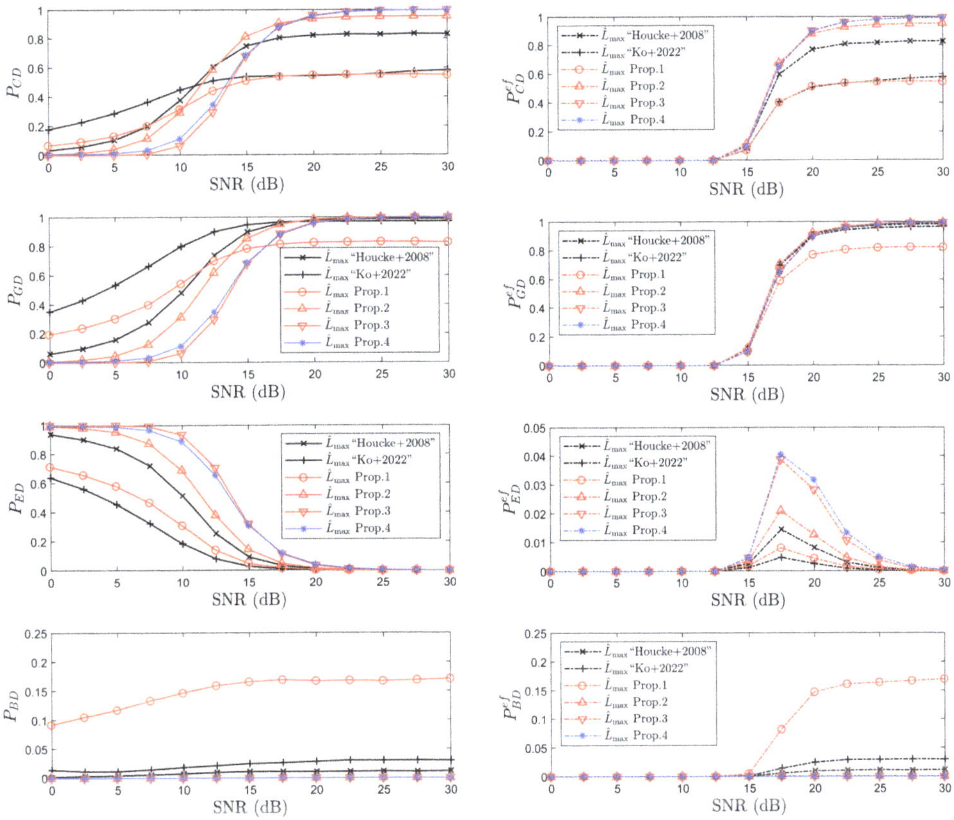

(a) (b)

Figure 7. Detection probability and error-free detection probability for $L_{\max}$ with 16QAM for Street Crossing NLOS (126 km/h, $L_{\max} = 7$, CR = 1/2, $M = 100$): (**a**) detection probability vs. SNR (dB) and (**b**) error-free detection probability vs. SNR (dB).

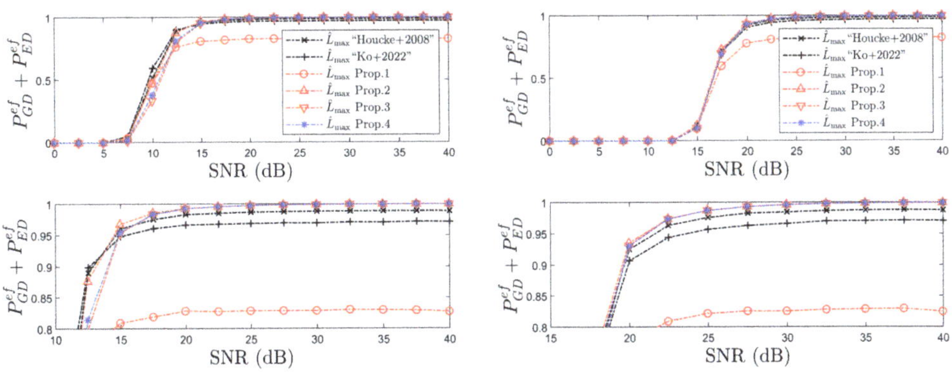

(a) (b)

Figure 8. Error-free GD+ED probability for $L_{\max}$ for Street Crossing NLOS (126 km/h, $L_{\max} = 7$, CR = 1/2, $M = 100$): (**a**) QPSK and (**b**) 16QAM.

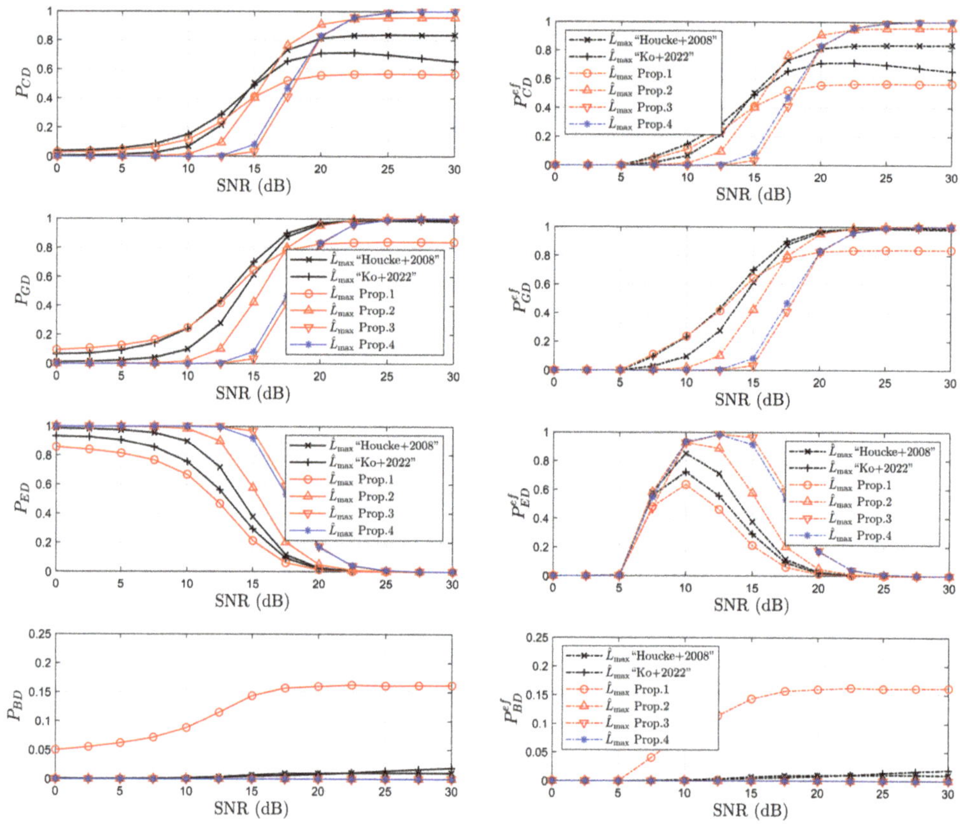

(**a**) (**b**)

Figure 9. Detection probability and error-free detection probability for L_{max} with QPSK for Highway LOS (252 km/h, $L_{max} = 6$, CR = 1/2, $M = 100$): (**a**) detection probability vs. SNR (dB) and (**b**) error-free detection probability vs. SNR (dB).

For 'Highway NLOS with 252 km/h', from Figures 12–17, it can be observed that the three proposed schemes (i.e., red lines corresponding to Prop.2–Prop.4) outperform the two existing schemes (black lines). Furthermore, the proposed schemes (Prop.2–Prop.4) are shown to achieve the MSE bounds (blue lines) of 'TDLS+Ideal(L_{max})' and 'NCCE+Ideal(L_{max})'. In Figures 12a and 13a, the NCCE combined with the proposed scheme exhibits saturated AMSE performance. Note that at a low SNR (i.e., SNR = 10 dB), 'NCCE+Ideal(L_{max})' yields a lower MSE than 'TDLS+Ideal(L_{max})' due to the assistance of the estimated CFRs for virtual subcarriers. Conversely, at a high SNR (i.e., SNR = 30 dB), NCCE schemes tend to exhibit saturated MSE performance for virtual subcarriers, allowing TDLS schemes to outperform their NCCE counterparts. As shown in Figures 14c and 15c, the MSE difference between NCCE and TDLS is the most pronounced under these conditions. Moreover, this disparity becomes more significant near the edges of the virtual subcarrier band and in higher-order modulation scenarios such as 16QAM, as observed by comparing Figures 16c and 17c.

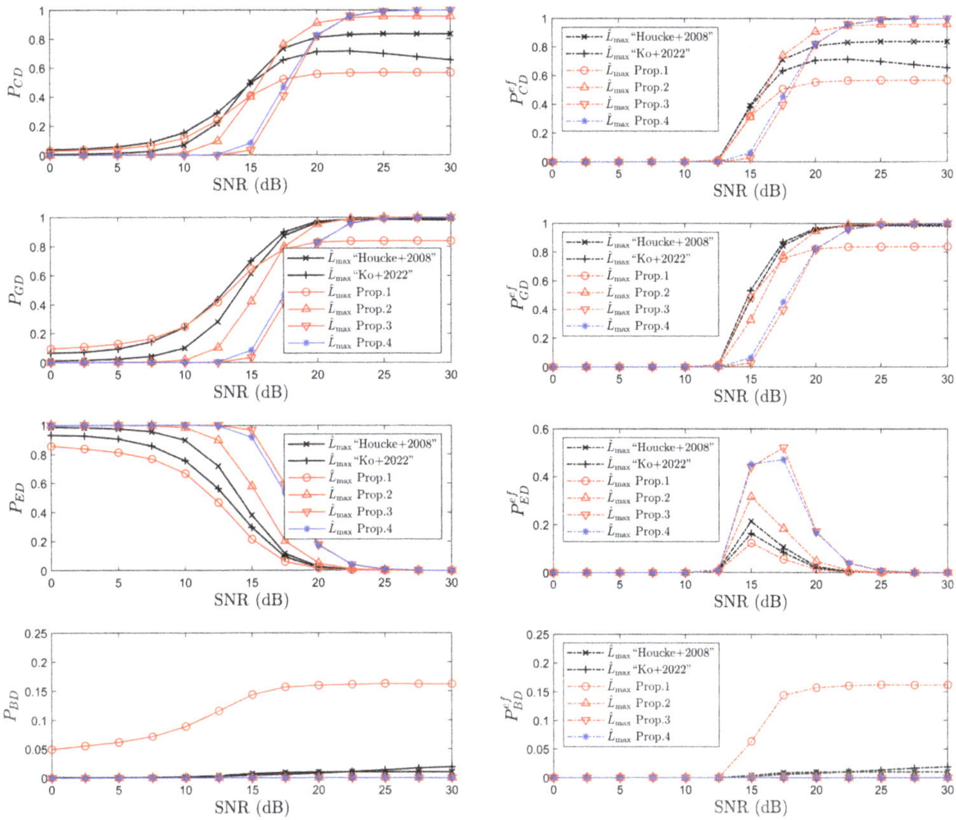

(**a**) (**b**)

Figure 10. Detection probability and error-free detection probability for L_{max} with 16QAM for Highway LOS (252 km/h, L_{max} = 6, CR = 1/2, M = 100): (**a**) detection probability vs. SNR (dB) and (**b**) error-free detection probability vs. SNR (dB).

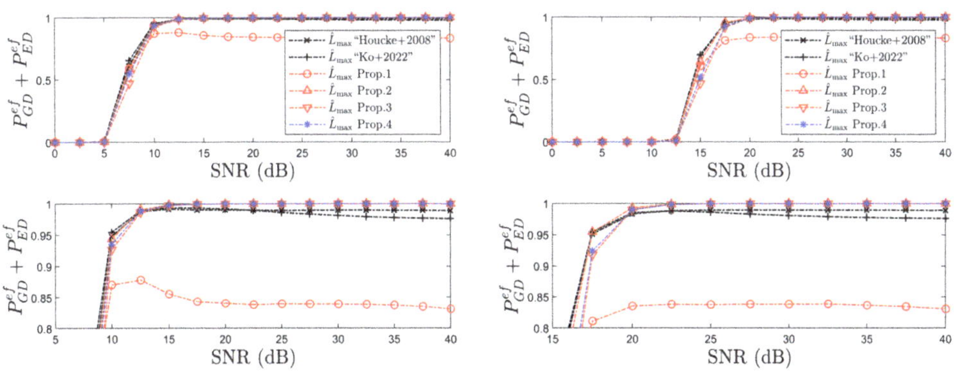

(**a**) (**b**)

Figure 11. Error-free GD+ED probability for L_{max} for Highway LOS (252 km/h, L_{max} = 6, CR = 1/2, M = 100): (**a**) QPSK and (**b**) 16QAM.

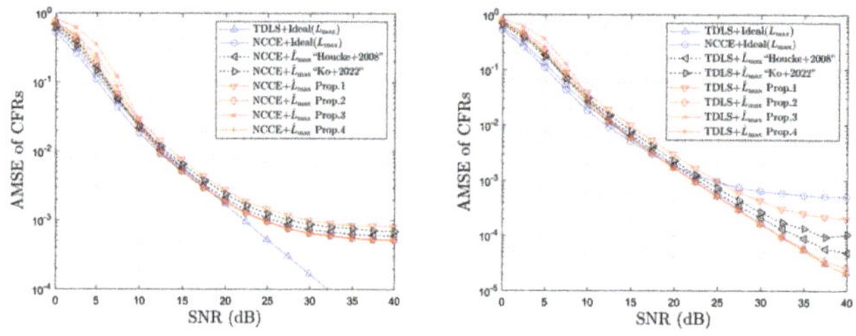

(**a**) (**b**)

Figure 12. CFR AMSE comparison for QPSK for Highway NLOS (252 km/h, $L_{\mathrm{max}} = 8$, CR = 1/2, $M = 100$): (**a**) NCCE and (**b**) TDLS.

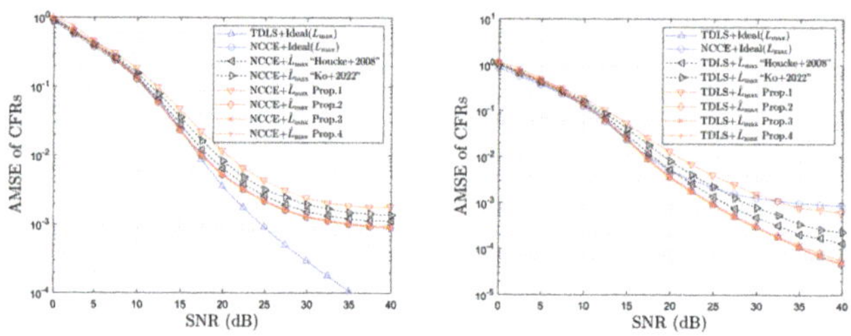

(**a**) (**b**)

Figure 13. CFR AMSE comparison for 16QAM for Highway NLOS (252 km/h, $L_{\mathrm{max}} = 8$, CR = 1/2, $M = 100$): (**a**) NCCE and (**b**) TDLS.

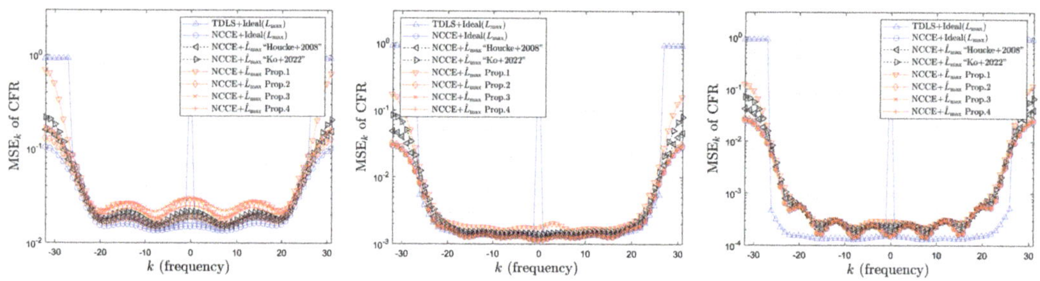

(**a**) (**b**) (**c**)

Figure 14. CFR MSE comparison for QPSK and NCCE for Highway NLOS (252 km/h, $L_{\mathrm{max}} = 8$, CR = 1/2, $M = 100$): (**a**) SNR = 10 dB, (**b**) SNR = 20 dB, and (**c**) SNR = 30 dB.

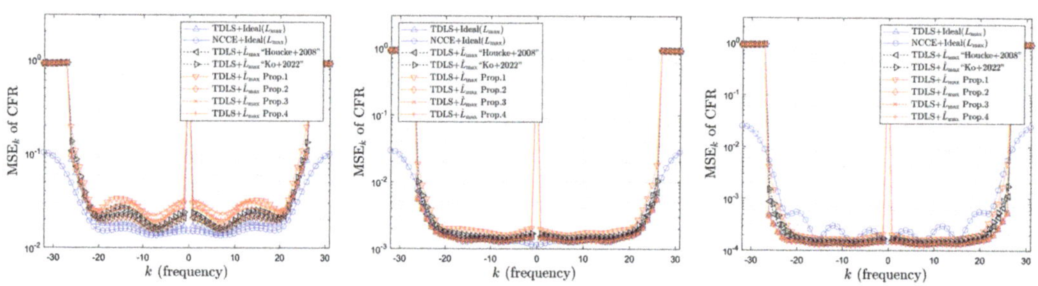

(**a**) (**b**) (**c**)

Figure 15. CFR MSE comparison for QPSK and TDLS for Highway NLOS (252 km/h, $L_{\mathrm{max}} = 8$, CR = 1/2, $M = 100$): (**a**) SNR = 10 dB, (**b**) SNR = 20 dB, and (**c**) SNR = 30 dB.

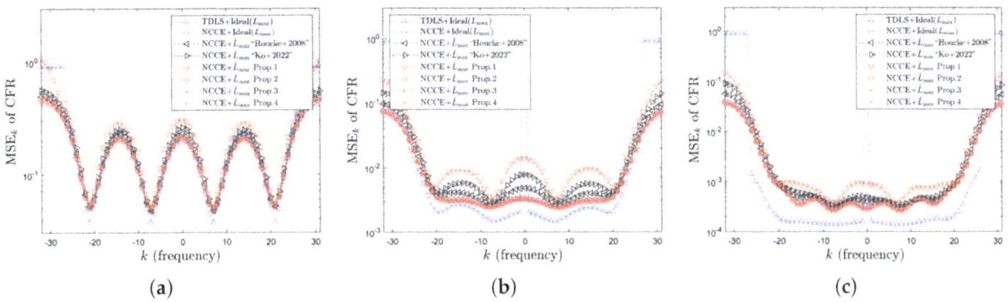

Figure 16. CFR MSE comparison for 16QAM and NCCE for Highway NLOS (252 km/h, $L_{\max} = 8$, CR = 1/2, $M = 100$): (**a**) SNR = 10 dB, (**b**) SNR = 20 dB, and (**c**) SNR = 30 dB.

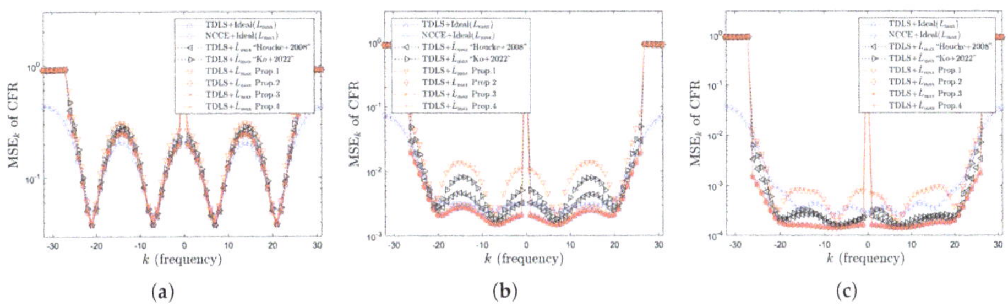

Figure 17. CFR MSE comparison for 16QAM and TDLS for Highway NLOS (252 km/h, $L_{\max} = 8$, CR = 1/2, $M = 100$): (**a**) SNR = 10 dB, (**b**) SNR = 20 dB, and (**c**) SNR = 30 dB.

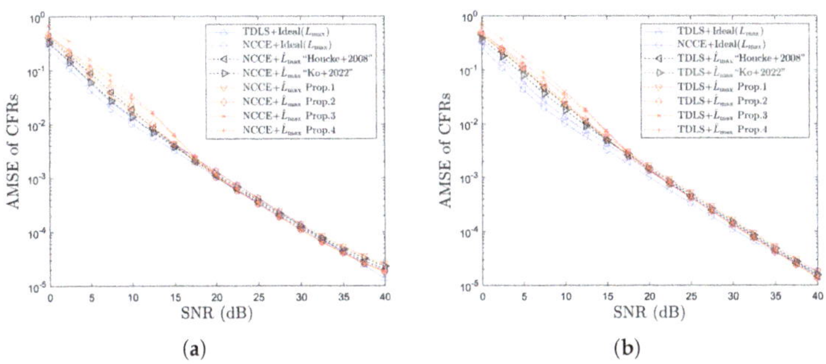

Figure 18. CFR AMSE comparison for QPSK for Street Crossing NLOS (126 km/h, $L_{\max} = 7$, CR = 1/2, $M = 100$): (**a**) NCCE and (**b**) TDLS.

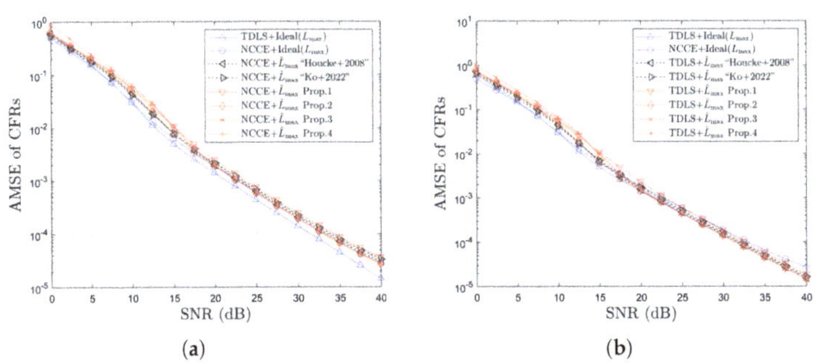

Figure 19. CFR AMSE comparison for 16QAM for Street Crossing NLOS (126 km/h, $L_{\max} = 7$, CR = 1/2, $M = 100$): (**a**) NCCE and (**b**) TDLS.

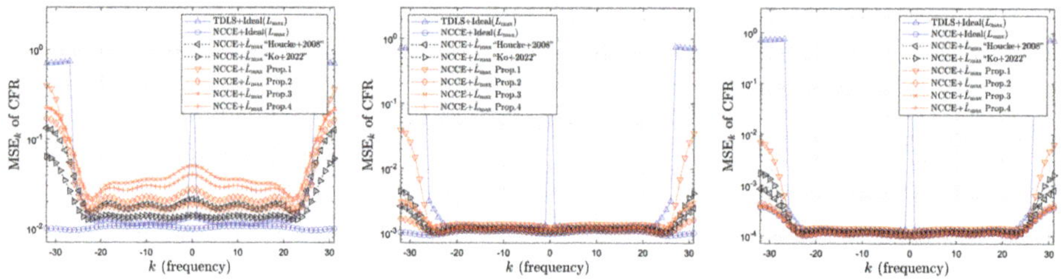

(a) (b) (c)

Figure 20. CFR MSE comparison for QPSK and NCCE for Street Crossing NLOS (126 km/h, $L_{\text{max}} = 7$, CR = 1/2, $M = 100$): (**a**) SNR = 10 dB, (**b**) SNR = 20 dB, and (**c**) SNR = 30 dB.

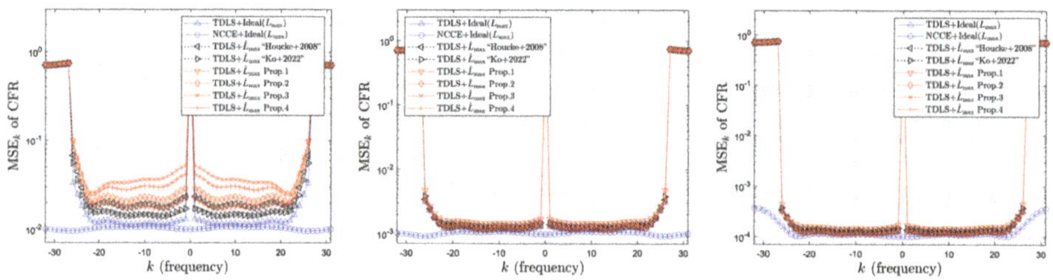

(a) (b) (c)

Figure 21. CFR MSE comparison for QPSK and TDLS for Street Crossing NLOS (126 km/h, $L_{\text{max}} = 7$, CR = 1/2, $M = 100$): (**a**) SNR = 10 dB, (**b**) SNR = 20 dB, and (**c**) SNR = 30 dB.

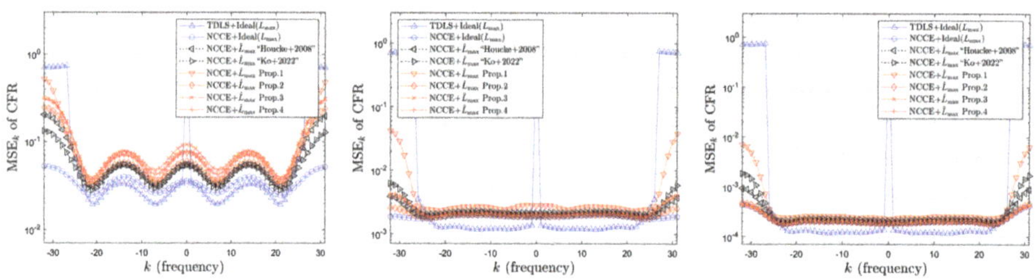

(a) (b) (c)

Figure 22. CFR MSE comparison for 16QAM and NCCE for Street Crossing NLOS (126 km/h, $L_{\text{max}} = 7$, CR = 1/2, $M = 100$): (**a**) SNR = 10 dB, (**b**) SNR = 20 dB, and (**c**) SNR = 30 dB.

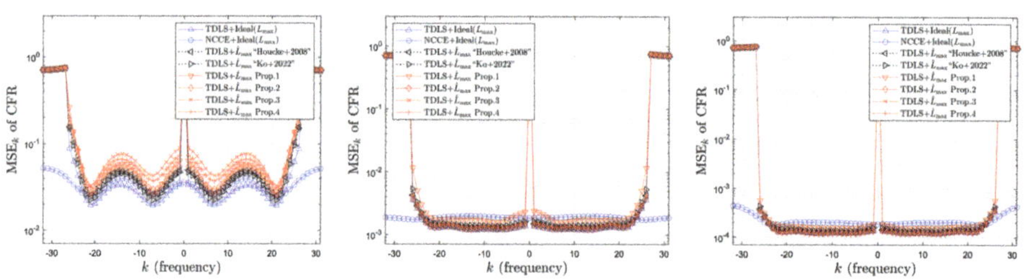

(a) (b) (c)

Figure 23. CFR MSE comparison for 16QAM and TDLS for Street Crossing NLOS (126 km/h, $L_{\text{max}} = 7$, CR = 1/2, $M = 100$): (**a**) SNR = 10 dB, (**b**) SNR = 20 dB, and (**c**) SNR = 30 dB.

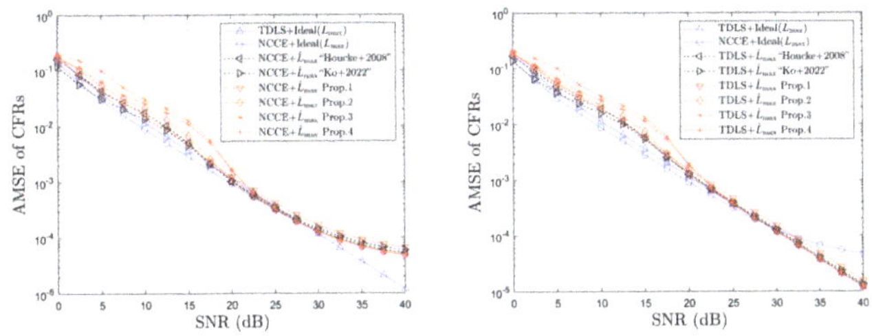

(a) (b)

Figure 24. CFR AMSE comparison for QPSK for Highway LOS (252 km/h, $L_{\max} = 6$, CR = 1/2, $M = 100$): (**a**) NCCE and (**b**) TDLS.

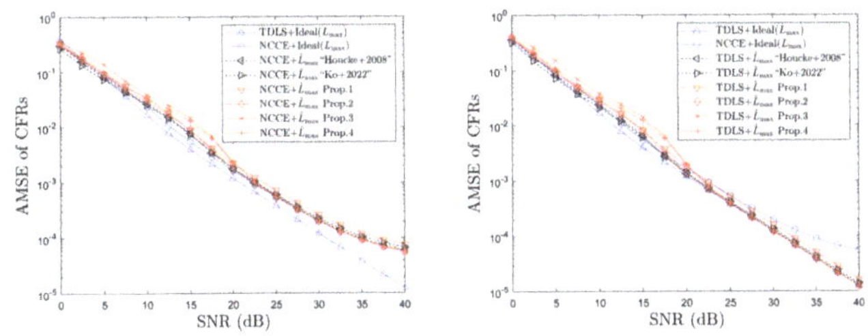

(a) (b)

Figure 25. CFR AMSE comparison for 16QAM for Highway LOS (252 km/h, $L_{\max} = 6$, CR = 1/2, $M = 100$): (**a**) NCCE and (**b**) TDLS.

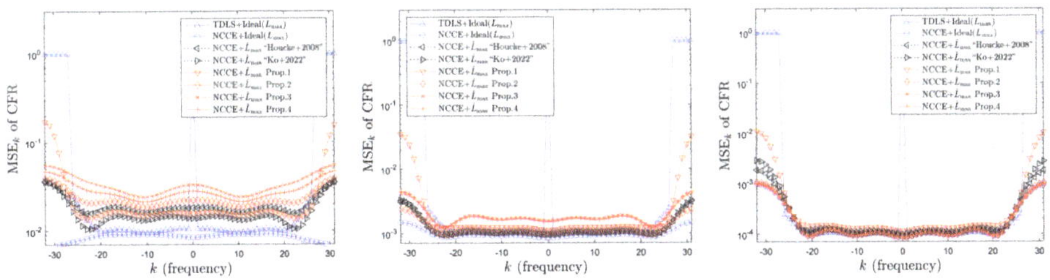

(a) (b) (c)

Figure 26. CFR MSE comparison for QPSK and NCCE for Highway LOS (252 km/h, $L_{\max} = 6$, CR = 1/2, $M = 100$): (**a**) SNR = 10 dB, (**b**) SNR = 20 dB, and (**c**) SNR = 30 dB.

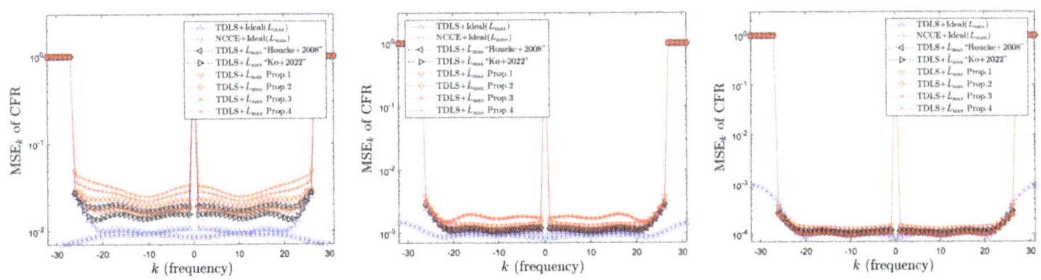

(a) (b) (c)

Figure 27. CFR MSE comparison for QPSK and TDLS for Highway LOS (252 km/h, $L_{\max} = 6$, CR = 1/2, $M = 100$): (**a**) SNR = 10 dB, (**b**) SNR = 20 dB, and (**c**) SNR = 30 dB.

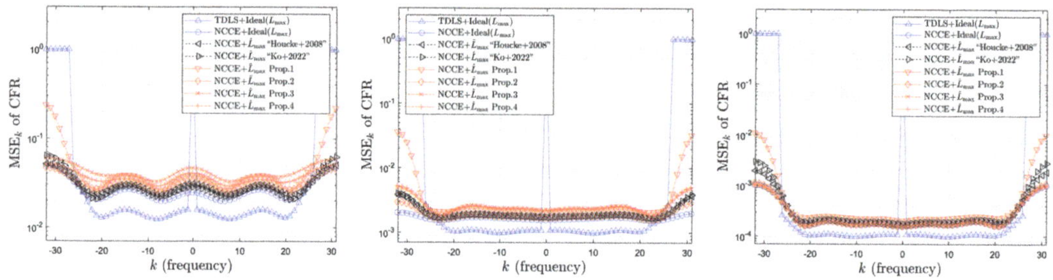

(a) (b) (c)

Figure 28. CFR MSE comparison for 16QAM and NCCE for Highway LOS (252 km/h, $L_{\max} = 6$, CR = 1/2, $M = 100$): (**a**) SNR = 10 dB, (**b**) SNR = 20 dB, and (**c**) SNR = 30 dB.

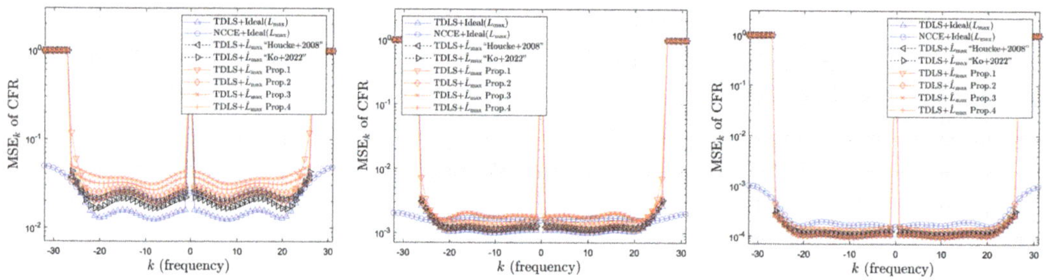

(a) (b) (c)

Figure 29. CFR MSE comparison for 16QAM and TDLS for Highway LOS (252 km/h, $L_{\max} = 6$, CR = 1/2, $M = 100$): (**a**) SNR = 10 dB, (**b**) SNR = 20 dB, and (**c**) SNR = 30 dB.

The AMSE and the kth MSE performances are generally observed to be similar in both the Street Crossing NLOS (see Figures 18–23) and Highway LOS (see Figures 24–29) environments. In general, Prop.2 demonstrates performance that is comparable or superior to the two existing schemes across all three channel scenarios. However, Prop.3 and Prop.4 may exhibit inferior AMSE and kth MSE performance when the SNR falls below a certain threshold in both the Street Crossing NLOS and Highway LOS scenarios. Consequently, the robustness of the Prop.2 scheme is evident across various channel conditions.

4.3. Simulation Results for Error Rates

Figures 30 and 31 illustrate the PER performance comparison of various CE schemes in the 'Highway NLOS' scenario at a velocity of 252 km/h for QPSK and 16QAM, respectively. Figures 32 and 33 present the PER performance comparison in the 'Street Crossing NLOS' scenario at 126 km/h for QPSK and 16QAM, respectively. Figures 34 and 35 show the PER performance in the 'Highway LOS' scenario at 252 km/h for QPSK and 16QAM, respectively. Figures 30a–35a illustrate the error rate performance of the NCCE scheme. Figures 30b–35b present the performance of the TDLS scheme. Figures 30c–35c provide a comparative analysis of both NCCE and TDLS schemes.

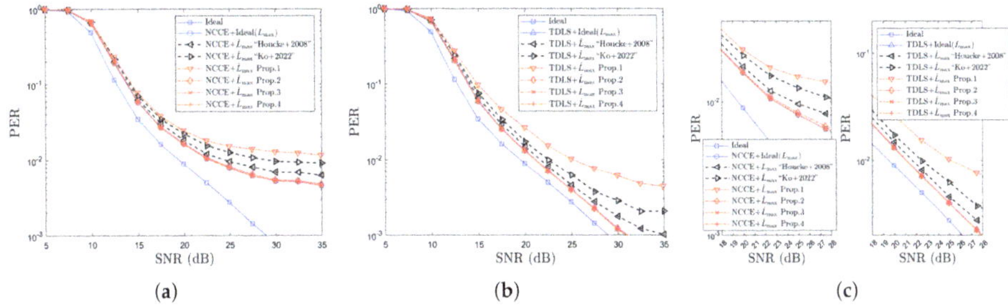

Figure 30. Error performance comparison for QPSK for Highway NLOS (252 km/h, $L_{\text{max}} = 8$, CR = 1/2, $M = 100$): (**a**) NCCE, (**b**) TDLS, and (**c**) NCCE vs. TDLS.

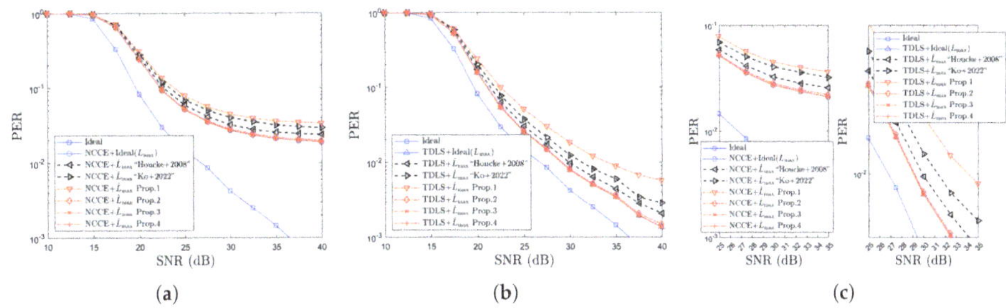

Figure 31. Error performance comparison for 16QAM for Highway NLOS (252 km/h, $L_{\text{max}} = 8$, CR = 1/2, $M = 100$): (**a**) NCCE, (**b**) TDLS, and (**c**) NCCE vs. TDLS.

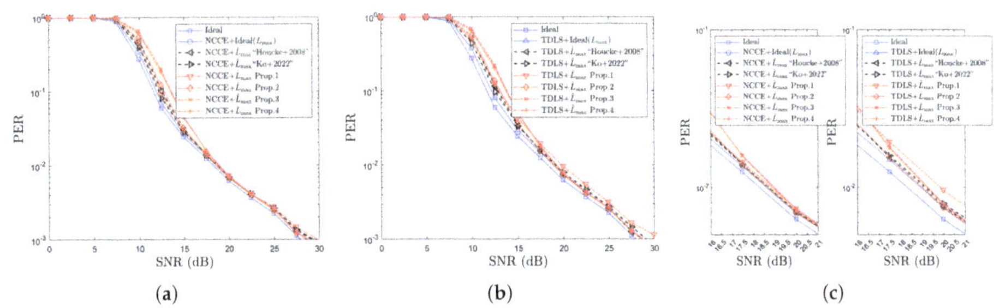

Figure 32. Error performance comparison for QPSK for Street Crossing NLOS (126 km/h, $L_{\text{max}} = 7$, CR = 1/2, $M = 100$): (**a**) NCCE, (**b**) TDLS, and (**c**) NCCE vs. TDLS.

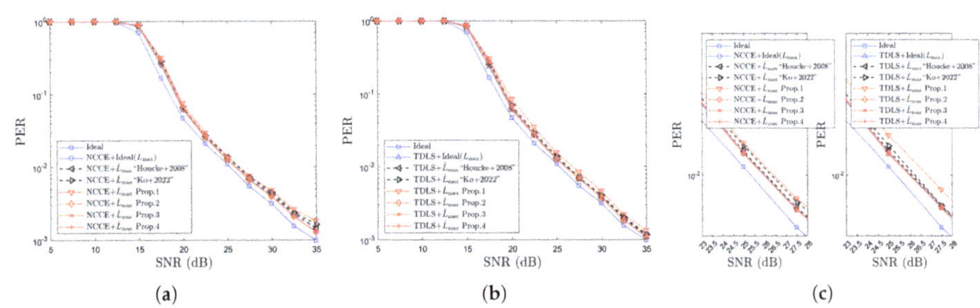

Figure 33. Error performance comparison for 16QAM for Street Crossing NLOS (126 km/h, $L_{\text{max}} = 7$, CR = 1/2, $M = 100$): (**a**) NCCE, (**b**) TDLS, and (**c**) NCCE vs. TDLS.

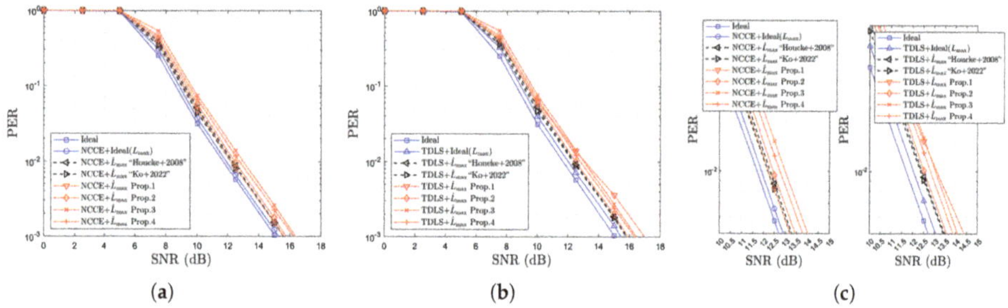

Figure 34. Error performance comparison for QPSK for Highway LOS (252 km/h, $L_{\max} = 6$, CR = 1/2, $M = 100$): (**a**) NCCE, (**b**) TDLS, and (**c**) NCCE vs. TDLS.

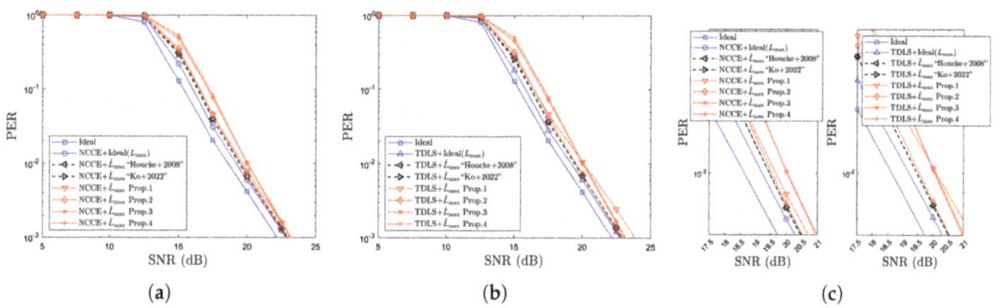

Figure 35. Error performance comparison for 16QAM for Highway LOS (252 km/h, $L_{\max} = 6$, CR = 1/2, $M = 100$): (**a**) NCCE, (**b**) TDLS, and (**c**) NCCE vs. TDLS.

In the 'Highway NLOS' scenario, Figure 30 shows the error rate performance comparison for QPSK with respect to $L_{\max}$ estimation schemes within both NCCE and TDLS. Figure 31 illustrates the error rate performance comparison for 16QAM with respect to $L_{\max}$ estimation schemes under both NCCE and TDLS frameworks. In Figures 30 and 31, it can be observed that the three proposed schemes (i.e., red lines corresponding to Prop.2–Prop.4) outperform the two existing schemes (black lines) and closely approach the error performance bounds (blue lines) of 'TDLS+Ideal($L_{\max}$)' and 'NCCE+Ideal($L_{\max}$)'. As mentioned earlier, NCCE results in higher MSE near the virtual subcarriers, as shown in Figures 14c–16c, leading to a saturated AMSE of approximately 10^{-3} in Figures 12a and 13a. Consequently, Figures 30a and 31a reveal a packet error floor in the case of NCCE. At PER $= 10^{-2}$ for QPSK, it is confirmed that the proposed schemes (Prop.2–Prop. 4) achieve SNR gains of 1.0 dB and 0.5 dB over the scheme in [6] under NCCE and TDLS, respectively. For 16QAM, PER $= 10^{-2}$ cannot be achieved with NCCE. However, the 0.8 dB SNR gain is observed with TDLS. The SNR gains achieved by Prop.2–Prop.4 over the $\hat{L}_{\max}$–based approach in [6] are summarized in Table 4. It is worth noting that the SNR gain can be achieved without additional computational complexity by applying $g(l)$ in (17). In other words, this gain can be obtained by appropriately adjusting the bias term B_u in (22).

In both the Street Crossing NLOS and Highway LOS scenarios, no packet error floor is observed, even in the case of NCCE (see Figures 32a–35a). It is noteworthy that in the Street Crossing NLOS scenario, both $L_{\max}$ and the velocity are lower than those in the Highway NLOS scenario. In this relatively more favorable channel environment, Prop.2–Prop.4 exhibit ER performance comparable to that of the existing methods (see Figures 32c and 33c). On the other hand, the Highway LOS scenario has a smaller $L_{\max}$ compared with the Highway NLOS scenario. Figures 34c and 35c illustrate that Prop.2 achieves ER performance similar to that of the existing methods.

By comparing the ER performance across the three channel environments, it can be inferred that the adaptive selection of $g(l)$ is feasible. Specifically, Prop. 2 to Prop. 4 are more effective in challenging channel conditions where L_{max} is relatively large (i.e., the Highway NLOS scenario), while Prop. 2 alone is sufficient in environments characterized by a shorter L_{max} (i.e., the Street Crossing NLOS and Highway LOS scenarios). Consequently, as discussed in Section 4.2, the robustness of the Prop.2 scheme is evident across various channel conditions.

Table 4. SNR gain (dB) of Prop.2–Prop.4 over $\hat{L}_{max}$ in [6].

Environment	CE Scheme	At PER = 10^{-2}	From
Highway NLOS, QPSK	NCCE	1.0 dB	Figure 30c
Highway NLOS, QPSK	TDLS	0.5 dB	Figure 30c
Highway NLOS, 16QAM	TDLS	0.8 dB	Figure 31c

5. Conclusions

In this paper, an enhanced maximum access delay time estimation method for OFDM systems operating over multipath fading channels has been proposed. By formulating a novel LLR based on the correlation properties introduced by the CP, the proposed scheme effectively approximates the ML estimation of MADT. The generality of the proposed framework has been analytically validated by showing that the existing method represents a special case within it.

Comprehensive simulation results demonstrate that the proposed estimator significantly improves MADT detection performance in terms of correct and good detection probabilities. Furthermore, when applied to the NCCE and TDLS techniques, the proposed MADT estimator enhances overall channel estimation accuracy, approaching the performance of ideal MADT scenarios. These results highlight the practical effectiveness of the proposed approach in realistic vehicular communication environments such as IEEE 802.11p/OFDM systems.

Future research directions include the development of a channel length estimation technique that fully exploits the CP-based correlation properties of the noise estimator, thereby enhancing the robustness of channel parameter estimation. Additionally, as demonstrated in this work, the NCCE technique shows inferior performance compared with TDLS in terms of PER. Therefore, another promising research direction is to design an improved NCCE channel estimation algorithm that can overcome this limitation and further close the performance gap.

As another promising direction for future research, the estimation of the maximum delay spread and the number of channel taps using deep learning approaches, particularly long short-term memory (LSTM) architectures, merits comprehensive investigation. Given the LSTM network's ability to capture long-range temporal dependencies, it is well-suited for learning channel characteristics from sequential data such as channel state information (CSI) or pilot signal observations. In dynamic V2X communication environments, where the channel properties vary rapidly and non-linearly, LSTM-based models could be trained to estimate either the effective tap count as a regression task or the underlying channel model as a classification task. This unified framework would enable the simultaneous inference of both the tap structure and the channel type, thereby enhancing the adaptability and robustness of wireless communication systems under diverse and time-varying propagation conditions.

Author Contributions: Conceptualization, K.K. and S.L.; methodology, S.L.; software, K.K.; validation, K.K. and S.L.; formal analysis, K.K.; investigation, K.K. and S.L.; resources, K.K.; data curation, K.K.; writing—original draft preparation, K.K.; writing—review and editing, S.L.; visualization, K.K.; supervision, S.L.; project administration, K.K. and S.L.; funding acquisition, K.K. and S.L. All authors have read and agreed to the published version of the manuscript.

Funding: This work was supported by the National Research Foundation of Korea (NRF) grant funded by the Korean government (MSIT) (NRF-2020R1A2C1005260, NRF-2021R1A2C2012558).

Data Availability Statement: Data are contained within the article.

Conflicts of Interest: The authors declare no conflicts of interest.

Abbreviations

The following abbreviations are used in this manuscript:

AMSE	average mean square error
AWGN	additive white Gaussian noise
BD	bad detection
CD	correct detection
CDP	constructed data pilots
CE	channel estimation
CFR	channel frequency response
C-ITS	cooperative intelligent transportation systems
CIR	channel impulse response
CLT	central limit theorem
CP	cyclic prefix
CSI	channel state information
GD	good detection
GI	guard interval
ED	erroneous detection
ISI	inter-symbol interference
LLR	log-likelihood ratio
LS	least square
LSTM	long short-term memory
MAC	medium access control
MADT	maximum access delay time
ML	maximum likelihood
MSE	mean square error
NCCE	noise-canceling channel estimation
FFT	fast Fourier transform
IDFT	inverse discrete Fourier transform
IFFT	inverse fast Fourier transform
OFDM	orthogonal frequency division multiplexing
PDF	probability density function
QAM	quadrature amplitude modulation
QPSK	quadrature phase shift keying
SNR	signal-to-noise ratio
TRFI	time-domain reliability test and frequency-domain interpolation
TDLS	time-domain least square
V2X	vehicle-to-everything
WAVE	wireless access in vehicular environments
WSSUS	wide-sense stationary uncorrelated scattering

References

1. Zhao, Z.; Cheng, X.; Wen, M.; Jiao, B.; Wang, C.X. Channel Estimation Schemes for IEEE 802.11p Standard. *IEEE Intell. Transp. Syst. Mag.* **2013**, *5*, 38–49. [CrossRef]
2. *IEEE Std. 1609.0-2013*; IEEE Guide for Wireless Access in Vehicular Environments (WAVE)-Architecture. IEEE: Piscataway, NJ, USA , 2014.
3. Bourdoux, A.; Cappelle, H.; Dejonghe, A. Channel Tracking for Fast Time-Varying Channels in IEEE802.11p Systems. In Proceedings of the 2011 IEEE Global Telecommunications Conference—GLOBECOM 2011, Houston, TX, USA, 5–9 December 2011; pp. 1–6.
4. Lim, S. A New Channel Estimation Scheme Using Virtual Subcarriers based on Successive Interpolation in IEEE 802.11p WAVE Systems. In Proceedings of the 2021 International Conference on Convergence Content (ICCC 2021), Jeju, Republic of Korea, 19–20 December 2021; pp. 441–442.
5. Han, S.; Park, J.; Song, C. Virtual Subcarrier Aided Channel Estimation Schemes for Tracking Rapid Time Variant Channels in IEEE 802.11p Systems. In Proceedings of the 2020 IEEE 91st Vehicular Technology Conference (VTC2020-Spring), Antwerp, Belgium, 25–28 May 2020; pp. 1–5.
6. Socheleau, F.; Aissa-El-Bey, A.; Houcke, S. Non data-aided SNR estimation of OFDM signals. *IEEE Commun. Lett.* **2008**, *12*, 813–815. [CrossRef]
7. Wang, H.; Lim, S.; Ko, K. Improved Maximum Access Delay Time, Noise Variance, and Power Delay Profile Estimations for OFDM Systems. *KSII Trans. Internet Inf. Syst.* **2022**, *16*, 4099–4113.
8. Lim, S.; Wang, H.; Ko, K. Power-Delay-Profile-Based MMSE Channel Estimations for OFDM Systems. *Electronics* **2023**, *12*, 510. [CrossRef]
9. Kim, Y.; Oh, J.; Shin, Y.; Mun, C. Time and frequency domain channel estimation scheme for IEEE 802.11p. In Proceedings of the 17th International IEEE Conference on Intelligent Transportation Systems (ITSC), Qingdao, China, 8–11 October 2014; pp. 1085–1090.
10. Kwak, K.; Lee, S.; Kim, J.; Hong, D. A New DFT-Based Channel Estimation Approach for OFDM with Virtual Subcarriers by Leakage Estimation. *IEEE Trans. Wirel. Commun.* **2008**, *7*, 2004–2008. [CrossRef]
11. Kim, K.; Hwang, H.; Choi, K.; Kim, K. Low-Complexity DFT-Based Channel Estimation with Leakage Nulling for OFDM Systems. *IEEE Commun. Lett.* **2014**, *18*, 415–418. [CrossRef]
12. Cho, Y.; Son, H. Enhanced DFT-Based Channel Estimator for Leakage Effect Mitigation in OFDM Systems. *IEEE Commun. Lett.* **2019**, *23*, 875–879. [CrossRef]
13. Ko, K.; Wang, H. Noise-Canceling Channel Estimation Schemes Based on the CIR Length Estimation for IEEE 802.11p/OFDM Systems. *Electronics* **2024**, *13*, 1110. [CrossRef]
14. Ko, K.; Park, M.; Hong, D. Performance Analysis of Asynchronous MC-CDMA systems with a Guard Period in the form of a Cyclic Prefix. *IEEE Trans. Commun.* **2006**, *2*, 216–220.
15. Park, M.; Ko, K.; Park, B.; Hong, D. Effects of asynchronous MAI on average SEP performance of OFDMA uplink systems over frequency-selective Rayleigh fading channels. *IEEE Trans. Commun.* **2010**, *2*, 586–599.
16. Kwak, K.; Lee, S.; Min, H.; Choi, S.; Hong, D. New OFDM Channel Estimation with Dual-ICI Cancellation in Highly Mobile Channel. *IEEE Trans. Wirel. Commun.* **2010**, *10*, 3155–3165.
17. Cui, T.; Tellambura, C. Power delay profile and noise variance estimation for OFDM. *IEEE Commun. Lett.* **2006**, *1*, 25–27.
18. van de Beek, J.; Sandell, M.; Borjesson, P. ML estimation of time and frequency offset in OFDM systems. *IEEE Trans. Signal Process* **1997**, *7*, 1800–1805.
19. Popoulis, A.; Pillai, S. *Probability, Random Variables, and Stochastic Processes*, 4th ed.; McGraw Hill: Boston, MA, USA, 2002.
20. Kahn, M. IEEE 802.11 Regulatory SC DSRC Coexistence Tiger Team V2V Radio Channel Models. Doc.: IEEE 802.11-14, February 2014. Available online: https://mentor.ieee.org/802.11/dcn/14/11-14-0259-00-0reg-v2v-radio-channel-models.ppt (accessed on 22 October 2022).

electronics

MDPI

Article

Research on an Algorithm for High-Speed Train Positioning and Speed Measurement Based on Orthogonal Time Frequency Space Modulation and Integrated Sensing and Communication

Jianli Xie [1,*], Yong Hao [1], Cuiran Li [1] and Huiqin Wang [2]

[1] School of Electronics and Information Engineering, Lanzhou Jiaotong University, Lanzhou 730070, China; 12221829@stu.lzjtu.edu.cn (Y.H.); licr@mail.lzjtu.cn (C.L.)

[2] School of Computer and Communication, Lanzhou University of Technology, Lanzhou 730050, China; whq1222@lut.edu.cn

* Correspondence: xiejl@mail.lzjtu.cn

Abstract: The Doppler effect caused by the rapid movement of high-speed rail services has a great impact on the accuracy of train positioning and speed measurement. Existing train positioning algorithms require a large number of trackside equipment and sensors, resulting in high construction and maintenance costs. Aiming to solve the above two problems, this article proposes a train positioning algorithm based on orthogonal time–frequency space (OTFS) modulation and integrated sensing and communication (ISAC). Firstly, based on the OTFS, the positioning and speed measurement architecture of communication awareness integration is constructed. Secondly, a two-stage estimation (TSE) algorithm is proposed to estimate the delay Doppler parameters of HST. In the first stage, a low-complexity coarse grid search is used, and in the second stage, a refined off-grid search is used to obtain the delay Doppler parameters. Then, the time difference of arrival/frequency difference of arrival (TDOA/FDOA) algorithm based on multiple base stations is used to locate the target, the weighted least square method is used to calculate the location, and the Cramér–Rao lower bound (CRLB) for positioning and speed measurement is derived. The simulation results demonstrate that, compared to GNSS/INS and OFDM radars, the algorithm exhibits enhanced positioning and speed measurement accuracy.

Keywords: OTFS; ISAC; parameter estimation; time difference of arrival; high-speed rail communication

1. Introduction

Train positioning technology plays a crucial role in ensuring the efficiency and safety of railway operations. Train positioning technology based on a single measurement method, such as track circuit positioning, odometer positioning, radar speed measurement positioning, etc., cannot meet the requirements of the positioning accuracy and the positioning continuity [1].

The current mainstream method for train positioning is combination positioning technology, among which the global navigation satellite system (GNSS) and inertial navigation system (INS) can provide accurate position information for trains. However, when a train passes through environments such as bridges, mountains and tunnels, the complex and changeable railway environment along the line can easily interfere with the transmission of GNSS signals, leading to a decline in positioning performance [2]. An inertial navigation system is an autonomous navigation system. Although it has a strong anti-interference ability, it is limited by the characteristics of its own micro inertial devices, and its position and attitude error information gradually accumulates, making it difficult to provide long-term accurate positioning information [3].

In the high-speed train (HST) positioning and speed measurement system, millimeter wave radar and light detection and ranging (LIDAR) can work in various environments, because they are not affected by light and weather conditions [4]. However, the construction and maintenance costs of high-precision radar systems and related equipment are high, requiring regular maintenance and calibration. Therefore, utilizing existing trackside base stations of the railway communication system to achieve low-cost and high-precision HST positioning and speed measurement is an unresolved problem.

Integrated sensing and communication (ISAC) is a new type of information processing technology that integrates communication and sensing functions into the same system, offering significant advantages in terms of equipment cost and size, system performance and spectrum utilization [5]. ISAC breaks the working mode of traditional communication or positioning systems, achieving both high-quality communication and high-precision positioning in one system [6]. In the HST scenarios, the ISAC modules can be added to the existing trackside base stations to realize the high-precision train positioning function. Compared with the current mainstream train positioning method, the ISAC-based positioning and speed measurement method has the following advantages: ISAC allows the communication and perception subsystems to share the same hardware, while realizing different functions, so it does not need to install additional sensors beside the track. Using the existing base station for reconstruction can realize the train positioning function and reduce the construction and maintenance costs. In the research on target sensing using ISAC systems, many studies have proposed corresponding solutions to improve the accuracy of target state sensing. In [7], the radial velocity of vehicle motion is obtained by using the echo signal reflected from the vehicle through a matched filter and the tracking error of the vehicle is minimized by using an extended Kalman filter (EKF). In [8], a novel and flexible orthogonal frequency division multiplexing (OFDM)-based sensing framework is developed, where the range-Doppler maps (RDMs) obtained using the cyclic cross-correlation (CCC) algorithm provide better sensing performance than the RDMs obtained based on the conventional point-by-point segmentation in the low-signal-to-noise ratio (SNR) region. In [9], to improve the accuracy of vehicle state sensing, a novel joint solution algorithm is proposed, which utilizes the structural features of the channel model, such as random channel fading, multipath propagation interference and the Doppler effect, and then employs downlink multicarrier signal echo sensing to derive the vehicle position and channel state and jointly estimate the vehicle speed.

With the gradual maturity of ISAC technology based on multicarrier waveforms, the application of the orthogonal time–frequency space (OTFS) modulation technique has attracted extensive attention from scholars. The core idea of OTFS is to use transforms to characterize the time–frequency bi-selective channels in the delay Doppler (DD) domain, so as to obtain a sparse channel that maintains good communication performance in high-mobility scenarios [10]. The OFDM is widely used in current wireless communication systems; the orthogonality between subcarriers is destroyed due to serious Doppler effect, which also makes OFDM modulation unable to carry out efficient and reliable communication in high-speed mobile environments [11]. The OTFS is more suitable than OFDM for communication transmission in high-speed mobility environments.

Research on OTFS-ISAC is currently on the rise [12]. OTFS is characterized by mapping data symbols and pilots into a DD domain grid in a specific way, where the size of the DD grid depends on the available bandwidth and signal duration, which is consistent with the concept of fast and slow time in radar signal processing [13,14]. By sensing the multipath delay and Doppler parameters of the target in the DD domain, the corresponding target distance and velocity can also be directly obtained. Therefore, the integration of OTFS and ISAC has been widely investigated.

In [15], regarding the theoretical performance indicators that OTFS-ISAC can achieve, the authors evaluate the effectiveness of OTFS modulation for ISAC transmission, proving that OTFS can maintain the same range and velocity estimation accuracy as frequency modular continuous wave (FMCW) radar. In [16], a scheme based on the maximum likelihood (ML) function is proposed in the OTFS-ISAC system based on multiple-input multiple-output (MIMO), which performs joint target detection and parameter estimation in discovery mode and high-resolution parameter estimation in tracking mode. Numerical results show that the proposed method can reliably detect multiple targets while approaching the Cramér–Rao lower bound (CRLB) of the corresponding parameter estimation problem.

Some studies consider the application of OTFS-ISAC in multi-base-station regimes but do not take into account HST application scenarios. Reference [17] discusses the problem of three-dimensional positioning of a single target in an ISAC system. Firstly, a joint position estimation model of direction of arrival (DOA) and time difference of arrival (TDOA) is established, and then an efficient two-step weighted least squares (TWLS) solution is derived. When the measurement error is large, the TWLS solution will reduce the positioning accuracy. Reference [18] proposes an integrated design approach in the delay Doppler domain, bypassing the time–frequency dimension, and significantly reduces the time–frequency resources required by the radar through a multi-station collaborative radar architecture. Reference [19] starts with the modulation principle of OTFS and delves deeper, layer by layer, to analyze the OTFS-ISAC system architecture under two different radar modes: monostatic radar and bistatic radar. It briefly discusses the communication and perception performance of OTFS-ISAC relative to OFDM. Reference [20] proposed a novel Doppler positioning scheme in HST scenarios, which arranges access points (APs) connected to base stations on the trackside and utilizes fractional Doppler estimation in OTFS to achieve higher positioning and velocity estimation accuracy, with lower requirements for bandwidth and hardware.

Based on the above analysis, this article proposes an HST positioning and velocity measurement algorithm based on OTFS-ISAC. The main contributions of this article are as follows:

1. Different from the existing high-speed train positioning and speed measurement methods, this paper proposes a communication perception integrated positioning model based on OTFS. The proposed algorithm does not require the deployment of additional trackside sensors and can achieve stable positioning and speed measurement of high-speed trains using only existing trackside base stations.
2. In order to improve the accuracy of target parameter estimation, a two-stage parameter estimation (TSE) algorithm is proposed, which uses the propagation characteristics of the DD domain to estimate parameters in the grid, and can accurately estimate the Doppler and delay parameters of high-speed trains.
3. Through the cooperation of multiple base stations, the information of TDOA and frequency difference of arrival (FDOA) are used, and then weighted least squares (WLS) methods are used to locate and measure the speed of high-speed trains.

The rest of this article is organized as follows: The communication-sensing integrated positioning model based on OTFS is designed in Section 2. Section 3 proposes the two-stage estimation (TSE) algorithm, which utilizes the propagation characteristics of the DD domain to accurately estimate the Doppler and time-delay parameters of the targets in the grid. In Section 4, multiple base stations collaborate to utilize TDOA and FDOA information and then use WLS for positioning and speed measurement of HSTs. We complete the simulation comparison experiment in Section 5. Finally, the conclusion of this article is presented in Section 6.

2. Train Positioning Model Based on OTFS-ISAC

2.1. Scene Modeling

Consider the OTFS-ISAC positioning scenario shown in Figure 1, where three base stations are located next to the railroad and equipped with integrated sensing modules for transmitting and receiving OTFS-ISAC signals. The HST, as the sensing target and the communication receiver, is equipped with a receiver on the roof to receive and demodulate the downlink communication signals sent by the base stations. The antenna arrays of the base stations are arranged in parallel with the running line of the HST, but the position and speed information of the HST are unknown. The base station has two functions: one is to communicate with the communication receiver, and the other is to accomplish the target speed measurement and distance measurement. When the HST moves forward at the speed of v, the base station transmits OTFS-ISAC signals, the communication receiver on the roof demodulates the received signals and obtains the communication signals, and at the same time, the HST reflects the integrated signals as a target, so that the reflected echo signals are received by the base station [21]. The integrated sensing module in the base station first uses a two-stage parameter estimation algorithm to estimate the parameters of the target and obtain the delay and Doppler parameters, and then uses TDOA/FDOA to locate and measure the speed of the target.

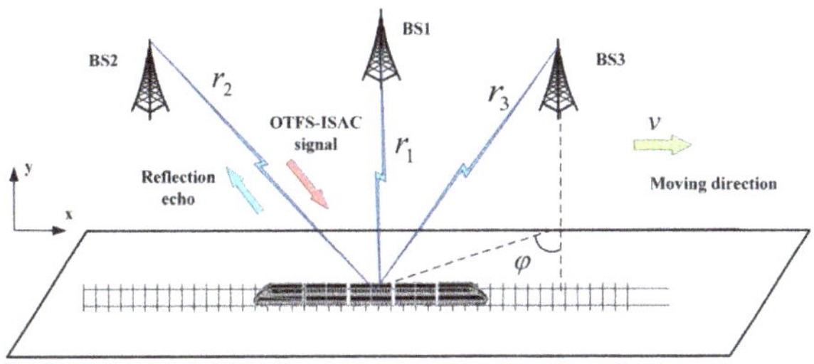

Figure 1. Schematic diagram of OTFS-ISAC positioning and speed measurement in the HST scenario.

2.2. Design of ISAC Signal Based on OTFS

As shown in Figure 2, the OTFS- ISAC transceiver adopts a configuration similar to that of a monostatic radar, consisting of an integrated transmitter and receiver. The system senses the target and estimates relevant parameters while communicating with the target. That is, the OTFS sensing integrated signal generated by the transmitter can achieve dual functions of transmission and perception simultaneously. For communication, the OTFS-ISAC transceiver can communicate bidirectionally with the user or the receiver at the target end. In terms of sensing, the active downlink sensing method is adopted. The integrated signal sent by the transmitter is reflected by the target, and the receiver processes the echo signal to complete the detection of target perception. After receiving the echo signal $r_R(t)$, it needs to be restored to the DD domain for speed and distance parameter estimation of the target.

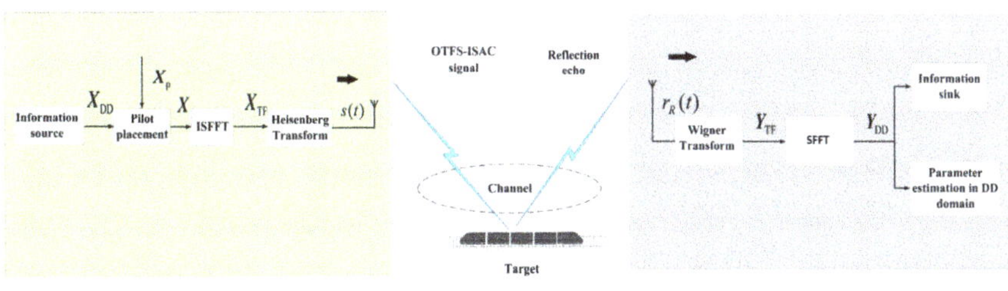

Figure 2. Block diagram of OTFS-ISAC system.

2.2.1. Transmitter

At the communication transmitter, the transmit modulation symbol vector $x \in \mathbb{C}^{1\times MN}$ of length MN is placed in the two-dimensional plane of the DD domain and arranged into a two-dimensional matrix X_{DD}, where N and M represent the number of OTFS symbols and the number of subcarriers, respectively. The $X_{DD}[l,k] \in X_{DD}$ represents the transmission modulation signal at the grid index of the k_{th} Doppler domain and the l_{th} delay domain in the two-dimensional plane grid of the delay Doppler domain. k and l are the Doppler and time delay grid indices for $l = 0, 1, \cdots, M-1, k = 0, 1, \cdots, N-1$. Therefore, X_{DD} is called the transmit modulation symbol matrix of the DD domain. In this article, we consider the superimposed frequency pilot scheme, in which the frequency pilot symbols

X_p are superimposed on X_{DD} in the delay Doppler domain, that is, $X = X_{DD} + X_p$, where the arrangement of the frequency pilot symbols can be expressed as

$$X_p[l,k] = \begin{cases} \sqrt{MN\sigma_p^2}, & (l,k) = (l_p,k_p) \\ 0, & others \end{cases} \tag{1}$$

where $0 \le l_p \le M-1$ and $0 \le k_p \le N-1$ denote the position of the frequency pilot symbol in the delay Doppler domain, and the average energy of the frequency pilot symbols can be defined as

$$E\left\{\left|X_p[l,k]\right|^2\right\} = \sigma_p^2 \tag{2}$$

The OTFS pilot design is shown in Figure 3, where data symbols are arranged on the entire delay Doppler plane, and non-zero pilot symbols are only placed in one delay Doppler grid (l_p, k_p), which is superimposed onto the data symbol $X_{DD}[l_p, k_p]$. Zero-padding is performed for pilots at other delay Doppler grids. Compared with the scheme of arranging protection areas around the embedded pilot (EP) [22], there is no dedicated grid for the pilot arrangement in the superimposed pilot scheme, which can reduce pilot overhead and improve spectral efficiency.

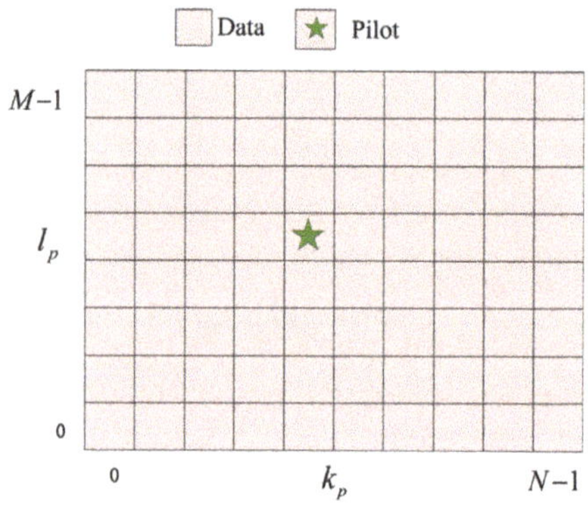

Figure 3. Delay Doppler domain frequency guidance design.

Then, the transmitter processes X and maps it to the time domain in two steps. In the first step, $X_{TF}[n,m]$ is mapped to the time–frequency domain using the inverse symplectic finite Fourier transform (ISFFT), as

$$X_{TF}[n,m] = \frac{1}{\sqrt{NM}} \sum_{k=0}^{N-1} \sum_{l=0}^{M-1} X_{DD}[k,l] e^{j2\pi(\frac{nk}{N} - \frac{ml}{M})} \tag{3}$$

where $m = 0, 1, \cdots, M-1$, $n = 0, 1, \cdots, N-1$ denote the frequency and time grid indexes, respectively. In the second step, the time-domain baseband signal of OTFS-ISAC can be obtained using the Heisenberg transform, as

$$s(t) = \sum_{n=0}^{N-1} \sum_{m=0}^{M-1} X_{TF}[n,m] g_{tx}(t - nT) e^{j2\pi m\Delta f(t-nT)} \tag{4}$$

where k and l are the Doppler and time delay grid indices, m and n are the frequency and time grid indices, the duration of each OTFS symbol is set to T, the carrier spacing is set to Δf, and g_{tx} represents the pulse shaping filter at the transmitter. After the baseband signal is upconverted, the OTFS-ISAC signal is sent through the transmit antenna.

2.2.2. Receiver

The transmitter generates and transmits the OTFS signal; the receiver on the roof of the HST receives the communication signal and demodulates it, while the receiver at the base station receives the echo signal reflected by the target and extracts the target parameter information. In the integrated communication and perception scenario, radar sensing typically operates in high-frequency bands such as sub-6 GHz and terahertz bands, which are dominated by line-of-sight transmission. Therefore, the channel between the base station transmitter and the target is considered to be a single direct channel. This article considers a time-delay Doppler channel with perception targets (or paths), as

$$h(\tau,\mu) = \sum_{i=1}^{P} h_i \delta(\tau - \tau_i)\delta(\mu - \mu_i) \tag{5}$$

where τ_i and μ_i are the delay and Doppler parameters of the i_{th} target, h_i represents the complex channel gain, and $\delta(\cdot)$ represents the Dirichlet function. After the transmitted OTFS signal passes through the DD domain radar channel, the received echo signal $r_R(t)$ at the receiver can be expressed as

$$r_R(t) = \iint h_R(\tau,\mu) s(t-\tau) e^{j2\pi\mu(t-\tau)} d\tau d\mu + \omega(t) \tag{6}$$

where $h_R(\tau,\mu) = h(\tau,\mu)e^{j2\pi f_c \tau}$ represents the equivalent DD domain radar sensing channel, and $\omega(t)$ represents additive Gaussian white noise. After the received signal $r(t)$ undergoes analog-to-digital (A/D) and serial–parallel conversion, the time-domain baseband received signal is obtained, that is, $r(t) = r_R(t)e^{-j2\pi f_c t}$; the specific expression is shown in (7).

$$r(t) = \iint h(\tau,\mu) s(t-\tau) e^{j2\pi\mu t} d\tau d\mu \tag{7}$$

Then, OTFS demodulation is performed on the $r(t)$. Firstly, the Wigner transform is used to enter the time–frequency (TF) domain to obtain the received modulation symbol matrix $Y_{TF} \in \mathbb{C}^{M\times N}$, which is expressed as

$$Y_{TF}[n,m] = \int_{\infty}^{\infty} r(t) g_{rx}(t-nT) e^{-j2\pi m\Delta f(t-nT)} \tag{8}$$

where g_{rx} stands for the receiver pulse shaping filter. Then, the received signal in the TF domain is transformed into DD domain through the symplectic finite Fourier transform (SFFT), as

$$Y_{DD}[k,l] = \frac{1}{\sqrt{MN}} \sum_{n=0}^{N-1} \sum_{m=0}^{M-1} Y_{TF}[n,m] e^{-j2\pi(\frac{nk}{N} - \frac{ml}{M})} \tag{9}$$

where $Y_{DD}[k,l] \in Y_{DD}$ denotes the received modulation signal in the k_{th} Doppler domain and l_{th} time-delay domain at the grid index on the DD domain. So far, the transformation of the received signal from the TF domain to the DD domain has been completed.

3. Target Parameter Estimation

Traditional radar signal processing algorithms often perform parameter estimation in the time–frequency domain, while OTFS radar can process signals in the DD domain and directly obtain the target's moving velocity by obtaining the target's Doppler frequency shift information [23]. This feature can be used to search on a two-dimensional (2D) grid in the DD domain, gradually narrowing down the area containing the true values of Doppler and delay parameters. In this section, the TSE algorithm is used to estimate the parameters of the target. In the first stage, a uniform grid is used for search to obtain rough estimates. Then, we continue off-grid search in the area established in the first stage. The proposed TSE algorithm process is described below.

At the first stage, in order to reduce the uncertainty area, a grid parameter search is performed on the discrete grid of the delay Doppler plane, as

$$\Lambda_i = \left\{ \left(\frac{l}{M\Delta f}, \frac{k}{NT} \right), l = 0\ldots, M-1, k = -\frac{N}{2}, \ldots, \frac{N}{2} - 1 \right\} \tag{10}$$

When the values of delay and Doppler are integer multiples of the delay resolution and Doppler resolution, $\tau_i = \frac{l_i}{M\Delta f}$, $\mu_i = \frac{k_i}{NT}$, the received signal $Y[l,k]$ in the time-delay Doppler domain can be approximated as a two-dimensional circular shift in the transmitted signal [24], as

$$Y[l,k] \approx \sum_{i=1}^{P} e^{j2\pi(\frac{l-l_i}{M})\frac{k_i}{N}} \alpha_i(l,k) X[[l-l_i]_M, [k-k_i]_N] + \widetilde{W}[l,k] \tag{11}$$

where $\alpha_i(l,k) = \begin{cases} 1 & l_i \leq l \leq M \\ \frac{N-1}{N} e^{-j2\pi(\frac{[k-k_i]_N}{N})} & 0 \leq l \leq l_i \end{cases}$ represents the linear phase shift in the Doppler domain caused by the cyclic displacement of time-domain samples. $\widetilde{W}[l,k]$ denotes the noise. Then, the maximum estimate obtained is

$$(\hat{l}_i, \hat{k}_i) = \underset{(\hat{l}_i,\hat{k}_i)\in\Lambda}{\text{argmax}} \left| (\Gamma_i x)^H \left(y - \sum_{j=1}^{i-1} \Gamma_j x(\hat{\tau}_j, \hat{\mu}_j) \right) \right|^2 \tag{12}$$

where $\Gamma_i x$ represents the useful signal and $\Gamma_j x$ represents the interfering signal. By using this formula, the index of time delay Doppler in the grid can be obtained. At this point, the estimated time delay $\hat{\tau}_i$ is set between $\frac{\hat{l}_i-1}{M\Delta f}$ and $\frac{\hat{l}_i+1}{M\Delta f}$; the estimated Doppler $\hat{\mu}_i$ is set between $\frac{\hat{k}_i-1}{NT}$ and $\frac{\hat{k}_i+1}{NT}$. Therefore, the lower limit and upper limit of the uncertainty area are determined. The search area of the second stage is

$$\Lambda_i = \left\{ (\tau,\mu), \frac{\hat{l}_i-1}{M\Delta f} \leq \tau \leq \frac{\hat{l}_i+1}{M\Delta f}, \frac{\hat{k}_i-1}{NT} \leq \mu \leq \frac{\hat{k}_i+1}{NT} \right\} \tag{13}$$

At the second stage, it is necessary to perform an off-grid search on the region Λ_i established in the first stage and solve the two-dimensional maximization problem in (14)

$$(\hat{\tau}_i, \hat{\mu}_i) = \arg \max_{(\tau_i,\mu_i)\in\Lambda_i} \left| (\Gamma_i x)^{\mathrm{H}} y_i \right|^2 \tag{14}$$

where $y_i = y - \sum_{j=1}^{i-1} \Gamma_j x(\hat{\tau}_j, \hat{\mu}_j)$; the meaning is to subtract the interference term $\Gamma_j x$ from the received signal y. Using the golden section, which belongs to the interval shrinkage method, the interval containing the best Doppler and delay solutions is gradually shrunk until the length of the interval is zero, and the shrinkage ratio $\rho \in [0,1]$ is set to $\frac{\sqrt{5}-1}{2}$. Based on the estimated delay and Doppler, the target distance and velocity can be calculated as $\hat{d}_i = \frac{\hat{\tau}_i c}{2}$ and $\hat{v}_i = \frac{\hat{\mu}_i c}{2 f_c}$.

When estimating the azimuth information in the angular domain, the maximum likelihood estimation algorithm is utilized to estimate the angular domain information [25], as

$$\hat{\varphi} = \underset{\varphi}{\text{argmax}}\ p(\mathrm{r}_R(t)|\varphi) \tag{15}$$

where φ represents the included angle between the connecting line between the base station and HST and the relative movement velocity v_0. After completing the parameter estimation using the two-stage parameter estimation algorithm, the Doppler and delay values of the HST can be obtained by each base station, respectively, which are converted into TDOA/FDOA information after differentiation and are processed for the localization and velocimetry solving in the next chapter.

The proposed two-stage parameter estimation algorithm is shown in Algorithm 1.

Algorithm 1: Proposed two-stage parameter estimation algorithm.

Input: Received vector y, transmit vector x.
Output: Estimated time-delay Doppler parameters $(\hat{\tau}, \hat{\mu})$.
Initialization: $y_i = y - \sum_{j=1}^{i-1} \Gamma_j x(\hat{\tau}_j, \hat{\mu}_j), \rho = \frac{\sqrt{5}-1}{2}$;
for $i = 1 : P$ **do**
Calculating (12) yields the time-delay Doppler index $(\hat{l}_i, \hat{k}_i)$;
$a_u = \frac{\hat{l}_i - 1}{M\Delta f}, a_l = \frac{\hat{l}_i + 1}{M\Delta f}, b_u = \frac{\hat{k}_i + 1}{NT}, b_l = \frac{\hat{k}_i - 1}{NT}$;
repeat
    $I_a = a_u - a_l, I_b = b_u - b_l$;
    $a_1 = a_l + (1-\rho)I_a, a_2 = a_l + \rho I_a$;
    $b_1 = b_l + (1-\rho)I_b, b_2 = b_u + \rho I_b$;
    $f_{11} = \left|(\Gamma_i(a_1, b_1)x)^H y_i\right|^2, f_{12} = \left|(\Gamma_i(a_1, b_2)x)^H y_i\right|^2$;
    $f_{21} = \left|(\Gamma_i(a_2, b_1)x)^H y_i\right|^2, f_{22} = \left|(\Gamma_i(a_2, b_2)x)^H y_i\right|^2$;
    switch $\max\{f_{11}, f_{12}, f_{21}, f_{22}\}$ **do**
        case1 $f_{11}, a_u = a_2, b_u = b_2$;
        case2 $f_{12}, a_u = a_2, b_l = b_1$;
        case3 $f_{21}, a_l = a_1, b_u = b_2$;
        case4 $f_{22}, a_l = a_1, b_l = b_1$;
    end
until Stopping criteria: $\hat{\tau}_i = \frac{a_l + a_u}{2}, \hat{\mu}_i = \frac{b_l + b_u}{2}$;
end

4. Radar Positioning and Speed Measurement

In general, the process of mobile target localization can be roughly divided into two steps. The first step is to obtain measurement information for position estimation. Secondly, based on the measurement information, the positioning algorithms are used to estimate the position and velocity of the target. Due to the less strict synchronization requirements between transceivers, TDOA/FDOA is often used as measurement information for positioning. The most common method in TDOA/FDOA positioning and solving algorithms is WLS, which has the characteristics of easy implementation and high computational efficiency. Therefore, the WLS method is selected in this article for positioning and speed measurement calculation.

4.1. Positioning Method Based on TDOA/FDOA

TDOA and FDOA are two common positioning methods that calculate the distance and velocity between the nodes by measuring the arrival time difference and frequency difference of signals, thus achieving positioning and speed measurement. There are three types of TDOA/FDOA positioning scenarios: (1) fixed base station to mobile target; (2) mobile base station to fixed target; (3) mobile base station to mobile target. These positioning scenarios all utilize the measured FDOA information and the position and velocity information between the base stations to form a nonlinear equation system, and then obtain the position and velocity information of the target by solving the equation system. For FDOA positioning, Doppler shift can be described as

$$\mu = \frac{f_0}{c} v_0 \cos \varphi \tag{16}$$

where v_0 is the relative velocity between the target and the base station; f_0 is the frequency of the transmitted signal.

This article considers a 2D localization and speed measurement scenario with an HST as the target, assuming that there are three fixed base stations with known positions and one target, one of which serves as a reference base station. The position and velocity vectors of the target to be estimated are $p = [x, y]^{\mathrm{T}}$ and $u = [\dot{x}, \dot{y}]^{\mathrm{T}}$, respectively. The position of the base station is $s = [x_i, y_i]^{\mathrm{T}}$. In this article, the superscript " $\cdot$ " notation is used to denote the corresponding rate of change, the right superscript "o" denotes the true value, and T denotes the transpose of the matrix. The actual distance between the target and the i_{th} base station is

$$r_i^o = \|s^o - p\| = \sqrt{(s^o - p)^{\mathrm{T}}(s^o - p)} \tag{17}$$

The TDOA measurement equation is

$$r_{i1}{}^o = r_i{}^o - r_1{}^o \tag{18}$$

Combining (17) with (18) and squaring them yields a specific set of TDOA equations, the expansion of which can be expressed as

$$r_{i1}^{o2} - s_i^{\mathrm{T}} s_i + s_1^{\mathrm{T}} s_1 = 2 r_{i1}^o r_1{}^o - 2(s_i - s_1) u^o \tag{19}$$

The time-difference equation can only estimate the position of the target, and for velocity estimation, it depends on the frequency-difference equation; taking the derivative of time in Equation (17) yields the relationship between the true distance change rate and the target position parameters, as

$$\dot{r}^o{}_i = \frac{(\dot{u}^o - \dot{s}_i)^{\mathrm{T}}(u - s_i)}{r_i{}^o} \tag{20}$$

The Doppler difference of the signal received by each base station and the reference base station can be written as

$$\mu_{i1}^o = \mu_i - \mu_1 = \frac{v_i f_0}{c} - \frac{v_1 f_0}{c} \tag{21}$$

To estimate the speed of the target, the FDOA equation can be obtained by taking the time derivative of (17), as

$$2(\dot{r}_{i1}^o r_{i1}^o + \dot{r}_{i1}^o r_i^o + r_{i1}^o \dot{r}_1^o) = 2(\dot{s}_i^{\mathrm{T}} - \dot{s}_1^{\mathrm{T}} s_1 - (\dot{s}_i - \dot{s}_1)^{\mathrm{T}} u^o - (s_i - s_1)^{\mathrm{T}} \dot{u}^o) \tag{22}$$

where $\dot{r}_{i1}^o$ represents the range different rate (RDR) for the variation in distance difference between the base stations. It can be found that TDOA and FDOA are closely related in the process of establishing the equations, and the relationship between them needs to be considered in order to realize the target localization and speed measurement.

Different from the ideal situation, in practical applications, when the base station receives the reflected signal for target positioning, it can only obtain the observed values of the time difference and frequency difference data, which contain measurement noise. The measured values of the distance difference vector and the distance difference rate of the change vector can be expressed as

$$\begin{cases} r = r^o + n_t \\ \dot{r} = \dot{r}^o + n_\mu \end{cases} \tag{23}$$

where n_t and n_μ are the measurement noise vectors of time and frequency differences, obeying the Gaussian distribution with zero mean and variance σ^2, and the covariance matrices are Q_t and Q_μ, respectively. The two measurements r and $\dot{r}$ are merged to form the localization parameter measurement vector $\beta = [r^{\mathrm{T}}, \dot{r}^{\mathrm{T}}]$, and the covariance matrix of the error vector between the true value and the measured value is $Q_\beta = E[\Delta\beta\Delta\beta^{\mathrm{T}}]$.

4.2. Weighted Least Squares Solutions

Least squares (LS) is a commonly used algorithm for solving localization problems, which essentially requires minimizing the sum of squares of estimation errors. The WLS assigns different weights to the observed values during the calculation process, essentially weighting the various items of the model to eliminate heteroscedasticity in the original mathematical model before parameter optimization. The positioning calculation process of WLS for TDOA/FDOA is as follows:

Firstly, define the variables to be measured $\theta^o = \left[u^{o\mathrm{T}}, r_1^o, \dot{u}^{o\mathrm{T}}, \dot{r}_1^o\right]^{\mathrm{T}}$, which include the unknown position of the target $u^{o\mathrm{T}}$, the velocity $\dot{u}^{o\mathrm{T}}$, the distance between the target and the reference base station r_1^o and the rate of change in the distance to the main base station $\dot{r}_1^o$. These four variables are uncorrelated. At this point, by replacing the observed values of (19) and (22) with the true values, the error vector of the positioning model obtained by combining them is

$$\lambda = h - G\theta \tag{24}$$

where

$$h = \begin{bmatrix} r_{21}^2 - s_2^{\mathrm{T}} s_2 + s_1^{\mathrm{T}} s_1 \\ r_{31}^2 - s_3^{\mathrm{T}} s_3 + s_1^{\mathrm{T}} s_1 \\ \vdots \\ 2(\dot{r}_{21} r_{21} - \dot{s}_2^{\mathrm{T}} s_2 + \dot{s}_1^{\mathrm{T}} s_1) \\ 2(\dot{r}_{31} r_{31} - \dot{s}_3^{\mathrm{T}} s_3 + \dot{s}_1^{\mathrm{T}} s_1) \end{bmatrix} \tag{25}$$

$$G = \begin{bmatrix} (s_2 - s_1)^{\mathrm{T}} & r_{21} & 0^{\mathrm{T}} & 0 \\ (s_3 - s_1)^{\mathrm{T}} & r_{31} & 0^{\mathrm{T}} & 0 \\ \vdots & \vdots & \vdots & \vdots \\ (\dot{s}_2 - \dot{s}_1)^{\mathrm{T}} & \dot{r}_{21} & (s_2 - s_1)^{\mathrm{T}} & r_{21} \\ (\dot{s}_3 - \dot{s}_1)^{\mathrm{T}} & \dot{r}_{31} & (s_3 - s_1)^{\mathrm{T}} & r_{31} \end{bmatrix} \tag{26}$$

where 0^{T} is a 1×3 row vector and the least squares estimate of θ^o is

$$\theta = \left[u^{\mathrm{T}}, r_1, \dot{u}^{\mathrm{T}}, \dot{r}_1\right]^{\mathrm{T}} = (G^{\mathrm{T}} W G)^{-1} G^{\mathrm{T}} W h \tag{27}$$

The weighting matrix in (27) is defined as follows:

$$W = B^{-\mathrm{T}} Q^{-1} B^{-1} \tag{28}$$

In the above equation, $B = \begin{bmatrix} B_t & O \\ B_\mu & B_t \end{bmatrix}$ and $B_t = 2diag\{r_2^o, r_3^o\}$ are the diagonal matrices of the distance between the base stations, $B_\mu = 2diag\left\{\dot{r}_2^0, \dot{r}_3^0\right\}$ is the diagonal matrix of the rate of change in the distance, and $Q = \begin{bmatrix} Q_t & O \\ O & Q_\mu \end{bmatrix}$ is the covariance matrix of the time–frequency difference data. Solve (27) to obtain the position and speed of the target; use the solution to update the weight matrix W, and substitute it into (27) again to complete the position estimation of the target.

5. Simulation Results and Performance Analysis

5.1. Performance Assessment Indicators

5.1.1. Root Mean Square Error

The performance of positioning and speed measurement can be evaluated and analyzed using root mean square error (RMSE), which is defined as

$$\text{RMSE} = \sqrt{\frac{1}{\text{n}}\sum_{i=1}^{m}(y_i - \hat{y}_i)^2} \tag{29}$$

The RMSE reflects the deviation between the estimated value and the true value of target speed measurement and positioning, which is easy to calculate and understand. The smaller the RMSE value, the more accurate the parameter estimation.

5.1.2. CRLB Performance Analysis

The CRLB is the lower bound of the unbiased estimator variance. To evaluate the radar performance of the integrated sensing algorithm proposed in this article, this section provides a performance analysis of CRLB. The true values of the position and speed of the HST to be located are given as $\theta^o = \left[u^{o\text{T}}, r_1^o, \dot{u}^{o\text{T}}, \dot{r}_1^o\right]^\text{T}$, the true value of the base station position is given as p_{BS}^o, the total unknown vector is defined as $\phi^o \left[\theta^{o^\text{T}}, p_{BS}^{o\text{T}}\right]^\text{T}$, and the total observation vector is defined as $z = \left[\beta^\text{T}, p_{BS}^{o\text{T}}\right]^\text{T}$. Firstly, calculate the joint probability density function (which has been taken as the natural logarithm) of the observation vector z under the given ϕ^o, as

$$\begin{array}{l} \ln f(z|\phi^o) = \ln f(\beta|\theta^{o^\text{T}}, p_{BS}^o) + \ln f(p_{BS}|\theta^{o^\text{T}}, p_{BS}^o) = \\ K - \frac{1}{2}[\beta - \beta^o]^\text{T} Q_\beta^{-1}[\beta - \beta^o] - \frac{1}{2}[p_{BS} - p_{BS}^o]^\text{T} Q_p^{-1}[p_{BS} - p_{BS}^o] \end{array} \tag{30}$$

Then, derive the Fisher Information Matrix (FIM) as follows:

$$\text{FIM}(\phi) = -\text{E}\left(\frac{\partial \ln^2 p(z|\phi^o)}{\partial\phi^o \partial\phi^{o\text{T}}}\right) = \text{E}\left[\left(\frac{\partial \ln p(z|\phi^o)}{\partial\phi^o}\right)\left(\frac{\partial \ln p(z|\phi^o)}{\partial\phi^o}\right)^\text{T}\right] \tag{31}$$

Based on the relationship that the CRLB is the reciprocal of the FIM, the CRLB of the parameter ϕ^o to be estimated can be obtained, as

$$\text{CRLB}(\phi^o) = [FIM(\phi^o)]^{-1} = \begin{bmatrix} X & Y \\ Y^\text{T} & Z \end{bmatrix}^{-1} \tag{32}$$

where X, Y, Z, respectively, denote

$$\begin{cases} X = \left(\frac{\partial\beta^o}{\partial\theta^{o\text{T}}}\right)^\text{T} Q_\beta^{-1}\left(\frac{\partial\beta^o}{\partial\theta^{o\text{T}}}\right) \\ Y = \left(\frac{\partial\beta^o}{\partial\theta^{o\text{T}}}\right)^\text{T} Q_\beta^{-1}\left(\frac{\partial\beta^o}{\partial p_{BS}^{o\text{T}}}\right) \\ Z = \left(\frac{\partial\beta^o}{\partial p_{BS}^{o\text{T}}}\right)^\text{T} Q_\beta^{-1}\left(\frac{\partial\beta^o}{\partial p_{BS}^{o\text{T}}}\right) + Q_p^{-1} \end{cases} \tag{33}$$

where Q_β and Q_p represent the covariance matrices of the error vectors between the true and measured values about β and p_{BS}, respectively. In practical applications, the main focus is on the position and velocity parameters of the target θ^o, and the corresponding matrix $\text{CRLB}(\theta^o)$ represents the 6×6 sub-matrix on the upper left of $\text{CRLB}(\phi)$, so the CRLB of θ^o can be expressed as

$$\text{CRLB}(\theta^o) = (X - YZ^{-1}Y^\text{T})^{-1} \tag{34}$$

The horizontal axis in Figure 4 represents the power of noise, expressed in logarithmic form and measured in dB. The vertical axis represents the RMSE of position or velocity estimation, measured in meters or m/s. From Figures 4 and 5, it can be seen that the accuracy of position estimation is higher than that of velocity estimation. When the measurement noise is below 4 dB, the RMSE for target localization can match the CRLB. However, as the noise level continues to increase, the position error also increases significantly. The RMSE for target speed measurement can reach CRLB when the measurement noise is below 0 dB. As the noise level continues to increase, the speed measurement error also increases significantly.

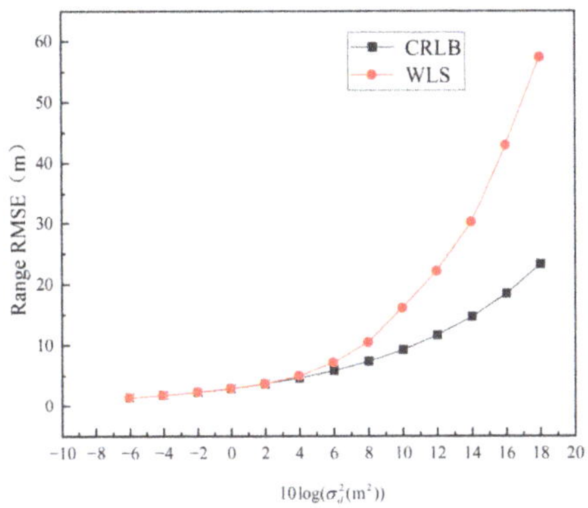

Figure 4. Comparison between RMSE of position estimation and CRLB.

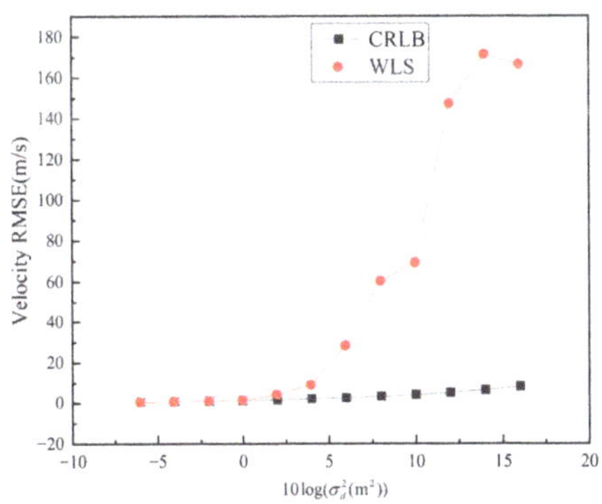

Figure 5. Comparison between RMSE of velocity estimation and CRLB.

5.2. Radar Performance Analysis

In order to verify the effectiveness of the proposed scheme, this section will simulate and validate the radar communication integrated signal based on OTFS and compare the algorithm proposed in this article with the multiple signal classification (MUSIC) algorithm and the CCC algorithm [8]. The simulation parameters are shown in Table 1. Table 2 shows the actual position coordinates of the three base stations and Table 3 shows the estimated position and velocity of the targets. Under the experimental parameters set in the following three tables, the simulation platform of this experiment is Matlab 2023a. The running environment is Intel(R) Core (TM) i7-13700F (2.1GHz), NVIDIA RTX4070Ti.

Table 1. Simulation parameters.

Parameters	Value	Parameters	Value
Carrier frequency/GHz	3	Base station pacing/m	500
Symbolic average energy	1	Number of iterations	100
Number of subcarriers	64	Radar SNR/dB	5
Subcarrier spacing	32	Cyclic prefix length	128
HSR speed/(m/s)	0–139	Distance/m	30

Table 2. True position of the base stations.

Name	x_i/m	y_i/m
Base station 1	0	0
Base station 2	2000	40
Base station 3	4000	60

Table 3. Estimated position and velocity of the target.

Number	r_i/m	$\hat{r}_i$/m	v_i/m/s	$\hat{v}_i$/m/s
1	30	29.98	97.22	97.86
2	40	39.98	111.11	110.92
3	100	100.16	138.89	139.53

Figure 6 shows the echograms of the radar signals received by the three base stations at an SNR of 10 dB. The x and y coordinates of the spike position correspond to the speed and distance of the target, respectively, and the z-axis represents the amplitude of the radar signals, with the spike position z = 0 dB. Usually, closer targets or higher-power radar systems produce stronger echo signals, which are manifested by higher spike amplitudes. By using the parameter estimation algorithm in Section 3 to process radar signal echoes, the time delay and Doppler information of the target can be obtained. Based on the arrival time difference and Doppler information obtained from each base station, the positioning and velocity estimation algorithm proposed in Sections 4.1 and 4.2 is used to estimate the position and velocity of the target.

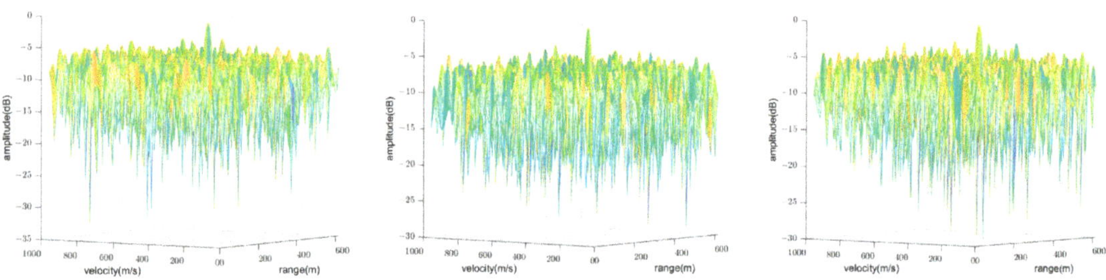

(**a**) Base Station 1 (**b**) Base Station 2 (**c**) Base Station 3

Figure 6. Radar signal echo diagram.

We compare the MUSIC algorithm and CCC algorithm with the algorithm proposed in this article when the SNR varies from 0 dB to 10 dB under the condition of an HST movement speed of 139 m/s. The RMSE is used to measure the radar detection performance and is represented by the vertical axis. The simulation comparison results are shown in Figure 7. Under the same SNR of 10 dB, the speed estimation of the perception algorithm in this article can achieve an accuracy of decimeters per second, which has significant advantages compared to the other two algorithms.

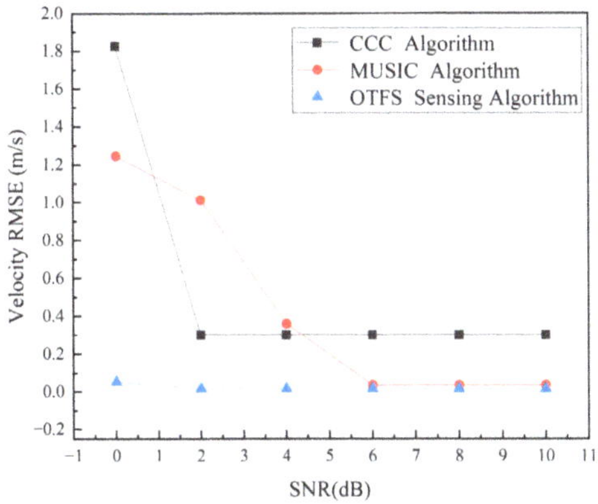

Figure 7. Velocity tracking error of algorithms under different SNR conditions.

Figure 8 shows the RMSE of distance estimation for each algorithm under different SNR conditions. It can be observed that the three algorithms are less affected by noise, but the two-stage estimation algorithm used in this article achieves millimeter-level distance estimation accuracy with an SNR of 10 dB.

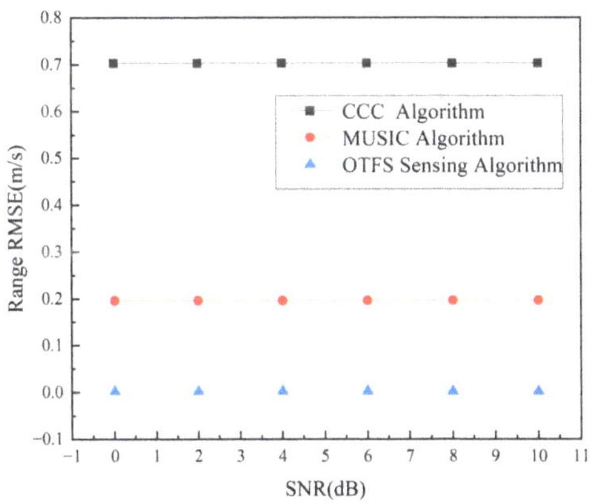

Figure 8. Distance estimation error of algorithms under different SNR conditions.

Next, the proposed algorithm is compared with the widely used GNSS/INS combined positioning and speed measurement algorithm in train positioning. The data fusion part of GNSS/INS uses EKF. The two algorithms are compared in the scenario where the train runs at a speed of 0–139 m/s. As shown in Figure 9, in the stationary state of the train, the measurement error of both positioning methods for train speed is 0, and with the increase in speed, the RMSE of GNSS/INS for speed increases from 0 to 7.71. The RMSE of the OTFS-ISAC perception algorithm in this article is maintained around 1, with lower estimation bias and significant advantages compared to GNSS/INS.

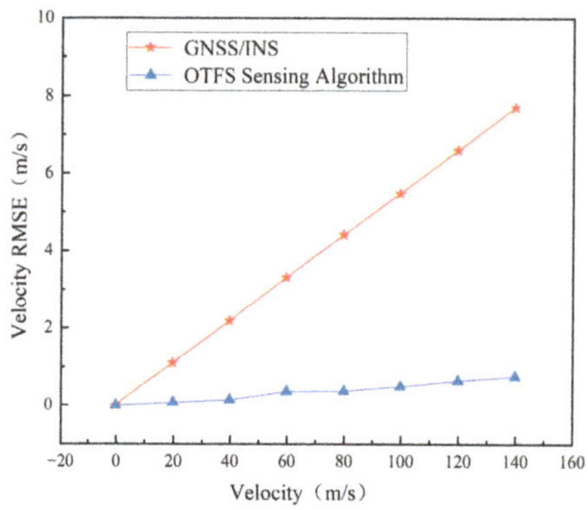

Figure 9. Comparison of proposed algorithms with GNSS/INS at different speeds.

5.3. Complexity and Time Performance Analysis

This section briefly analyzes the complexity and time performance analysis of the proposed TSE algorithm. The algorithm realizes the super-resolution estimation of the sensing target parameters. The estimation result can be a non-integer multiple of the system delay and Doppler resolution. The proposed TSE algorithm helps to reduce the complexity of the system, because it excludes the dictionary matrix that needs a lot of calculation in the second step of refined search. The specific operation is to re-establish a new smaller area Λ_i to conduct off-grid search after the first stage of rough estimation, thus greatly reducing the search interval and ultimately reducing the complexity of the algorithm. At the same time, the algorithm is conducted in the time domain, so it does not need to convert to the time-delay Doppler domain for parameter estimation. The complexity of the TSE algorithm in this article is $O(MN\log(MN))$.

In addition, this article compared the time of the proposed algorithm with the CCC algorithm [8] and two-dimensional fast Fourier transform (2D-FFT) algorithm in a single run, and recorded the running time of each algorithm five times in a row. It can be seen that the running time of the proposed TSE algorithm is relatively short. The comparison results are shown in Table 4.

Table 4. Comparison of running times of different algorithms.

Parameter Estimation Algorithm	Running Time (Unit: Second)				
TSE algorithm	0.234	0.221	0.202	0.210	0.212
CCC algorithm [8]	0.764	0.721	0.795	0.807	0.744
2D-FFT algorithm	0.537	0.587	0.609	0.629	0.688

6. Conclusions

Based on the modulation principle of OTFS and the characteristics of signal processing in the DD domain, OTFS is integrated with communication perception technology and applied in HST speed measurement and positioning scenarios. This article proposes an OTFS-based integrated sensing positioning and speed measurement algorithm for HST scenarios. Firstly, the composition structure of the radar communication integrated system based on OTFS modulation is analyzed, and the system scenario is analyzed. System parameters that meet the performance indicators of the radar and communication system were designed, and the TSE algorithm was used to estimate the speed and distance parameters of the HST. Then, the train is located and measured using TDOA/FDOA, and the CRLB for positioning and speed measurement of this method is derived. The comparative experimental results show that the algorithm proposed in this article has better positioning and speed measurement performance compared to OFDM radar and GNSS/INS combined positioning algorithms.

In terms of radar operating mode, this article only uses the simpler monostatic radar scheme. Therefore, the next step of work will consider expanding to the bistatic radar system to adapt to complex HST positioning and speed measurement scenarios with higher detection range and accuracy. In future research, the OTFS-based communication perception integrated positioning and speed measurement algorithm will play an important role in HST autonomous driving, natural disaster monitoring, and improving the safe transportation capacity of HST.

Author Contributions: Conceptualization, J.X., Y.H., C.L. and H.W.; methodology, J.X., Y.H., C.L. and H.W.; validation, J.X. and Y.H.; writing—original draft preparation, Y.H.; writing—review and editing, J.X. and Y.H.; supervision, J.X. and C.L. All authors have read and agreed to the published version of the manuscript.

Funding: This research was funded by the National Natural Science Foundation of China (62161016).

Data Availability Statement: Data are contained within the article.

Conflicts of Interest: The authors declare no conflicts of interest. The funders had no role in the design of the study; in the collection, analyses, or interpretation of data; in the writing of the manuscript; or in the decision to publish the results.

References

1. Mo, Z.; An, H. Research on Comprehensive Autonomous Positioning Technology of New Train Control System. *J. China Railw. Soc.* **2022**, *44*, 56–64.
2. Liu, D.; Jiang, W.; Cai, B. Train Localization Method for Tightly-coupled GNSS/INS Integrated Strategy Based on Simplified Robust UKF. *J. China Railw. Soc.* **2023**, *45*, 62–71.
3. Wang, Z.; Wang, Y.; Li, W. Research on BDS Information Fusion Algorithm of JingZhang High Speed Intelligent EMU. *J. Dalian Jiaotong Univ.* **2023**, *44*, 101–106.
4. Wang, Z.; Yu, G.; Zhou, B.; Wang, P.; Wu, X. A train positioning method based-on vision and millimeter-wave radar data fusion. *IEEE Trans. Intell. Transp. Syst.* **2021**, *23*, 4603–4613. [CrossRef]
5. Zhu, L.; He, D.; Zhong, Z. Application of Radar Communication Integration in Intelligent Railway. *China Railw.* **2021**, *1*, 17–24.
6. Wen, J.; Yi, X.; Han, K.; Chen, L. A New Paradigm of High-Precision Localization: Integrated Communication and Localization Technology and Application. *Mob. Commun.* **2023**, *47*, 17–25+109.
7. Liu, F.; Yuan, W.; Masouros, C.; Yuan, J. Radar-assisted predictive beamforming for vehicular links: Communication served by sensing. *IEEE Trans. Wirel. Commun.* **2020**, *19*, 7704–7719. [CrossRef]
8. Wu, K.; Zhang, J.A.; Huang, X.; Guo, Y.J. Integrating low-complexity and flexible sensing into communication systems. *IEEE J. Sel. Areas Commun.* **2022**, *40*, 1873–1889. [CrossRef]
9. Zhou, B.; Ma, S. Simultaneous vehicular location and velocity detection towards 6G integrated communication and sensing. *J. Commun.* **2023**, *44*, 81–92.
10. Zhang, K.; Yuan, W.; Li, S.; Liu, F.; Gao, F.; Fan, P.; Cai, Y. Radar sensing via otfs signaling: A delay doppler signal processing perspective. In Proceedings of the ICC 2023-IEEE International Conference on Communications, Rome, Italy, 28 May–1 June 2023; pp. 6429–6434.
11. Yuan, W.; Wei, Z.; Li, S.; Yuan, J.; Ng, D.W.K. Integrated sensing and communication-assisted orthogonal time frequency space transmission for vehicular networks. *IEEE J. Sel. Top. Signal Process.* **2021**, *15*, 1515–1528. [CrossRef]
12. Wei, Z.; Li, S.; Yuan, W.; Schober, R.; Caire, G. Orthogonal time frequency space modulation—Part I: Fundamentals and challenges ahead. *IEEE Commun. Lett.* **2022**, *27*, 4–8. [CrossRef]
13. Bondre, A.S.; Richmond, C.D. Dual-use of OTFS architecture for pulse Doppler radar processing. In Proceedings of the 2022 IEEE Radar Conference (RadarConf22), New York, NY, USA, 21–25 March 2022; pp. 1–6.
14. Wang, X.; Xiaogang, T.; Yiqing, Z.; Chong, Z.; Zhengang, P. Survey of channel estimation method in delay-Doppler domain for OTFS. *J. Commun.* **2022**, *43*, 188–201.
15. Gaudio, L.; Kobayashi, M.; Caire, G.; Colavolpe, G. On the effectiveness of OTFS for joint radar parameter estimation and communication. *IEEE Trans. Wirel. Commun.* **2020**, *19*, 5951–5965. [CrossRef]
16. Dehkordi, S.K.; Gaudio, L.; Kobayashi, M.; Caire, G.; Colavolpe, G. Beam-space MIMO radar for joint communication and sensing with OTFS modulation. *IEEE Trans. Wirel. Commun.* **2023**, *22*, 6737–6749. [CrossRef]
17. Liu, R.Y.; Liu, R.; Li, J.; Pan, Z. Three-Dimensional Joint Localization for Integrated Sensing and Communication Systems. *Mob. Commun.* **2024**, *48*, 90–94.
18. Wang, J.; Ma, H.; Liang, X.; Li, Y.; Chen, L. Research on Integrated Radar Communication Technology Based on Delay-Doppler Modulation. *J. Signal Process.* **2023**, *39*, 1047–1059.
19. Sun, C.; Li, L.; ZHANG, H. OTFS-enabled integrated sensing and communication techniques for next-generation V2X networks. *Chin. J. Eng.* **2023**, *45*, 1674–1683.

20. Wu, Y.; Jia, H.; Nakawaki, Y.; Pan, Z.; Shimamoto, S. Doppler-Only Wireless Positioning for High-Speed Railway Based on Fractional Doppler Estimation. In Proceedings of the 2024 IEEE Wireless Communications and Networking Conference (WCNC), Dubai, United Arab Emirates, 21–24 April 2024; pp. 1–6.
21. Zhao, S.; Du, H.; Liu, Z.; Yu, J. OTFS Radar Communication Integrated Multiplexed Waveform Design. *J. Signal Process.* **2024**, *40*, 1239–1248.
22. Raviteja, P.; Phan, K.T.; Hong, Y. Embedded pilot-aided channel estimation for OTFS in delay–Doppler channels. *IEEE Trans. Veh. Technol.* **2019**, *68*, 4906–4917. [CrossRef]
23. Xiong, L.; Liu, X.; Gong, J.; Peng, M. Orthogonal Time Frequency Space Modulation-based Sensing Signal Processing Algorithms. *Radio Commun. Technol.* **2023**, *49*, 100–109.
24. Wu, Y.; Han, C.; Chen, Z. DFT-spread orthogonal time frequency space system with superimposed pilots for terahertz integrated sensing and communication. *IEEE Trans. Wirel. Commun.* **2023**, *22*, 7361–7376. [CrossRef]
25. Du, Z.; Liu, F.; Yuan, W.; Masouros, C.; Zhang, Z.; Xia, S.; Caire, G. Integrated sensing and communications for V2I networks: Dynamic predictive beamforming for extended vehicle targets. *IEEE Trans. Wirel. Commun.* **2022**, *22*, 3612–3627. [CrossRef]

electronics

MDPI

Article

AIDETECT2: A Novel AI-Driven Signal Detection Approach for beyond 5G and 6G Wireless Networks

Bibin Babu [1], Muhammad Yunis Daha [1], Muhammad Ikram Ashraf [2], Kiran Khurshid [3] and Muhammad Usman Hadi [1,*]

[1] School of Engineering, Ulster University, Belfast BT15 1AP, UK; babu-b2@ulster.ac.uk (B.B.); daha-my@ulster.ac.uk (M.Y.D.)
[2] Nokia, 02610 Espoo, Finland; ikram.ashraf@nokia.com
[3] Department of Computer and Software Engineering, College of Electrical and Mechanical Engineering, National University of Sciences and Technology (NUST), Islamabad 44000, Pakistan; kiran.khurshid@ceme.nust.edu.pk
* Correspondence: m.hadi@ulster.ac.uk

Abstract: Artificial intelligence (AI) is revolutionizing multiple-input-multiple-output (MIMO) technology, making it a promising contender for the coming sixth-generation (6G) and beyond-fifth-generation (B5G) networks. However, the detection process in MIMO systems is highly complex and computationally demanding. To address this challenge, this paper presents an optimized AI-based signal detection method known as AIDETECT-2 which is based on feed forward neural network (FFNN) for MIMO systems. The proposed AIDETECT-2 network model demonstrates superior efficiency in signal detection in comparison with conventional and AI-based MIMO detection methods, particularly in terms of symbol error rate (SER) at various signal-to-noise ratios (SNR). This paper thoroughly explores various signal detection aspects using FFNN, including the design of system architecture, preparation of data, training processes of the network model, and performance evaluation. Simulation results show that the proposed model demonstrates a significant performance improvement ranging between 13.75% to 99.995% better SER compared to the best conventional method and also achieved between 56.52% to 97.69 better SER compared to benchmark AI-based MIMO detectors at 20 dB SNR for given MIMO scenarios respectively. It also presented the computational complexity analysis of different conventional and AI-based MIMO detectors. We believe that this optimized AI-based network model can serve as a comprehensive guide for deploying deep-learning (DL) neural networks for signal detection in the forthcoming 6G wireless networks.

Keywords: AIDETECT2; beyond 5G networks; MIMO detection; 6G

1. Introduction

We are now in the beyond-5G (B5G) era, also called sixth-generation (6G) networks, where millions of internet-connected devices compete for connectivity every day, and this number keeps increasing. This surge puts immense pressure on network resources, especially regarding energy efficiency, power consumption, reliability, and latency. Meeting these ever-growing network demands has become a significant challenge [1]. To address these needs, multiple-input-multiple-output (MIMO) technology, in combination with artificial intelligence (AI), plays a vital role in the development of 6G and B5G networks. MIMO technologies have brought remarkable improvements in wireless communication, especially when compared to older generation technologies such as single-input multiple-output (SIMO), multiple-input single-output (MISO), and single-input single-output (SISO) in terms of throughput, channel selection, spatial diversity, and spectral efficiency [2]. When the number of transmitting and receiving antennas is large, typically from tens to hundreds, these systems are referred to as massive multiple-input-multiple-output (Ma-MIMO) systems. Integrating AI with traditional networking techniques significantly transforms how we manage and enhance networks. AI can swiftly process large amounts of data, anticipate network performance, and adjust settings accordingly. This holds great potential for improving network reliability and speed. AI techniques are increasingly valued for tackling intricate and difficult problems with large search spaces [3,4]. With AI, network operators can detect and resolve issues before they impact users. Additionally, AI can optimize network resource utilization based on real-time usage, enhancing efficiency and reducing congestion. AI systems continuously learn and adjust their parameters to meet evolving network demands.

1.1. Motivation

The conventional wireless network technology is segmented into various working blocks. However, the delay between these blocks has made it challenging to achieve highly reliable and low-latency networks. Hence, nowadays, researchers are trying to integrate AI with each block of the conventional network technology, so that the working blocks will get replaced by AI or machine learning (ML) blocks as shown in Figure 1. Our study focuses on the red hierarchical part. In the next phase, multiple working blocks will be replaced by a single AI block. In the vision the future 6G network is expected to have end-to-end learning, i.e., both transmitter and receiver side will get replaced by a single block of AI as shown in Figure 1. This paper explores the development of AI-based signal detection for MIMO systems. To address the growing needs of wireless networks and the computational demands of signal detection in MIMO systems, AI integration is a promising solution. In our exploration of different models, we implemented AIDETECT with PatternNet, as well as Long Short-Term Memory (LSTM) networks and Convolutional Neural Networks (CNNs). While these models were evaluated, we encountered difficulties in achieving the desired balance between output performance and computational complexity. LSTMs and CNNs, despite their advanced capabilities, proved to be computationally intensive. As a result, we developed a customized feed-forward neural network (FFNN) that strikes a more effective balance between computational efficiency and performance. Also, given the significant data load in our MIMO system, FFNN has been chosen for its ability to capture long-term dependencies and handle sequential data input [5].

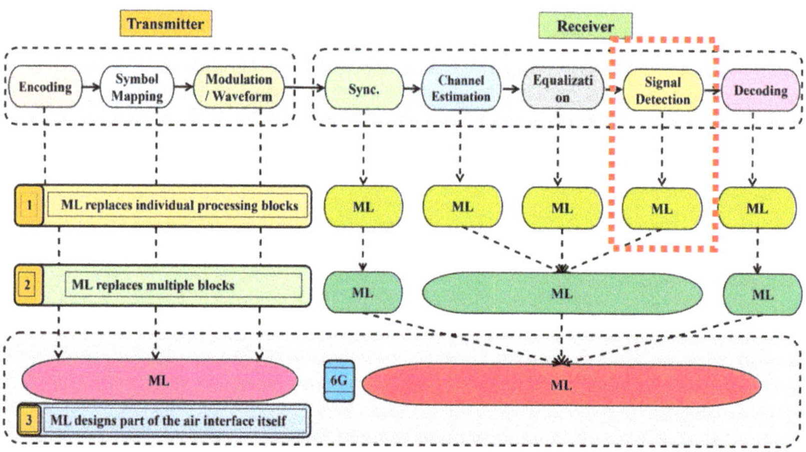

Figure 1. Mathematical architecture of MIMO system model. The red dotted box signifies that the proposed work entails this block.

1.2. Contributions

- In this paper, we present an optimized novel AI-enhanced signal detection model known as AIDETECT2. The proposed model employs an FFNN to enhance reliability and reduce the latency of wireless communication systems.

- The new AIDETECT2 model surpasses the performance of our previously proposed AIDETECT model, as well as the benchmarked AI models and traditional detectors.
- The performance evaluation of the proposed model is conducted in terms of both, symbol error rate (SER) and computational complexity.

By incorporating cutting-edge neural network techniques and refining the signal detection process, our model significantly boosts detection accuracy and computational efficiency. This advancement not only establishes a new benchmark in AI-powered signal detection but also paves the way for future developments in wireless communication, showcasing the transformative potential of AI in this domain. The paper is organized as follows: The system model is explained in Section 2. Section 3 reviews related work on signal detection in MIMO systems, focusing on the core aspects of MIMO systems and exploring various conventional methods and existing AI models. The problem formulation and data preparation are outlined in Section 4. This section details the framework and challenges of the system model used in the study, including assumptions, parameters, and configurations for MIMO systems. Section 5 introduces the proposed deep-learning (DL) model for signal detection in MIMO systems, specifically the novel AIDETECT2 scheme. This section describes the architecture and innovative aspects of AIDETECT2, highlighting how it builds upon previous models to improve performance. Theoretical underpinnings and algorithmic steps are elaborated to provide a comprehensive understanding of the scheme. In Section 6, we discuss the simulation setup used to evaluate the proposed models and the simulation results. This section details the simulation environment, including software tools, hardware specifications, and the parameters set for simulations. Also it presents the findings from the simulations and experiments, including a comparative analysis of the proposed models against existing techniques, highlighting performance improvements and any observed limitations. Following this, Section 6 also presents the optimization of the proposed model, and addresses the computational complexities of the signal detection models. The computational complexity analysis breaks down the complexity of each component, including the LSTM model, AIDETECT, AIDETECT2, and conventional methods. Finally, conclusions are drawn in Section 7. This section summarizes the key findings and contributions of the paper, reflecting on the effectiveness of the proposed models and their potential impact on MIMO signal detection. Future research directions are suggested, including possible enhancements and new areas of application.

2. System Model

We examine a system having N transmitting antennas and M receiving antennas, in which every antenna on both sides can communicate with each antenna on the other side. This results in $N \times M$ communication channels. Noise is considered for each receiving antenna. The transmitting symbols are generated using quadrature amplitude modulation of constellation size 16 (16-QAM) [6,7]. System performance is evaluated within a signal-to-noise ratio (SNR) range of 0 to 20 dB, with 5 dB step size. Figure 2 illustrates the general model for a MIMO system containing M receiving antennas and N transmitting antennas, which includes transmitted signal vector ($\boldsymbol{x}$), received signal vector ($\boldsymbol{y}$), channel matrix vector ($\boldsymbol{H}$), and noise vector ($\boldsymbol{n}$) [8]. The proposed system's development is based on the following equation:

$$\boldsymbol{y} = \boldsymbol{H}\boldsymbol{x} + \boldsymbol{n} \tag{1}$$

In this equation, $\boldsymbol{y}$ represents the signal that is received, $\boldsymbol{H}$ denotes the channel coefficient matrix vector, $\boldsymbol{x}$ signifies the signal vector which is transmitted, and $\boldsymbol{n}$ is the noise vector [9]. The model that was developed estimates the transmitted signal as $\hat{\boldsymbol{x}}$ using Equation (1).

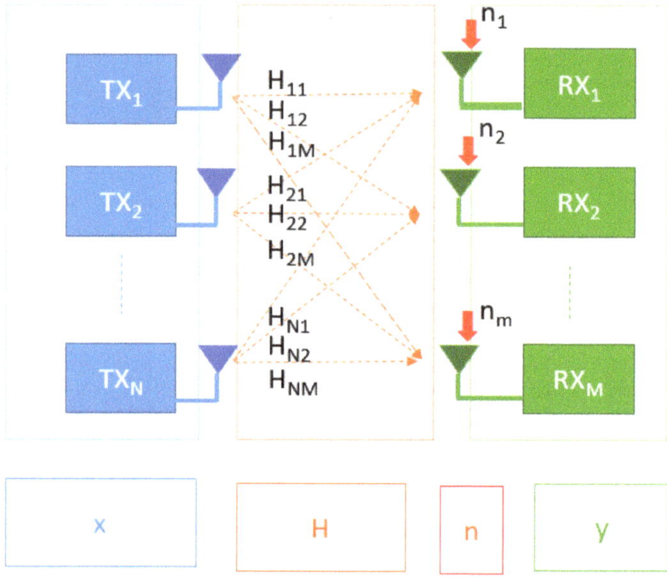

Figure 2. Mathematical architecture of MIMO system model.

3. Signal Detection in MIMO Systems

Multiple-input multiple-output, abbreviated as MIMO, is a wireless communication technology that employs numerous antenna elements at base stations to improve both spectral efficiency and energy [10]. This technology increases coverage at cell edges by spatially targeting transmissions toward users, resulting in enhanced signal strength as users distance themselves from the base station. Spatial multiplexing allows MIMO to communicate with several user devices simultaneously, enhancing overall spectral efficiency and throughput. Furthermore, when combined with millimeter-wave frequencies, MIMO boosts signal power [11]. During the signal detection phase in wireless networks, MIMO sustains a high SNR without altering bandwidth, unlike conventional approaches. Although MIMO has been successful in addressing these challenges, the growing number of connected devices introduces significant issues, such as fading, signal loss, and interference within networks. Various traditional signal detection techniques are available, including the commonly used minimum mean square error (MMSE) detector, maximum likelihood detector (MLD) and the zero-forcing (ZF) detectors [12].

3.1. Conventional MIMO Detectors

3.1.1. Zero-Forcing Detector

The zero-forcing detector effectively minimizes interference in the signal received by the receiver. It accomplishes this by designing a linear filter, which nullifies other user's interference within the network, effectively removing the presence of interference from the received signal at the same frequencies. This process allows the receiver to distinguish the target signal from the interference, thereby enhancing the accuracy of signal detection [12]. The ZF algorithm employs following equation to detect the transmitted signal $\hat{x}$.

$$\hat{x} = \boldsymbol{H}^{\dagger}\boldsymbol{y} = \boldsymbol{H}^{\mathrm{H}}(\boldsymbol{H}\boldsymbol{H}^{\mathrm{H}})^{-1}\boldsymbol{H}^{\mathrm{H}}\boldsymbol{y} \quad (2)$$

In this equation, $\boldsymbol{H}$ represents the channel matrix vector and $\boldsymbol{y}$ denotes the received signal matrix vector [13–15]. The pseudo-inverse of $\boldsymbol{H}$ is represented by $\boldsymbol{H}^{\dagger}$.

3.1.2. Minimum Mean Square Error Algorithm

The minimum mean square error detector is utilized in wireless communication systems for signal detection as it aims to reduce the mean square error between the actual signal detection response and the estimated signal response, thus reducing the effects of interference and noise on the signal received. In contrast to the ZF algorithm, which targets interference at specific frequencies, the MMSE detector employs a statistical method for signal detection, accounting for system interference and noise [16]. The detector employs particular equations to detect $\hat{x}$.

$$\hat{x} = \boldsymbol{H}^{\mathrm{H}}\boldsymbol{H} + \sigma^2 E_s^{-1}\boldsymbol{H}^{\mathrm{H}}\boldsymbol{y} \tag{3}$$

where $\boldsymbol{H}$ represents the channel matrix vector and $\boldsymbol{y}$ signifies the matrix of the received signal vector [13–15]. E_s stands for the mean symbol energy.

3.1.3. Maximum Likelihood Detector

The maximum likelihood detector is utilized in wireless communication systems to detect signals by identifying values that increase the likelihood of receiving the given signal, given the parameters that are estimated. This method aims to pinpoint the most probable parameter set that could have generated the data that is observed. Specifically, channel estimation seeks to determine the channel characteristics that best explain the received signal. Unlike the MMSE and ZF detectors, which focus on reducing noise and interference, the MLD aims to identify the most likely parameters that describe the signal based on data that is observed [17].

$$\hat{x} = \underset{x \in A^P}{\operatorname{argmin}} ||\boldsymbol{y} - \boldsymbol{H}\boldsymbol{x}||^2 \tag{4}$$

Here, $\boldsymbol{H}$ represents the channel matrix vector, $\boldsymbol{x}$ denotes the transmitted signal matrix vector, and $\boldsymbol{y}$ indicates the received signal matrix vector. A^P is the finite set of all possible sequences of 16-QAM, where each sequence forms a vector with P components [12]. z_i is the ith component of estimated signal.

3.2. *AI Based MIMO Detectors*

3.2.1. LSTM Model

The LSTM model is a recently developed DL-based signal detection model using LSTM neural network. It has demonstrated superior performance compared to traditional methods [8]. However, the increased computational complexity posed challenges when scaling to higher models. Mathematical representation is

$$\begin{aligned}\hat{x} = \arg\max \Bigg(\text{softmax}\Bigg(W_4 \cdot \text{Dropout}_{0.2}\Bigg(\text{LSTM}_{16}\{W_3, b_3, \\ \text{LSTM}_{32}[W_2, b_2, \text{LSTM}_{64}(W_1, b_1, F)]\}\Bigg) + b_4\Bigg)\Bigg)\end{aligned} \tag{5}$$

where F is $\arg\min_{x \in X^{N_t}} f(\mathbf{x}, (\mathbf{y}_{\mathrm{Re}}, \mathbf{H}_{\mathrm{Re}}), (\mathbf{y}_{\mathrm{Im}}, \mathbf{H}_{\mathrm{Im}}))$, which is a cost function, $\hat{x}$ denotes the estimated signal, W_i, b_i are weights bias of ith hidden layer, and $\boldsymbol{H}$, $\boldsymbol{y}$ and $\boldsymbol{x}$ represent the coefficients of channel vector, received signal vector, and transmitted signal vector respectively. $\text{Dropout}_{0.2}$ is the dropout layer having 0.2 dropout rate and LSTM_n is the LSTM layer having n number of neurons. softmax denotes the softmax function. The softmax function is a neural function that transforms a vector of raw scores into a probability distribution over potential classes. The argmax operation is then applied to choose the index of the highest value in the softmax output, indicating the predicted class label [18].

3.2.2. AIDETECT

AIDETECT is a recently introduced DL based model for signal detection within wireless networks [19]. The exceptional performance of this model serves as a benchmark for our proposed model. This model employs a 3-layer optimized pattern recognition DL network as shown in Figure 3. Mathematical representation of the model is given by the following equation:

$$\hat{x} = \arg\max\left(\text{softmax}(W_4 \times \sigma_3\{W_3 \times \sigma_2[W_2 \times \sigma_1(W_1 \times F + b_1) + b_2] + b_3\} + b_4)\right) \quad (6)$$

Here $\hat{x}$ denotes the estimated signal, W_i, b_i, σ_i are weight, bias, clipped ReLU activation function of ith hidden layer, and F is $\arg\min_{x \in X^{N_t}} f(\mathbf{x}, (\mathbf{y}_{\text{Re}}, \mathbf{H}_{\text{Re}}), (\mathbf{y}_{\text{Im}}, \mathbf{H}_{\text{Im}}))$, which is a cost function. $\boldsymbol{H}$, $\boldsymbol{y}$ and $\boldsymbol{x}$ represents the coefficients of channel vector, received signal vector and transmitted signal vector respectively. The softmax activation function maps a vector of raw scores to a normalized probability distribution, enabling classification. The argmax function then identifies the class label with the highest predicted probability [19].

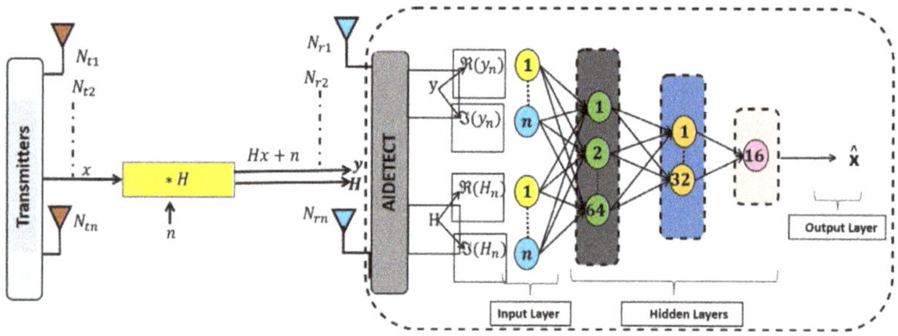

Figure 3. Schematic representation of AIDETECT.

4. Problem Formulation and Data Preparation

This part provides a discussion on the formulation of the problem and preparation of data in the MIMO systems.

4.1. Problem Formulation

Algorithms based on mathematical models encounter several limitations, such as limited scalability and high complexity [20]. To avoid these issues and enhance the utilization of MIMO systems present in wireless communications, incorporating AI-based methods, such as ML, offers a promising solution. ML algorithms can indeed be beneficial in the development of ultra-reliable and low-latency communication (URLLC) networks [21]. By optimizing channel allocation based on traffic patterns, AI can significantly improve overall spectral efficiency and network performance. The primary goal here is to develop an AI-enhanced signal detection model for MIMO systems in wireless communication. Instead of traditional detection methods, we designed a neural network method to detect the transmitted signal. Following the development of the neural network, we trained it utilizing multiple transmitted symbols and subsequently evaluated the system's performance with various symbols. The efficiency of the developed system needs to be contrasted with that of conventional methods and other AI models.

4.2. Data Preparation

Initially, symbols are produced based on the modulation scheme and constellation size. Random indices for the *ntrain* and *ntest* number of symbols are generated, which are then used to form the transmitted symbol vector $\boldsymbol{x}$ for training and testing. *ntrain* and *ntest* are the size of training testing dataset. The channel vector $\boldsymbol{H}$ is also generated and normalized so that the magnitude of its elements follows a Rayleigh distribution having a scaling parameter of 1. Rayleigh distribution is a mathematical model representing the probability of various measurement values influenced by several random factors [22]. The vector $\boldsymbol{y}$ for the received signal is created from the previously generated vectors for a selected SNR range, which is multiple times the size of vector $\boldsymbol{x}$. Feature extraction involves obtaining the real and imaginary parts from vectors $\boldsymbol{y}$ and $\boldsymbol{H}$. The targets correlating to the feature data developed are generated accordingly.

5. Proposed AIDETECT2 Scheme for MIMO Detection

The proposed model begins by generating transmitted signals, channel coefficients, and noise. Using these inputs, the received signal vector $\boldsymbol{y}$ is computed for various signal-to-noise ratios. Importantly, the system generates all necessary data internally, without relying on any pre-trained data from external sources. This internally generated data is then used to create training and testing datasets for the neural network, as mentioned above in the data preparation section. After the neural network is implemented and integrated into the system, it undergoes training and testing to ensure that the proposed system can efficiently detect signals. We have developed an AI model for signal detection using an optimized FFNN model, which is then integrated with the proposed system to form AIDETECT2. After the training and testing of the neural network, the network can detect the signals efficiently from $\boldsymbol{y}$ and $\boldsymbol{H}$, as shown in Figure 4.

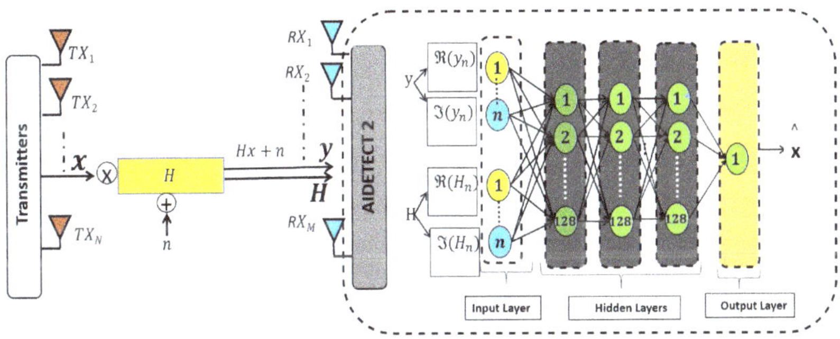

Figure 4. AIDETECT2 Block diagram.

Previously, a recently developed AI-based model AIDETECT fell short of performance expectations. To address these shortcomings, we developed a tailored FFNN model AIDETECT2. Drawing inspiration from the AIDETECT, our customized lightweight FFNN aims to reduce complexity while enhancing overall performance. AIDETECT has utilized patternet neural network, which has three hidden layers consisting of 64, 32 and 16 neurons. The customized FFNN used in AIDETECT2 is employed for a sequence of regression tasks. Initially, the feature data is transformed into a sequential format and passed to a sequence input layer sized according to the number of rows in the feature data. This input then flows through three hidden layers. Each hidden layer includes a clipped ReLU layer for introducing non-linearity, aiding the neural network in pattern recognition. In the fully connected layers, each neuron connects to the previous layer, incorporating weights and biases through an activation function [23]. Following these three hidden layers, the data enters the output layer for regression tasks. Training employs the Adam optimizer with an initial learning rate set to 0.001 and a mini-batch size of 10 [24]. The following Figure 5 shows the hidden layer architecture of the AIDETECT-2 network model.

The mathematical representation of the proposed system is given by:

$$\hat{x} = \arg\max\left(W_4 \times \sigma_3[W_3 \times \sigma_2\{W_2 \times \sigma_1(W_1 \times F + b_1) + b_2\} + b_3] + b_4\right) \quad (7)$$

where $\hat{x}$ denotes the estimated signal vector, W_i, b_i, and σ_i are weight, bias, and the clipped ReLU activation function of ith hidden layer, and F is $\arg\min_{x \in X^{N_t}} f(\mathbf{x}, (\mathbf{y}_{\mathrm{Re}}, \mathbf{H}_{\mathrm{Re}}), (\mathbf{y}_{\mathrm{Im}}, \mathbf{H}_{\mathrm{Im}}))$, which is a cost function. $\boldsymbol{H}$, $\boldsymbol{y}$ and $\boldsymbol{x}$ represent the coefficients of channel vector, received signal vector, and transmitted signal vector, respectively.

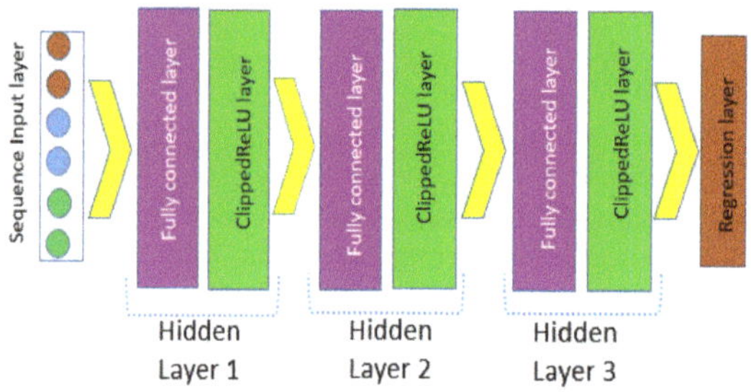

Figure 5. Block diagram and architecture of AIDETECT2 neural network Model.

The following Algorithm 1 shows the framework of the proposed AIDETECT2 MIMO detector. The output of the proposed system model is a predicted class with the highest probability. The model predicts the probability distribution across classes, and the class with the highest probability is selected as the output. This predicted output is then compared with the actual signal to evaluate the efficiency of the model, typically measured using metrics such as SER.

Algorithm 1: AIDETECT2: AI-based MIMO detector

Start;
Input: x, H
Output: $\hat{x}$
Data: Testing data set and Training data Set
Generate **x**, Generate **n**, Generate **H**
while $m = length(SNR)$ **do**
  $m = 1$
  Calculate **y**
  $m++$
end
end while
Feature Data = $\{\mathrm{Re}\{\tilde{y}\}\mathrm{Im}\{\tilde{y}\}\mathrm{Re}\{\tilde{H}\}\mathrm{Im}\{\tilde{H}\}\}$
while $m = length(SNR)$ **do**
  $m = 1$
  Concatenate $FeatureData$
  Concatenate $TargetData$
  $m++$
end
end while
Integrate AIDETECT2 Network Model

$$\hat{x} = \mathrm{AIDETECT2}\left(\arg\min_{\mathbf{x} \in X^{N_t}} f((\mathbf{y}_{\mathrm{Re}}, \mathbf{H}_{\mathrm{Re}}), (\mathbf{y}_{\mathrm{Im}}, \mathbf{H}_{\mathrm{Im}})) \right)$$

while $m = length(SNR)$ **do**
  Calculate SER
end
end while
End

AIDETECT2 represents a significant enhancement over its predecessor, AIDETECT, primarily due to a novel network model architecture. Notably, AIDETECT2 replaces the original network model with a customized FFNN design. This deliberate modification aims to reduce computational complexity and enhance the network's efficiency, particularly in terms of its ability to maintain robust performance across various SNRs. By leveraging this lightweight FFNN, AIDETECT2 aims for improved adaptability and accuracy in handling diverse signal conditions, which will make it a more effective solution. The comparison of different network architectures is provided in Table 1.

Table 1. Comparison of Neural Network Architectures.

Feature	LSTM	PatternNet-AIDETECT	AIDETECT2
Model Type	Recurrent	Feedforward	Feedforward
Layers	LSTM + Fully Connected	Custom Layers	Fully Connected + ClippedRelu
Activation Functions	Tanh, ReLU, Softmax	Sigmoid, Softmax	ReLU, Sigmoid
Input Type	Sequential	Pattern-based input	Sequential
Output Type	Sequence output	Classification output	Fixed-size output vector
Output Operation	Regression	Classification	Regression
Number of hidden layers	3	3	3
Number of neurons	64, 32, 16	64, 32, 16	128, 128, 128

6. Simulation Setup and Result Discussion

To replicate the proposed method, a custom simulator based on MATLAB was developed. This simulator runs on a stand-alone Intel i9-10900K CPU, @ 3.70 GHz with 128 GB RAM, and includes a 32 GB GPU for improved computational efficiency. Table 2 outlines the parameters used in the simulation setup.

Table 2. Simulation Parameters.

Parameters	Parameter Settings
Number of transmitters	2, 4, 8
Number of receivers	2, 4, 8
Modulation	QAM
Constellation size	16
Test data size ($ntest$)	1 Million
Train data size ($ntrain$)	1 Million
SNR Range	0:5:20 dB
Training Function	Adam
Number of neurons	128
Number of hidden layers	3
Maximum epochs	2000
Mini Batch Size	10
Initial learning rate	0.001

To train and evaluate the performance of the proposed deep learning model, simulations were conducted for both training data and testing data having a data size of 1 million (in terms of several symbols). QAM modulation with a constellation size of 16 was applied across different SNR points.

6.1. Result Discussion

To analyze the performance of the proposed AIDETECT-2 scheme is benchmarked against conventional (MMSE, ZF, and MLD) and AI-based detectors such as Variational Autoencoder-Enhanced Deep Neural Network-Based Detection (VAE-DNN-Det), LSTM, and AIDETECT MIMO detectors for different MIMO scenarios. The performance of the proposed AIDETECT-2 scheme is measured by conducting a simulation study based on 3 different MIMO simulating scenarios including 2 × 2, 4 × 4, and 8 × 8 MIMO systems with an equal number of transmitting and receiving antennas. Figure 6 presents the SER measurement of the 2 × 2 MIMO system for the proposed AIDETECT-2 MIMO detectors against benchmark conventional and AI-based MIMO detectors. Figure 6 simulation results show that the proposed AIDETECT-2 outperforms the conventional MIMO detectors and achieves between 98.31% to 95.56% better SER performance at 20 dB SNR. Similarly, simulation results also show that not only against conventional MIMO detectors but also, against AI-based MIMO detectors, the proposed AIDETECT-2 also achieves better SER performance ranging between 56.52% to 92.92% at 20 dB SNR respectively.

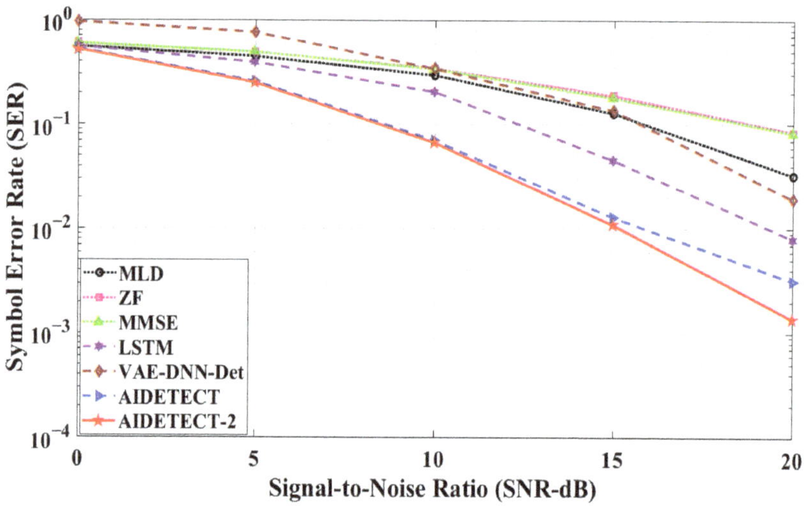

Figure 6. Comparison between outputs of conventional methods and AI models with AIDETECT2 for 2 × 2 MIMO systems.

We further extended the number of transmitting and receiving antennas and conducted a study for 4 × 4 MIMO scenarios as presented in Figure 7. Figure 8 simulation results show that the proposed AIDETECT-2 achieves excellent SER performance than conventional MIMO detectors and beat them with 99.914% to 99.995% better SER performance at 20 dB SNR. Compared with the AI-based MIMO detectors, the proposed AIDETECT-2 beat LSTM and VAE-DNN-Det MIMO detectors by a big margin and achieved 70% to 97.69% better SER performance than both AI-based MIMO detectors at 20 dB SNR. However, previously developed AIDETECT holds almost similar performance as AIDETECT-2 from 0-dB to 10-dB but after 10-dB the proposed AIDETECT-2 surpase the benchmark AIDETECT MIMO detector and achieves 88% better SER performance at 20 dB SNR.

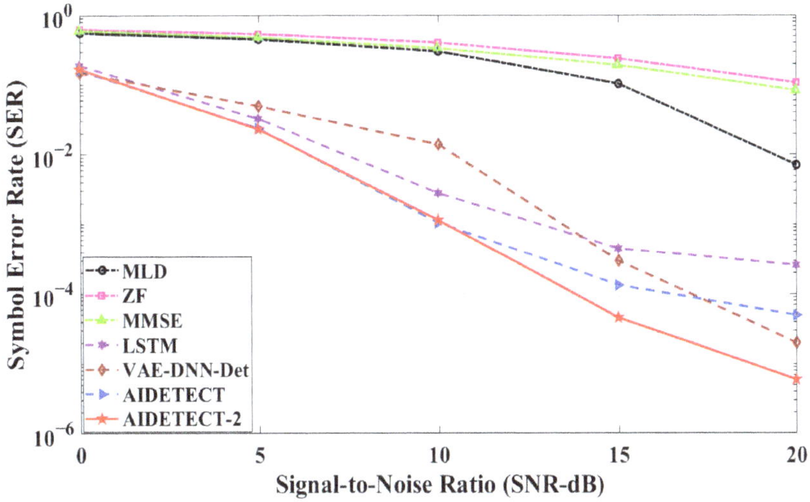

Figure 7. Comparison between outputs of conventional methods and AI models with AIDETECT2 for 4 × 4 MIMO systems.

Finally, in our last simulation scenario, we considered a bigger scenario of an 8 × 8 MIMO system with 8 transmitting and 8 receiving antennas as shown in Figure 8. Similar to Figures 6–8 also follows a similar trend showing the superiority of the proposed AIDETECT-2 against the benchmark conventional and AI-based MIMO detectors. Figure 8 results show that at 20 dB SNR, the proposed AIDETECT-2 achieves 99.4% to 99.7% better SER performance than conventional ZF and MMSE MIMO detectors. However, for a bit large 8 × 8 MIMO scenario, the optimal MLD MIMO detector achieves better SER performance than other AI-based MIMO detectors and holds very close SER performance (13.75% lower SER) than AIDETECT-2 at 20 dB SNR, but is still unable to beat the proposed AIDETECT-2 MIMO detection method. Figure 8 simulation results also show that the proposed AIDETECT-2 outperforms the benchmark AI-based MIMO detectors including AIDETECT and achieves between 78.5% to 89.2% better SER performance than all AI-based MIMO detectors at 20 dB SNR respectively.

We also analyze the training and validation graph based on root mean square error (RMSE) and loss for 8 × 8 MIMO system mode. Figure 9 shows the plotting of both loss and RMSE against the number of iterations in the model training and validation process. Figure 9 demonstrates how, as a result of the initial randomness of the model, RMSE begins high and rapidly drops in the first 200–300 iterations as the model learns, then steadily drops throughout fine-tuning. With a somewhat larger, steady value, the validation RMSE closely tracks the training RMSE, suggesting little overfitting and strong generalization.

Furthermore, in the case of loss scenario, Figure 9 shows how the loss increases in the first few hundred iterations as the model learns, but then rapidly drops as a result of the model's initial ignorance. Following this sharp reduction, the loss decreases more gradually when the model is fine-tuned. The validation loss closely follows the training loss, suggesting a similar trend in the form of small overfitting and good generalization.

In conclusion, the proposed model consistently outperformed the benchmark MIMO detectors including both conventional (MMSE, ZF, and MLD) and AI-based (LSTM, VAE-DNN-Det, and AIDETECT) MIMO detectors, and demonstrated higher noise resilience. These results suggest that AIDETECT-2 provides more reliable and accurate estimations, even under high interference conditions.

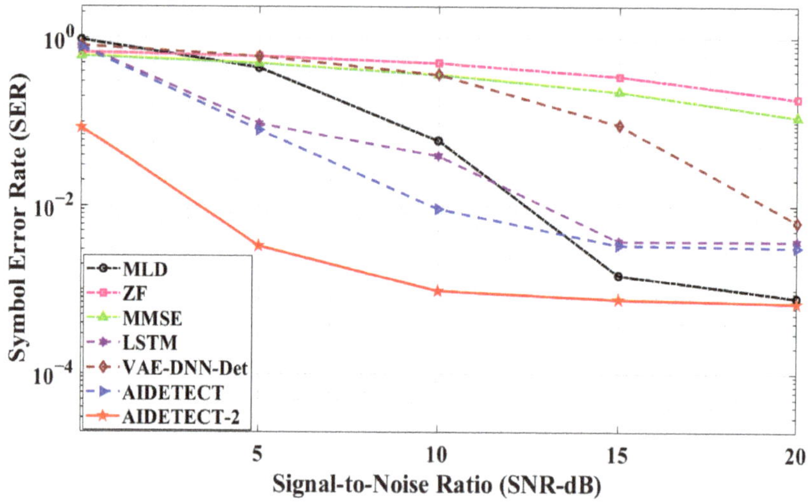

Figure 8. Comparison between outputs of conventional methods and AI models with AIDETECT2for 8 × 8 MIMO systems.

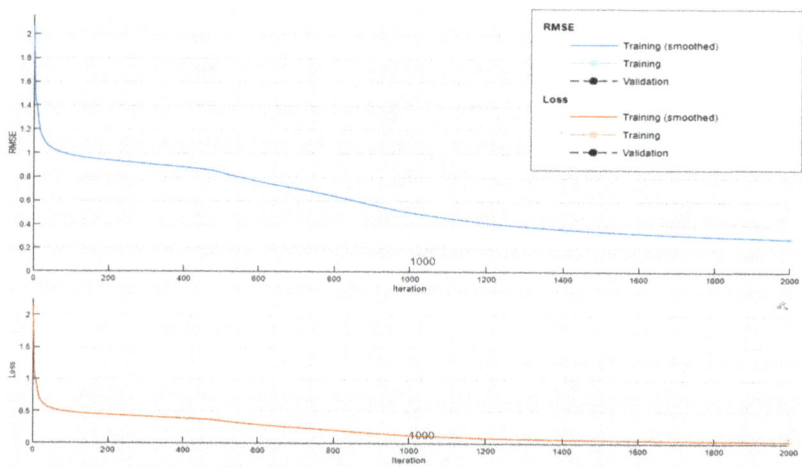

Figure 9. Training RMSE and training loss of AIDETECT2 for 8 × 8 MIMO system.

Furthermore, we also conducted an optimization study aimed at improving the performance of the proposed AIDETECT-2 network model based on the number of hidden layers for a 4 × 4 MIMO system focusing specifically on the proposed model's neural network architecture. The study involved evaluating the model's performance with varying numbers of hidden layers, specifically 2, 3, and 4 hidden layers as shown in Figure 10. Through experimentation and analysis, it was observed that AIDETECT-2 with 3 hidden layer model consistently outperformed both AIDETECT-2 with 2 hidden layer and AIDETECT-2 with 4 hidden layer configurations across key performance metrics. Consequently, the AIDETECT-2 network model based on FFNN with 3 hidden layers was selected as the optimal architecture for the proposed signal detection model. This decision was based on its superior ability to handle the complexity of MIMO systems, achieve higher detection accuracy, and maintain computational efficiency. The findings underscore the critical role of architectural design in optimizing neural network models for practical applications in communication systems.

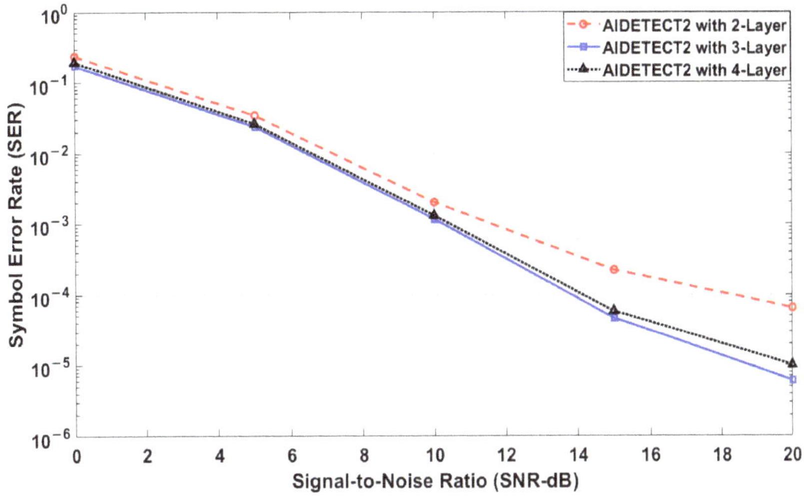

Figure 10. Comparison between outputs of AIDETECT2's neural network model for different number of hidden layers.

6.2. Computational Complexity Analysis

Computational complexity is an important parameter to measure the effectiveness of any MIMO detector. The complexity analysis is required to comprehend standard MIMO detectors' computational computing needs. In this section, we further analyze the computational complexity of different conventional (MMSE, ZF, and MLD) and AI-based (LSTM, VAE-DNN Det, AIDETECT, and AIDETECT-2) MIMO detectors respectively as presented in Table 3. The computational complexity of the proposed model on comparison with other model is plotted in Figure 11 in the form of number flops. Flop is the measure of number of floating point operations per second [25,26]. Where M is the number of receiving antennas and N is the transmitting antennas.

Table 3. Big-O Notation Comparison Table.

Model	Computational Complexity
ZF	$O(N^3 + N^2M)$
MMSE	$O(N^3 + N^2M)$
MLD	$O(M^N)$
VAE-DNN-Det	Varies widely
AIDETECT	$O(L^4)$
LSTM Model	$O(T(4a \cdot 64 + 4 \cdot 64^2 + 3 \cdot 64 + 4a \cdot 32 + 4 \cdot 32^2 + 3 \cdot 32 + 4a \cdot 16 + 4 \cdot 16^2 + 3 \cdot 16))$
AIDETECT2	$O(128a + 128^2 + 128m)$

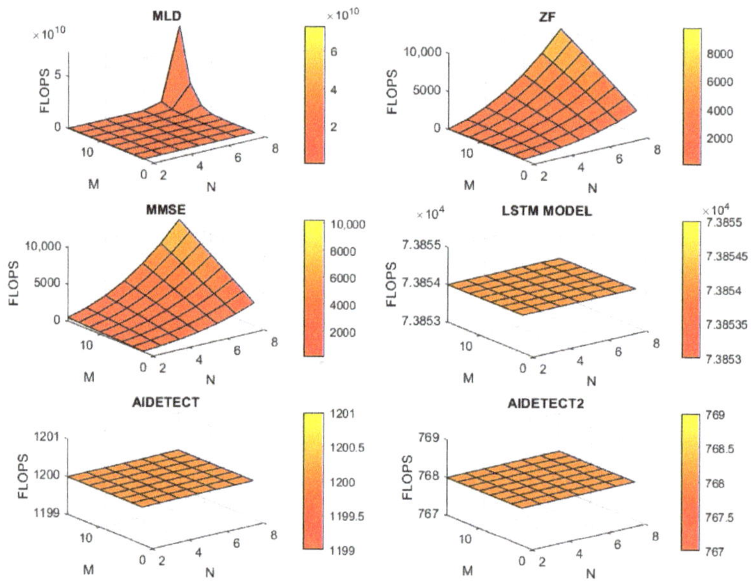

Figure 11. Computational complexity comparison in terms of flops.

6.2.1. Conventional Methods

Conventional signal detection algorithms such as ZF, MMSE, and MLD have distinct computational complexities. The ZF and MMSE algorithms involve matrix inversions, which have a complexity of $O(N^3)$, where N is the number of transmitting antennas. They also require matrix multiplications, adding $O(N^2M)$, where M is the number of receiving antennas. The ML algorithm has an exponential complexity $O(M^N)$ due to its exhaustive search through all possible transmitted vectors, making it impractical for large systems despite its optimal performance [27]. The LSTM model, with its multi-layer structure and sequence processing capabilities, exhibits a higher computational complexity compared to the FFNN model, which benefits from a simpler feed-forward architecture. The neural network models show significantly improved performance in signal detection over conventional algorithms such as ZF, MMSE, and MLD [28]. Despite their higher complexity, the adaptability and accuracy of AI-driven models make them suitable for advanced wireless communication systems like B5G and 6G networks.

6.2.2. VAE-DNN-DET Model

The computational complexity of the VAE-DNN-Det network model varies widely as does not provide its computational complexity exactly as the computational needs for training the VAE are determined by the design and size of the encoder and decoder networks, which include elements like layer count, neuron amount, and layer types. In addition, the sampling method for the latent variable space adds extra complexity. The size of the latent space and the architecture of the encoder network affect the difficulty of VAE inference.

6.2.3. LSTM Model

The computational complexity of the LSTM for the forward pass is determined by the operations in each LSTM cell, which include matrix-vector multiplications and element-wise operations. For an input dimension a and hidden dimensions of 64, 32, and 16, the complexity per time step per layer is $O(4ad + 4d^2 + 3d)$, where d represents the hidden dimension [29]. Summing up the layers and considering the sequence length T, the overall forward pass complexity is

$$O(T(4a \cdot 64 + 4 \cdot 64^2 + 3 \cdot 64 + 4a \cdot 32 + 4 \cdot 32^2 + 3 \cdot 32 + 4a \cdot 16 + 4 \cdot 16^2 + 3 \cdot 16)) \quad (8)$$

The backward pass doubles this complexity. Given the high number of epochs and the sequence data involved, the computational burden of the LSTM model is substantial but justified by its improved performance in signal detection.

6.2.4. AIDETECT

The patternet feed-forward network used in the benchmark model AIDETECT has three hidden layers with 64, 32, and 16 neurons. The computational complexity is largely influenced by the number of neurons in the layer. The total complexity of the model is given by

$$O\left(L^4\right) \quad (9)$$

where L is the number of hidden layers in the network [30].

6.2.5. AIDETECT2

The customized FFNN model used in AIDETECT2 has 3 layers with 128 units and uses the Clipped ReLU activation function. The complexity of the FFNN is largely influenced by the fully connected layers, each layer has a complexity of $O(n \cdot m)$, where n and m are the sizes of the input and output layers, respectively. For the developed architecture, each layer performs $128 \cdot n + 128 \cdot 128$ operations, summed across the layers. Therefore, the total complexity for a forward pass through the network is

$$O(128a + 128^2 + 128m) \quad (10)$$

where a is the input feature dimension. The regression layer at the end contributes less to the overall complexity. Given the fixed input size and number of layers, the computational complexity of the FFNN is generally lower than that of the LSTM and patternet.

7. Conclusions and Future Discussion

This study underscores the pivotal role of AI-driven methodologies in advancing beyond 5G wireless networks, with a focus on signal detection in MIMO systems. More specifically, this study presents AIDETECT-2, an AI-based signal detection technique designed for MIMO systems based on the FFNN network. Compared to conventional and other AI-based techniques, AIDETECT-2 greatly increases the efficiency of signal detection, especially in lowering the SER at various SNR points. The developed AI model AIDETECT2 for signal detection demonstrates superior performance compared to conventional methods and AI methods. Specifically, our proposed AIDETECT2 model surpasses the benchmarks set by traditional approaches, AIDETCT, LSTM Model, and VAE-DNN-Det. The simulation results indicate that the proposed model achieves a significant performance improvement, with SER improvements ranging from 13.75% to 99.995% better than the best conventional method and from 56.52% to 97.69 better than benchmark AI-based MIMO detectors at 20 dB SNR for given MIMO scenarios, respectively. This excellent performance with low complexity makes AIDETECT2 a more effective solution for signal detection in challenging environments. Despite these promising results, integrating multiple neural network architectures into a hybrid model could yield even better outcomes. Additionally, developing an AI-enhanced channel estimation model and integrating it with our current system could significantly impact the industry. This single, unified model would have the capability to estimate channels and detect signals more effectively.

Author Contributions: Conceptualization, M.U.H.; methodology, B.B. and M.Y.D.; software, B.B.; validation, B.B., M.Y.D., M.I.A. and M.U.H.; formal analysis, K.K. and M.U.H.; investigation, B.B.; resources, M.Y.D.; data curation, M.Y.D.; writing—original draft preparation, B.B.; writing—review and editing, B.B. and M.Y.D.; visualization, B.B.; supervision, M.U.H.; project administration, M.U.H. All authors have read and agreed to the published version of the manuscript.

Funding: This research received no external funding.

Data Availability Statement: The data that support the findings of this study were generated internally by the developed system. As no external data were used, there are no additional data available for sharing.

Conflicts of Interest: Muhammad Ikram Ashraf was employed by Nokia. The remaining authors declare that the research was conducted in the absence of any commercial or financial relationships that could be construed as a potential conflict of interest.

References

1. Vailshery, L.S. IoT Connected Devices Worldwide—Statista, 27 July 2023. [Online]. Available online: https://www.statista.com/statistics/1183457/iot-connected-devices-worldwide/ (accessed on 21 May 2024).
2. Wei, G.; Zhang, B.; Ding, G.; Zhao, B.; Wei, Y.; Guo, D. Massive MIMO-Based Distributed Signal Detection in Multi-Antenna Wireless Sensor Networks. *Sensors* **2007**, *20*, 2005. [CrossRef] [PubMed]
3. Chen, Y.; Tang, Y.; Jiang, B.; Zhao, Y.; Bao, J.; Tang, X. Efficient and Low-Complex Signal Detection with Iterative Feedback in Wireless MIMO-OFDM Systems. *Sensors* **2023**, *23*, 9798. [CrossRef]
4. Kim, K.; Lee, J.; Choi, J. Deep Learning Based Pilot Allocation Scheme (DL-PAS) for 5G Massive MIMO System. *IEEE Commun. Lett.* **2018**, *22*, 828–831. [CrossRef]
5. Akhtar, S.; Shah, M.A. Comparative Analysis of LSTM-FCN on Multiple Datasets. In Proceedings of the Competitive Advantage in the Digital Economy (CADE 2021), Online Conference, 2–3 June 2021; pp. 167–173.
6. Chinnusami, M.; Ravikumar, C.V.; Priya, S.B.M.; Arumainayagam, A.; Pau, G.; Anbazhagan, R.; Varma, P.S.; Sathish, K. Low Complexity Signal Detection for Massive MIMO in B5G Uplink System. *IEEE Access* **2023**, *11*, 91051–91059. [CrossRef]
7. Zhang, Y.; Li, Y.; Xue, T. SER Performance of Compact MIMO Systems with Receive Matching Networks in Nakagami-m Fading Channel. In Proceedings of the 2019 IEEE 19th International Conference on Communication Technology (ICCT), Xi'an, China, 16–19 October 2019; pp. 633–637.
8. Babu, B.; Daha, M.Y.; Hadi, M.U. Enhancing Signal Detection in 6G Networks through LSTM-based MIMO Technology. In Proceedings of the 2024 35th Irish Signals and Systems Conference (ISSC), Belfast, UK, 13–14 June 2024; pp. 1–6. [CrossRef]
9. Tan, X.; Ueng, Y.-L.; Zhang, Z.; You, X.; Zhang, C. A low-complexity massive mimo detection based on approximate expectation propagation. *IEEE Trans. Veh. Technol.* **2019**, *68*, 7260–7272. [CrossRef]
10. Kulshreshtha, G.; Chauhan, U. Signal-to-Interference-Noise Ratio (SINR) and Signal-to-Noise Ratio (SNR) Improvement in the Massive Multiple-Input Multiple-Output (MIMO) Systems. In Proceedings of the 2020 2nd International Conference on Advances in Computing, Communication Control and Networking (ICACCCN), Greater Noida, India, 18–19 December 2020; pp. 615–620.
11. MathWorks. Massive MIMO. Available online: https://uk.mathworks.com/discovery/massive-mimo.html (accessed on 14 March 2024).
12. Santos, I.; Murillo-Fuentes, J. Self and Turbo Iterations for MIMO Receivers and Large-Scale Systems. *IEEE Wirel. Commun. Lett.* **2019**, *8*, 1095–1098. [CrossRef]
13. Jojare, S.; Kant, N. SNR Performance of Various MIMO-OFDM Detectors using V-BLAST. *Int. J. Comput. Appl.* **2018**, *179*, 1–5. [CrossRef]
14. Chourasia, S.K.; Pandey, R. Extensive Survey on MIMO Technology using V-BLAST Detection Technique. *Int. J. Comput. Appl.* **2014**, *98*, 1–7.
15. Khurshid, K.; Khan, A.A.; Siddiqui, M.H.; Hadi, M.U.; Rashid, I.; Imran, M. Optimality of Linear MIMO Detection for 5G Systems via 1-Opt Local Search. *J. Electr. Eng. Technol.* **2021**, *16*, 1099–1108. [CrossRef]
16. Im, T.H.; Kim, J.; Yi, J.H.; Yun, S.; Cho, Y.S. MMSE-OSIC 2 signal detection for spatially multiplexed MIMO systems. In Proceedings of the VTC Spring 2008—IEEE Vehicular Technology Conference, Singapore, 11–14 May 2008; pp. 1468–1472.
17. Elghariani, A.; Zoltowski, M. Branch and Bound with M algorithm for near optimal MIMO detection with higher order QAM constellation. In Proceedings of the MILCOM 2012—2012 IEEE Military Communications Conference, Orlando, FL, USA, 29 October–1 November 2012; pp. 1–5.
18. Softmax. MathWorks, 2023. [Online]. Available online: https://uk.mathworks.com/help/deeplearning/ref/softmax.html#description (accessed on 5 August 2024).
19. Daha, M.Y.; Rafferty, J.; Ashraf, M.I.; Hadi, M.U. AIDETECT—AI-based Integratable Detection for Beyond 5G Networks. In Proceedings of the International Conference on Electronics, Computers, and Communication in the Modern Era (ICECCME), Online, 20–21 July 2023.

20. Khani, M.; Alizadeh, M.; Hoydis, J.; Fleming, P. Adaptive neural signal detection for massive MIMO. *IEEE Trans. Wirel. Commun.* **2020**, *19*, 5635–5648. [CrossRef]
21. Adnan, S.; Linbo, Z.; Dars, M.A.; Zahoor, M.I. Low complexity MIMO detection algorithm by combining modified OSIC and ML detection. In Proceedings of the 2016 8th IEEE International Conference on Communication Software and Networks (ICCSN), Beijing, China, 4–6 June 2016.
22. ScienceDirect. Rayleigh Distribution. [Online]. Available online: https://www.sciencedirect.com/topics/engineering/rayleigh-distribution (accessed on 5 August 2024).
23. "Patternnet" MATLAB Deep Learning Toolbox Documentation. [Online]. Available online: https://uk.mathworks.com/help/deeplearning/ref/patternnet.html (accessed on 14 May 2024).
24. MathWorks. Example Deep Learning Networks Architectures. Available online: https://uk.mathworks.com/help/deeplearning/ug/example-deep-learning-networks.html (accessed on 14 May 2024).
25. Omondi, G.; Olwal, T.O. Variational autoencoder-enhanced deep neural network-based detection for MIMO systems. *e-Prime—Adv. Electr. Eng. Electron. Energy* **2023**, *6*, 100335. [CrossRef]
26. He, H.; Wen, C.-K.; Jin, S.; Li, G.Y. Model-driven deep learning for mimo detection. *IEEE Trans. Signal Process.* **2020**, *68*, 1702–1715. [CrossRef]
27. Bai, L.; Choi, J. *Low Complexity MIMO Detection*; Springer Science & Business Media: Berlin, Germany, 2012.
28. Yu, A.; Jing, S.; Tan, X.; Wu, Z.; Yan, Z.; Zhang, Z.; You, X.; Zhang, C. Efficient successive over relaxation detectors for massive mimo. *IEEE Trans. Circ. Syst. I Regul. Pap.* **2020**, *67*, 2128–2139. [CrossRef]
29. Aslam, S.; Alam, F.; Hasan, S.F.; Rashid, M.A. A Machine Learning Approach to Enhance the Performance of D2D-Enabled Clustered Networks. *IEEE Access* **2021**, *9*, 15307–15316. [CrossRef]
30. Zhang, Z.; Li, Y.; Yan, X.; Ouyang, Z. A low-complexity amp detection algorithm with deep neural network for massive mimo systems. *Digit. Commun. Netw.* 2022, *in press*. [CrossRef]

electronics

MDPI

Article

High-Speed Rail Multiple-Input–Multiple-Output Channel Model Based on Reconfigurable Intelligent Surface System

Qiong Xu, Jianli Xie * and Zepeng Zhang

School of Electronic and Information Engineering, Lanzhou Jiaotong University, Lanzhou 730070, China; xuqiong@mail.lzjtu.cn (Q.X.); 13220079@stu.lzjtu.edu.cn (Z.Z.)
* Correspondence: xiejl@mail.lzjtu.cn

Abstract: This paper presents a geometry-based stochastic model (GBSM) for a reconfigurable intelligent surface (RIS)-assisted multiple-input–multiple-output (MIMO) system tailored to high-speed rail environments, addressing energy leakage at space lattice points caused by the time-dependent reception direction and differing antenna array resolutions. Initially, this study explores the spatial domain feature map and spreading antenna lobe characteristics from various RIS array configurations to reshape the polarization distribution and facilitate a virtual antenna array. Subsequently, the time-dependent reception direction and position are derived by integrating the dynamics of train operations. Finally, the received signal is aligned with the polarization direction to assess the signal reception efficiency and finalize the model output. Research findings indicate that the number of RIS elements, spacing between RIS elements, and mobile relay (MR) movement characteristics significantly influence the performance of the system. Compared to existing models, the proposed model proficiently captures the effects of time-dependent receiver angle properties and RIS array configuration.

Keywords: RIS; communication channel model; MIMO; signal leakage; reception efficiency

1. Introduction

Strengthening the construction of transportation systems and establishing an intelligent high-speed rail (HSR) system is sure to increase demand for wireless communications technology [1,2]. Studies under the European Union (EU) Shift2Rail Plan illustrate that advanced traffic management and remote control systems must rely on state-of-the-art communication systems [3]. In 2019, the World Radio Communication Conferences (WRC) introduced the WRC-19 standard for vehicle-to-ground communications in rail systems, specifying mandatory targets for rail communication infrastructure [4]. RISs have garnered significant attention due to their unique advantages such as low cost, passive operability, and ease of deployment. The RIS-assisted MIMO system [5–8] facilitates automatic reflective beam tuning with an ultra-large aperture, allowing for base stations (BSs) to operate with fewer antennas while maintaining quality of service (QoS) for users and mitigating the costs and power consumption associated with RF chain deployment in MIMO systems [4].

RIS technology introduces a range of new and promising functionalities for wireless communications, effectively mitigating communication channel impairments. Research on RISs primarily focuses on enhancing coverage, suppressing interference, and refining communication channel efficiency [9–15]; however, the effectiveness of such research relies heavily on precise channel modeling. In terms of path loss and wireless parameters, E. Basar et al. proposed a corresponding loss model based on reflection and scattering, grounded in the basic working principles of RIS [7]. E. Björnson et al. employed physical optics to derive far-field signal loss data based on this model [16,17]. Tang et al. developed signal loss models for different scenarios based on laboratory verification results, providing precise parameters for each model [18]. Danufane et al. introduced a physics based method to characterize the path loss of RIS components, taking into account the deployment conditions and size of RIS [19]. Assuming isotropic scattering, they utilized the correlation matrix for an RIS with half-wavelength spacing, demonstrating that the channel exhibits uncorrelated Rayleigh fading characteristics [20]. For the GBSM, E. Basar et al. formulated a unified narrowband channel model by effectively integrating the MIMO cluster model with RIS for both indoor and outdoor scenarios [21]. H. Jiang et al. segmented the channel into three sub-channels to analyze time-dependent characteristics, revealing that the three sub-channel domains have low correlation [22,23]. Sun et al. explored the quantifiable characteristics of a channel that uses RIS discrete phase control and determined that it does not alter the time correlation compared to using continuous phase shifts [24]. Guiqi Sun et al. examined the statistical characteristics of line-of-sight, single-bounce, and dual-bounce modes to model an RIS-assisted MIMO HSR system based on a 3D cylindrical model [25]. Lastly, Tao Zhou et al. proposed a GBSM for an RIS-assisted MIMO HSR system based on a spherical model and two elliptic cylindrical models, integrating the dynamic evolution of clusters in both the time and array domains [26].

Current research on RIS channel modeling predominantly focuses on static scenarios, and studies in high-speed rail channel modeling often assume that all signals are effectively received [21–26]. This assumption leads to an overly idealized portrayal of channel performance metrics. To generate and detect electromagnetic waves with varying angles of incidence, the spacing of RIS elements must be denser than $\lambda/2$. This denser array structure influences the spatial resolution of the RIS [5]. In practical situations, the rapid motion of the train alters the time-dependent receiver angle of the MR, causing the angles emitted by the RIS to not fully align with the MR's receiving range. The distinct spatial feature maps of Tx, RIS, and MR inevitably result in signal reception losses [27,28] due to energy leakage at the lattice points. Moreover, previous studies have identified additional challenges: (1) the coordinates of the transmitter (Tx) and MR are presumed to follow an ideal distribution, overlooking the variations in radiation direction maps of different arrays due to the MR's movement [22–25]; (2) the RIS phase adjustment is modeled under ideal propagation conditions using a simplistic phase offset method, neglecting the effects of the RIS array configuration on the channel characteristics [24–26]. Consequently, this paper introduces a non-smooth channel model for RIS-assisted MIMO wireless communication, based on the established operational conditions of trains:

- Initially, a physical model of the millimeter-wave MIMO channel assisted by RIS in HSR communication is constructed. A single-frequency network (SFN) is utilized to establish uniform frequency coverage, with RIS strategically placed between two Tx units to enhance communication. According to these conditions, a comprehensive channel response model that incorporates the polarization effect is introduced.
- This study explores the relationship between the radiative gain of the RIS composite array and its configuration, including variations in the spatial periodic spectrum and the deficiency of spatial base rank in the angular domain. These influencing factors are incorporated into the calculation of antenna weights for the horizontal and vertical polarization bases, subsequently deriving the virtual antenna array response of the RIS.
- Based on the train trajectory and speed, the departure and arrival angles of each line-of-sight (LOS) and non-line-of-sight (NLOS) propagation path are continuously updated, and then the angle signal is projected onto the corresponding array space to compute the polarization reception efficiency.
- Based on the relative positions of the Tx, RIS, and MR, the propagation and antenna polarization effects are integrated to facilitate the calculation of the complete channel response, and the statistical characteristics of the model are analyzed, including the time autocorrelation function (ACF) and the spatial cross-correlation function (CCF). This research demonstrates that the physical properties of the RIS, such as the number of RIS elements, the spacing of RIS elements, and the MR's movement speed,

significantly influence the channel characteristics; thus, the model effectively captures the dynamic variations in spatial angle and array configuration.

The remainder of this paper is structured as follows: Section 2 describes the channel model for the RIS-assisted communication system for rail systems. Section 3 delves into the virtual antenna array of the RIS and calculates the polarization reception efficiency. Section 4 presents numerical and simulation results. Lastly, Section 5 presents conclusions.

2. Channel Model of RIS-Assisted Communication System for Rail

2.1. Scenario Description

Considering the complexity of a typical millimeter-wave rail MIMO communication system, this study employs the HSR 30 GHz scenario as defined in 3GPP 38.913 [29], as illustrated in Figure 1. The high-speed deployment scenario prioritizes continuous coverage, focusing on providing a pleasant experience for passengers and maintaining reliability at high speeds. Consequently, an SFN is utilized in the millimeter-wave band, creating uniform frequency coverage by linking the remote radio head (RRH) with the baseband unit (BBU) within the same cell. Here, the RRH serves as the TX. To minimize penetration loss, the MR is typically positioned at the mid-point of the top of the train, where the TX connects to the MR via millimeter waves, and the MR then relays data to the users. Given the susceptibility of millimeter waves to obstruction, and their tendency to be intermittently blocked during high-speed movement, it is crucial to incorporate an RIS to bolster communication performance through its robust phase adjustment capability. To align with the same-frequency networking strategy and maximize the signal efficiency of RIS, it is strategically placed between neighboring RRHs and within the LOS range. When the LOS is unobstructed, the TX communicates directly with the MR. When the LOS is blocked, the transmission occurs through the RIS. After RIS reflection, two propagation paths are considered: the RIS-reflected signal directly reaching the MR and the RIS-reflected signal reaching the MR via a single scatterer.

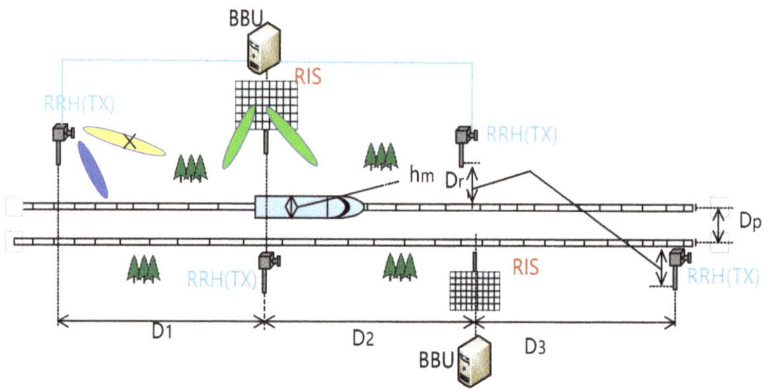

Figure 1. 30 GHz RIS-assisted high-speed rail deployment scenario.

2.2. Description of Model Parameters

In the rail system, owing to the high-speed motion of the MR, a global coordinate system (GCS) is established with the RIS as the reference, as depicted in Figure 2. The first reflection element in the lower left corner of the RIS is designated as the origin. To minimize the effect of the elevation angle on array resolution, both the TX and RIS are positioned at the same height, h_t, and the distance across the rail is maintained uniformly. The TX is stationed on a static platform, while the MR moves at a speed of v_u and in the direction of γ_u. The RIS is structured into N_H rows and N_V columns, spaced by d in both the horizontal and vertical directions. The coordinates of the element indexed by $n_r \in [1, \cdots, N_r]$ element is $u_{n_r} = [0, p_i d, q_i d]^T$, and (p_i, q_i) are the corresponding horizontal and vertical indices, respectively. The computation method is detailed in Equation (1), where 'mod' is used for rounding down, and $\Phi(pi, qi)$ denotes the corresponding phase shift [20].

$$p_i = \mathrm{mod}(n_r - 1, N_H), q_i = \lfloor n_r - 1/N_H \rfloor \tag{1}$$

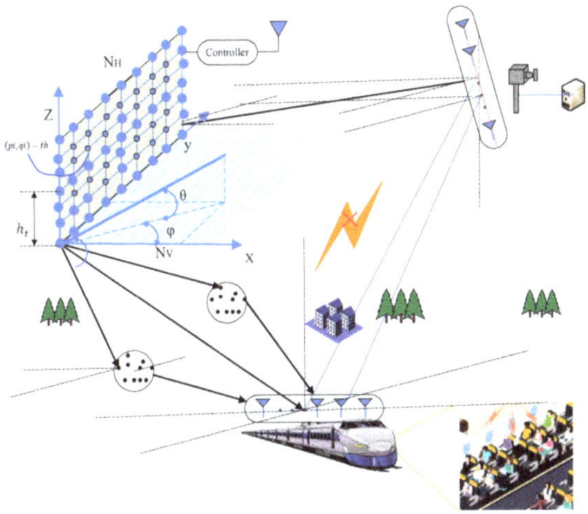

Figure 2. RIS-assisted high speed rail communication channel model based on 3D geometry.

The elevation angle of departure (EAOD) and azimuth angle of departure (AAOD) transmitted from the m_ith antenna element of the TX to the n_ith antenna element of the MR and to the RIS element indexed as (p_i,q_i) are $\left(\Theta_t^{m_i,n_i}, \varphi_t^{m_i,n_i}\right), \left(\Theta_{tr}^{m_i,p_i,q_i}, \varphi_{tr}^{m_i,p_i,q_i}\right)$. The Elevation angle of arrival (EAOA) and the azimuth angle of arrival (AAOA) received by the n_ith antenna element of the MR from the TX m_ith antenna element, from the RIS element indexed as (p_i,q_i) and from scatterer s_i are $\left(\Theta_u^{m_i,n_i}, \varphi_u^{m_i,n_i}\right), \left(\Theta_{ur}^{n_i,p_i,q_i}, \varphi_{ur}^{n_i,p_i,q_i}\right), \left(\Theta_{ur,s_i}^{n_i,}, \varphi_{ur,s_i}^{n_i}\right)$. The reflector element indexed as (p_i,q_i) receives the TX incident signal and retransmits it after phase adjustment, creating two distinct transmission paths: either direct or through the scatterer. The respective EAOA, AAOA, EAOD, and AAOD for these paths are $\left(\Theta_{r,in}^{m_i,p_i,q_i}, \varphi_{r,in}^{m_i,p_i,q_i}\right)$, $\left(\Theta_{r,out}^{p_i,q_i,n_i}, \varphi_{r,out}^{p_i,q_i,n_i}\right), \left(\Theta_{ur,s_i}^{n_i,}, \varphi_{ur,s_i}^{n_i}\right)$. The model assumes the presence of S scatterers between the RIS and the MR. The EAOA and AAOA of the incident signal received by scatterer s_i are $\left(\Theta_{si,in}^{p_i,q_i}, \varphi_{si,in}^{p_i,q_i}\right)$, and the EAOD and AAOD of the outgoing signal transmitted by scatterer s_i are $\left(\Theta_{si,out}^{p_i,q_i}, \varphi_{si,out}^{p_i,q_i}\right)$. The parameters $d_{mi,ni}$, $d_{mi,pi,qi}$, $d_{pi,qi,ni}$, $d_{pi,qi,si}$, and $d_{si,ni}$ represent the distances from TX to MR, TX to RIS, RIS to MR, RIS to scatterer (S), and scatterer to MR, respectively. A detailed description of these parameters is provided in Table 1.

Table 1. Parameters definition of the channel model.

Notation	Definition
$d_0, d_0^r, d_0^{rs}, d_0^{us}$	Tx-MR, RIS-MR, RIS-S, S-MR initial distance
M_T, N_U	Tx/MR number of antenna elements
δ_T, δ_T	Tx/MR antenna element space
h_t	Tx/RIS center height

Table 1. *Cont.*

Notation	Definition
$I(\Theta_t, \varphi_t)$ and $v_u(\Theta_u, \varphi_u)$	Tx antenna array elevation, azimuth angles and MR velocity vector
$\left(\Theta_t^{m_i,n_i}, \varphi_t^{m_i,n_i}\right), \left(\Theta_{tr}^{m_i,p_i,q_i}, \varphi_{tr}^{m_i,p_i,q_i}\right)$	EAOD, AAOD from Tx m_ith antenna element by LOS Path, RIS
$\left(\Theta_u^{m_i,n_i}, \varphi_u^{m_i,n_i}\right)\left(\Theta_{ur}^{n_i,p_i,q_i}, \varphi_{ur}^{n_i,p_i,q_i}\right)\left(\Theta_{ur,S_i}^{n_i}, \varphi_{ur,S_i}^{n_i}\right)$	EAOA, AAOA to MR n_ith antenna element by LOS path, RIS, scatterer
$\left(\Theta_{r,in}^{m_i,p_i,q_i}, \varphi_{r,in}^{m_i,p_i,q_i}\right)$	RIS (p_i, q_i) reflection element EAOA, AAOA
$\left(\Theta_{r,out}^{p_i,q_i,n_i}, \varphi_{r,out}^{p_i,q_i,n_i}\right)$	RIS (p_i, q_i) reflection element EAOA, AAOA
$\left(\Theta_{r,out}^{p_i,q_i,si}, \varphi_{r,out}^{p_i,q_i,si}\right)$	EAOA, AAOA from RIS (p_i, q_i) reflection element to S_i scatterer
S	Number of scatterers
$\left(\Theta_{si,in}^{p_i,q_i}, \varphi_{si,in}^{p_i,q_i}\right)$	EAOA, AAOA of S_i scatterer incident signal
$\left(\Theta_{si,out}^{p_i,q_i}, \varphi_{si,out}^{p_i,q_i}\right)$	EAOA, AAOA of S_i scatterer transmitted signal
$d_{mi,ni}, d_{mi,pi,qi}, d_{pi,qi,ni}, d_{pi,qi,si}, d_{si,ni}$ $d_{m,n}, d_{m,p,q}, d_{p,q,n}, d_{p,q,s}, d_{s,n}$	TX-MR, TX-RIS, RIS-MR, RIS-S, S-MR unit distance; center distance

2.3. Description of Model Parameters

The RIS-assisted MIMO system is underpinned by a widely used physical channel model that employs RIS for transmission when the TX-MR pathway is obstructed. Given the high signal loss characteristics of millimeter-wave frequencies, we disregard signals that have been reflected and scattered by the RIS more than once. The total channel response and TX-MR, TX-RIS, RIS-MR channel responses are represented as follows.

$$\boldsymbol{H}_m = \boldsymbol{H}_d + \boldsymbol{H}_u \Phi \boldsymbol{H}_r, \tag{2}$$

where

$$\boldsymbol{H}_d = h_d^{dl} \boldsymbol{a}_u(\boldsymbol{\Theta}_u, \boldsymbol{\varphi}_u) \boldsymbol{a}_t^H(\boldsymbol{\Theta}_t, \boldsymbol{\varphi}_t), \tag{3}$$

and

$$\boldsymbol{H}_r = h_{\mathrm{r}}^{\mathrm{dl}} \boldsymbol{a}_{\mathrm{r}}(\boldsymbol{\Theta}_{r,in}, \boldsymbol{\varphi}_{r,in}) \boldsymbol{a}_t^H(\boldsymbol{\Theta}_{tr}, \boldsymbol{\varphi}_{tr}), \tag{4}$$

and

$$\boldsymbol{H}_u = h_u^{los} \boldsymbol{a}_{\mathrm{u}}\left(\boldsymbol{\Theta}_u^{los}, \boldsymbol{\varphi}_u^{los}\right) \boldsymbol{a}_r^{\mathrm{H}}\left(\boldsymbol{\Theta}_{r,out}^{los}, \boldsymbol{\varphi}_{r,out}^{los}\right) + \frac{1}{\sqrt{SK_f}} \sum_{\updownarrow=1}^{S} h_u^{nlos} \boldsymbol{a}_{\mathrm{u}}\left(\boldsymbol{\Theta}_u^{nlos}, \boldsymbol{\varphi}_u^{nlos}\right) \boldsymbol{a}_r^H\left(\boldsymbol{\Theta}_{r,out}^{nlos}, \boldsymbol{\varphi}_{r,out}^{nlos}\right), \tag{5}$$

The variables $h_d^{dl}, h_r^{dl}, h_u^{los}, h_u^{nlos}$ represent the large-scale channel coefficients for the LoS and $\updownarrow$th NLoS path ($\updownarrow \epsilon S$, Sscatterers); K_f is the Rician factor that denotes the ratio of the energy between the LoS and NLoS channels. $\boldsymbol{a}_{\mathrm{t}}(\boldsymbol{\Theta}_t, \boldsymbol{\varphi}_t) \in \mathrm{C}^{M_T \times 1}$, $\boldsymbol{a}_{\mathrm{u}}(\boldsymbol{\Theta}_u, \boldsymbol{\varphi}_u) \in \mathrm{C}^{N_U \times 1}$, $\boldsymbol{a}_{\mathrm{r}}(\boldsymbol{\Theta}_{r,in}, \boldsymbol{\varphi}_{r,in}) \in \mathrm{C}^{N \times 1}$ describe the array responses of TX, MR, and RIS, while $\boldsymbol{\Phi} = \mathrm{diag}(\Theta_{piqi}) \in \mathrm{C}^{N \times N}$, (Θ, φ) are the elevation and azimuth angle of the corresponding paths.

The array responses for TX-MR transmission are detailed in Equations (6) and (7), and the array responses for TX-RIS and RIS-MR are calculated similarly.

$$\boldsymbol{a}_{\mathbf{t,los}}\left(\boldsymbol{\Theta}_t^{m_i}, \boldsymbol{\varphi}_t^{m_i}\right) = \left[\mathrm{e}^{-\mathrm{j}k(\Theta_t^{m_i}, \varphi_t^{m_i})^{\mathrm{T}} \delta_T}, \cdots, \mathrm{e}^{-\mathrm{j}k(\Theta_t^{m_i}, \varphi_t^{m_i})^{\mathrm{T}} \delta_T M_T}\right]^{\mathrm{T}}, \tag{6}$$

$$\boldsymbol{a}_{\mathbf{u,los}}\left(\boldsymbol{\Theta}_u^{n_i}, \boldsymbol{\varphi}_u^{n_i}\right) = \left[\mathrm{e}^{-\mathrm{j}k(\Theta_u^{n_i}, \varphi_u^{n_i})^{\mathrm{T}} \delta_U - \mathrm{j}2\pi v_1 t}, \cdots, \mathrm{e}^{-\mathrm{j}k(\Theta_u^{n_i}, \varphi_u^{n_i})^{\mathrm{T}} \delta_U M_U - \mathrm{j}2\pi v_{ni} t}\right]^{\mathrm{T}}, \tag{7}$$

where $\boldsymbol{k}(\boldsymbol{\Theta}, \boldsymbol{\varphi}) = \frac{2\pi}{\lambda}[\cos(\boldsymbol{\Theta})\cos(\boldsymbol{\varphi}), \cos(\boldsymbol{\Theta})\sin(\boldsymbol{\varphi}), \sin(\boldsymbol{\Theta})]^{\mathrm{T}}$ is the wave vector and v_{ni} signifies the Doppler shift for different antenna array elements; hence, Equation (7) captures the array response of different elements influenced by the Doppler shift. Consequently, the modifications in the path and those induced by the Doppler shift are reflected in the real-time phase characteristics. As the MR moves rapidly along the rail track, it inevitably alters the direction of polarization, and Equation (8) presents the comprehensive channel response model that incorporates the polarization effect [30].

$$h_{l,mi,ni}(t) = \sqrt{\frac{P_l}{S}} \sum_{si=1}^{S} \begin{bmatrix} F_{rx,ni,\Theta}\left(\Theta_{l,si}, \varphi_{l,si}\right) \\ F_{rx,ni,\varphi}\left(\Theta_{l,si}, \varphi_{l,si}\right) \end{bmatrix}^T \begin{bmatrix} F_{tx,mi,\Theta}(\Theta_{l,si}, \varphi_{l,si}) \\ F_{tx,mi,\varphi}(\Theta_{l,si,EOD}, \varphi_{l,si}) \end{bmatrix} \boldsymbol{M}\boldsymbol{H}_m, \tag{8}$$

where

$$\boldsymbol{M} = \begin{bmatrix} \exp\left(j\Phi_{l,si}^{\Theta\Theta}\right) & \sqrt{\kappa_{l,si}^{-1}}\exp\left(j\Phi_{l,si}^{\Theta\varphi}\right) \\ \sqrt{\kappa_{l,si}^{-1}}\exp\left(j\Phi_{l,si}^{\varphi\Theta}\right) & \exp\left(j\Phi_{l,si}^{\varphi\varphi}\right) \end{bmatrix}, \tag{9}$$

where l denotes a cluster, s_i denotes a ray within cluster l, S represents the number of rays (S scatterers), P_l is the power of cluster l, and m_i and n_i designate the antenna elements. $F_{tx,\Theta}$,$F_{tx,\varphi}$,$F_{rx,\Theta}$,$F_{rx,\varphi}$ illustrate the field patterns of TX and MR in both vertical and horizontal directions. M is the 2×2 polarization coupling matrix that describes the alteration in polarization during transmission from TX to MR. κ measures the cross-polarizability, which quantifies the separation between two polarization channels.

3. Virtual Model of RIS Array

3.1. Equivalent Baseband Channel Model

The effective channel is described as the inner product between the beamforming vectors at the transceiver and the guiding vectors of the physical channel. The beamforming vectors for the TX and MR are denoted by $f \in \mathbb{C}^{M_T \times 1}$, $w \in \mathbb{C}^{N^U \times 1}$. The following relationships are established:

$$\begin{aligned} \boldsymbol{H}_m &= \boldsymbol{w}^{\mathrm{H}}(\boldsymbol{H}_u\boldsymbol{\Phi}\boldsymbol{H}_r + \boldsymbol{H}_d)\boldsymbol{f} \\ &= h_r^{dl} g_{\mathrm{tr}}(\Omega_{tr}) \left[h_u^{los} g_{\mathrm{r}}(\Omega_r) g_{\mathrm{ur}}(\Omega_{ur}) + \frac{1}{\sqrt{LK_f}} \sum_{\updownarrow=1}^{S} h_u^{nlos} g_{\mathrm{r}}(\Omega_r) g_{\mathrm{ur}}(\Omega_{ur}) \right] + h_d^{dl} g_{\mathrm{tu}}(\Omega_{tu}) g_{\mathrm{ut}}(\Omega_{ut}), \end{aligned} \tag{10}$$

$\Omega = \left| \boldsymbol{a}_2(\boldsymbol{\Theta}_2, \boldsymbol{\varphi}_2)^{\mathrm{H}} \boldsymbol{a}_1(\Theta_1, \boldsymbol{\varphi}_1) \right|$ represents the spatial cosines of the two vectors, and $g_{\mathrm{tr}}(\Omega_{tr}) = \boldsymbol{a}_{\mathrm{t}}\left(\boldsymbol{\Theta}_{tr}^{m_{0,}}, \boldsymbol{\varphi}_{tr}^{m_0}\right)\boldsymbol{f}$ denotes the beamforming response of the TR-RIS transmitter. The variables $g_{\mathrm{ur}}(\Omega_{ur})$, $g_{\mathrm{tu}}(\Omega_{tu})$, $g_{\mathrm{ut}}(\Omega_{ut})$ are defined in a similar manner. Beyond the array gain, a phase adjustment factor is incorporated into the RIS, which expression in the spatial domain is $g_{\mathrm{r}}(\Omega_r) = \boldsymbol{a}_{\mathrm{r}}(\boldsymbol{\Theta}_{r,out}, \boldsymbol{\varphi}_{r,out}) \boldsymbol{\Phi} \boldsymbol{a}_{\mathrm{r}}(\boldsymbol{\Theta}_{r,in}, \boldsymbol{\varphi}_{r,in})$. All the responses exhibit a slow rate of change in the spatial domain [30].

3.2. RIS Reflective Array Response Model

To optimize the RIS signal convergence capability, the most effective approach is to increase the number of reflective elements and reduce the space between array elements, while maintaining a constant array area. Through examining the spatial beam map of the RIS reflective array, the impact of array structural changes on performance was analyzed.

$$\begin{aligned} g_{\mathrm{r}}(\Omega_r) &= \boldsymbol{a}_{\mathrm{r}}(\boldsymbol{\Theta}_{r,out}, \boldsymbol{\varphi}_{r,out}) \boldsymbol{\Phi} \boldsymbol{a}_{\mathrm{r}}(\boldsymbol{\Theta}_{r,in}, \boldsymbol{\varphi}_{r,in}) \\ &= \Phi(pi, qi) \exp\left[-j\left(\frac{N_H \Omega_{rH}}{2} + \frac{N_V \Omega_{rV}}{2} - \frac{\Omega_{rH}}{2} - \frac{\Omega_{rH}}{2} \right) \right] \times \frac{\sin\left(\frac{N_H \Omega_{rH}}{2}\right)}{N_H \sin\left(\frac{\Omega_{rH}}{2}\right)} \times \frac{\sin\left(\frac{N_V \Omega_{rV}}{2}\right)}{N_V \sin\left(\frac{\Omega_{rV}}{2}\right)}, \end{aligned} \tag{11}$$

The detailed proof is outlined in Appendix A. According to Equation (11), the spatial cosine is maximized at a value of 0. The RIS is rotated by $g_{\mathrm{r}}(\Omega_r)$ to align the spatial cosine to a value of 0. $\Phi(pi, qi)$ can be considered a two-dimensional discrete signal in the spatial domain, with a uniform spatial sampling period, d, in each direction. The spatial domain is analyzed using the discrete time Fourier transform (DTFT) [27,31].

$$\Phi(pi,qi) = \left(\frac{d}{2\pi}\right)^2 \int_0^{2\pi/d} \int_0^{2\pi/d} \omega(\mathrm{k},\mathrm{l}) e^{\mathrm{j}(dpik+dqil)} \mathrm{d}k\mathrm{d}l, \tag{12}$$

$$\Omega(k,l) = \sum_{pi}^{N_H} \sum_{qi}^{N_V} \Phi(pi,qi) e^{-\mathrm{j}(dpik+dqil)}, \tag{13}$$

$\Omega(k,l)$ is a two-dimensional periodic function in terms of k and l, with a period of $2\pi/d$. This is derived by substituting Equation (13) into Equation (11):

$$g_r(\Omega_r) = \frac{N_H N_V d^2}{4\pi^2} e^{j[\frac{d(1-N_H d)\Omega_{rH}}{2} + \frac{d(1-N_V d)\Omega_{rV}}{2}]} \int_0^{2\pi/d} \int_0^{2\pi/d} \Omega(k,l) e^{j(\frac{N_H+1}{2} + \frac{N_V+1}{2})} \frac{\sin[N_H(\Omega_{rH} - dk)/2]}{N_H \sin[(\Omega_{rH} - dk)/2]} \times \frac{\sin[N_V(\Omega_{rV} - dl)/2]}{N_V \sin[(\Omega_{rV} - dl)/2]}, \tag{14}$$

The detailed proof is outlined in Appendix B. From Equation (14), it is clear that the array gain response follows the Dirichlet function distribution and exhibits spatial periodicity. Precise beam steering can be achieved by appropriately adjusting the RIS phase to manipulate the spatial domain parameters (d,k). To examine the RIS radiation characteristics, it is crucial to consider the dimensions of power leakage and resolution. There is no issue of spectral leakage if the peak value is unique within the angular range of the train's travel, specifically if $(\Omega_{\mathrm{rH}} - dk)/2$ does not exceed the $[-\pi/2, \pi/2]$ interval. When (d,k) equals 0, it is determined that there is no spectral leakage when d is less than 0.5λ. Utilizing $N_H \times N_V$ reflective array aligns with a horizontal and vertical resolution of $2\pi/N_{H/V}$ (established on an orthogonal basis as described in Equation (16)). If $\Omega_{rH} \geq 2\pi$ is consistent along the entire railway trajectory, it guarantees complete coverage by the orthogonal basis. Substituting $\frac{2\pi d\cos(\Theta)\sin(\varphi)}{\lambda}$ and $\Theta = 0$, we find that $d \geq 0.5\lambda$. Thus, there exists a spatial rank deficiency when $d < 0.5\lambda$. This condition implies that as the train moves at high speeds and the spatial angle changes, it affects the signal reception efficiency.

$$\Omega = \begin{bmatrix} 1 \\ \exp(-\mathrm{j}2\pi d\Omega) \\ \vdots \\ \exp(-\mathrm{j}2\pi(n-1)d\Omega \end{bmatrix} \quad \begin{matrix} \delta_H = \left\{ \urcorner(0), \urcorner\left(\frac{1}{N_H}\right), \cdots, \urcorner\left(N_H - \frac{1}{N_H}\right) \right\} \\ \delta_V = \left\{ \urcorner(0), \urcorner\left(\frac{1}{N_V}\right), \cdots, \urcorner\left(N_V - \frac{1}{N_V}\right) \right\} \end{matrix}. \tag{15}$$

3.3. RIS Virtual Antenna Establishment

An RIS operates by weighting location samples through phase adjustment, effectively creating a virtual antenna with a consistent directional map in space. The previous literature [32] has suggested using uniformly arranged omnidirectional antennas to simulate RIS arrays; however, this method overlooks the passive operation of RISs and the variations in phase for different polarization directions. In this paper, based on the array radiation gain described in Equation (14), the directional map is composed of a horizontally polarized base and a vertically polarized base as outlined in Equation (16). Following the guidelines of Equation (17) and using omnidirectional radiating subarrays, the configuration of the virtual antenna array is fully established. Algorithm 1 outlines the generation process.

$$\boldsymbol{F} = \begin{pmatrix} \boldsymbol{F}^{[\Theta]}(\boldsymbol{\Theta}, \boldsymbol{\psi}) \\ \boldsymbol{F}^{[\varphi]}(\boldsymbol{\Theta}, \boldsymbol{\psi}) \end{pmatrix} = \begin{pmatrix} \frac{\sin(N_V \Omega_{rV}/2)}{N_V \sin(\Omega_{rV}/2)} \\ \frac{\sin(N_H \Omega_{rH}/2)}{N_H \sin(\Omega_{rH}/2)} \end{pmatrix}, \tag{16}$$

where

$$\boldsymbol{F}^{[\Theta]}(\boldsymbol{\Theta}, \boldsymbol{\psi}) = \sqrt{2} \frac{\sin(N_V \Omega_{rV}/2)}{N_V \sin(\Omega_{rV}/2)} / \sqrt{\left[\sum \left(\left(\frac{\sin\left(\frac{N_V \Omega_{rV}}{2}\right)}{N_V \sin\left(\frac{\Omega_{rV}}{2}\right)} \right)^2 /2 + \left(\frac{\sin\left(\frac{N_H \Omega_{rH}}{2}\right)}{N_H \sin\left(\frac{\Omega_{rH}}{2}\right)} \right)^2 /2 \right) w_{ele} \right]}, \tag{17}$$

and

$$F^{[\varphi]}(\Theta,\psi) = \sqrt{2}\frac{\sin(N_H\Omega_{rH}/2)}{N_H\sin(\Omega_{rH}/2)} / \sqrt{\left[\sum\left(\left(\frac{\sin\left(\frac{N_V\Omega_{rV}}{2}\right)}{N_V\sin\left(\frac{\Omega_{rV}}{2}\right)}\right)^2/2+\left(\frac{\sin\left(\frac{N_H\Omega_{rH}}{2}\right)}{N_H\sin\left(\frac{\Omega_{rH}}{2}\right)}\right)^2/2\right)w_{azi}\right]}, \quad (18)$$

Algorithm 1. RIS Virtual Antenna Generation

1. Generate the vertical direction ele grid ($0^{\circ} \sim 180^{\circ}$), horizontal direction azi_grid ($0^{\circ} \sim 360^{\circ}$);
2. Generate horizontally and vertically polarized radiation patterns based on Equations (17) and (18);
3. Calculate horizontal and vertical polarization angle weights:
wele = [diff(ele_grid)-μ] × sum(averagecos(diff(ele_grid)));
wazi = diff(azi_grid); wazi = wazi/sum(waz);
4. Calculate the antenna gain
$P = \left(F^{[\Theta]}\right)^2 + \left(F^{[\varphi]}\right)^2$, calculate the maximum value max(P) in P, power normalized:
$\mathrm{P} = \mathrm{P}./\max(\mathrm{P})$, $\mathrm{w} = \mathrm{wele}\times\mathrm{wazi}$, calculate the gain : $\mathrm{Gain} = 1/(\mathrm{sum}(\mathrm{P})\cdot w)$
5. Power normalization
$F^{[\Theta]} = F^{[\Theta]}\cdot\sqrt{Gain/\max(\mathbf{P})}; F^{[\varphi]} = F^{[\varphi]}\cdot\sqrt{Gain/\max(\mathbf{P})};$

$F^{[\Theta]}$ and $F^{[\varphi]}$ represent the vertical and horizontal polarimetric antenna responses, respectively. 'ele_grid' refers to the vertical lattice points, and 'azi_grid' to the horizontal lattice points. The symbol μ denotes the median value between two adjacent vertical points. 'wele' and 'wazi' are the vertical and horizontal polarimetric weights, respectively. 'P' signifies the normalized transmission power, and 'Gain' refers to the antenna gain.

3.4. Mobile Parameters Update

In the HSR system, the TX-RIS setup remains fixed; hence, updates are only considered for MR-related paths:

$$\text{TX-RIS-MR}: \ I_{pi,qi,ni}(\mathrm{t}) = \frac{d_{pi,qi,ni}(t)}{\left|d_{pi,qi,ni}(t)\right|}, \quad (19)$$

$$d_{pi,qi,ni}(t) = \begin{cases} d_{p,q,n}(t-1) + \int_0^{\triangle t} v_u(\Theta_v,\varphi_v)\cdot I_{pi,qi,ni}(t)\mathrm{d}t - u_{n_r} + n_i\delta_T\cdot\frac{v_u(\Theta_u,\varphi_u)}{|v_u(\Theta_u,\varphi_u)|} \\ d_0^r - u_{n_r} + n_i\delta_\mathrm{T}\cdot\frac{v_u(\Theta_u,\varphi_u)}{|v_u(\Theta_u,\varphi_u)|} \qquad t=0 \end{cases} \quad (20)$$

$$d_{mi,pi,qi,ni}(t) = d_{mi,pi,qi} + d_{mi,pi,qi} + d_{pi,qi,ni}(t), \quad (21)$$

$$\varphi_{ur}^{n_i,p_i,q_i}(t) = \mathrm{arctan2}(-d_{pi,qi,ni}(t,x), -d_{pi,qi,ni}(t,y)), \quad (22)$$

$$\Theta_{ur}^{n_i,p_i,q_i}(\mathrm{t}) = \arcsin\left(\frac{d_{pi,qi,ni}(t,z)}{\left|d_{pi,qi,ni}(t)\right|}\right), \quad (23)$$

$$\psi_{mi,pi,qi,ni}(t) = (2\pi/\lambda)\times\mathrm{mod}(d_{mi,pi,qi,ni}(\mathrm{t}),\lambda) \quad (24)$$

The TX-MR and TX-RIS-S-MR paths are calculated similarly. The relative position of the scatterer ΔS is determined through linear drift calculation.

3.5. Determine Received Signal Efficiency

The high-speed movement of the train results in the beams transmitted at different times not aligning perfectly with the receiving grid points. Given that the angle of the receiver exhibits a slow-varying process, the efficiency calculation is conducted in the spatial domain. This is achieved by dividing the angular domains of the RIS and the MR into vertical and horizontal sections.

$$\begin{aligned} \text{RIS}: \ ris_V(qi) &= \frac{2\pi}{\lambda}\left[-1,-1+\frac{2(pi)}{\Omega_{rV}},\cdots 1\right] \\ ris_H(pi) &= \frac{2\pi}{\lambda}\left[-1,-1+\frac{2(qi)}{\Omega_{rH}},\cdots 1\right], \end{aligned} \quad (25)$$

$$\begin{aligned} \text{MR}: \ mr_V(ni) &= \frac{2\pi}{\lambda}\left[-1,-1+\frac{2(n_i)}{N_U \sin(\Theta_u)},\cdots,1\right] \\ mr_H(ni) &= \frac{2\pi}{\lambda}\left[-1,-1+\frac{2(n_i)}{N_U \sin(\Theta_u)\cos(\varphi_u)},\cdots,1\right] \end{aligned} \quad (26)$$

The elevation and azimuth angles derived from the mobile update are projected onto the receiving grid points, and the grid point with the highest projection value is selected. The spatial distance between the actual spatial angle and the optimal receiving grid point is calculated and normalized to ascertain the polarized signal reception efficiency.

$$F^{[\Theta]}{}_{efficiency} = \left(\frac{\sin[N_U(mr_V(n_{opt}) - \sin\Theta_{ur})/2]}{N_U \sin[(mr_V(n_{opt}) - \sin\Theta_{ur})/2]}\right) / \int \left(\frac{\sin[N_U(mr_V(n_{opt}) - \sin\Theta_{ur})/2]}{N_U \sin[(mr_V(n_{opt}) - \sin\Theta_{ur})/2]}\right) \quad (27)$$

$$F^{[\varphi]}{}_{efficiency} = \left(\frac{\sin[N_U(mr_H(n_{opt}) - \sin\varphi_{ur}\cos\Theta_{ur})/2]}{N_U \sin[(mr_H(n_{opt}) - \sin\varphi_{ur}\cos\Theta_{ur})/2]}\right) / \int \left(\frac{\sin[N_U(mr_H(n_{opt}) - \sin\varphi_{ur}\cos\Theta_{ur})/2]}{N_U \sin[(mr_V(n_{opt}) - \sin\varphi_{ur}\cos\Theta_{ur})/2]}\right) \quad (28)$$

4. Numerical Results and Analysis

In this section, we present numerical results to assess the impact of the RIS array configuration on the performance of the proposed channel model. Given the absence of relevant measurement data, our analysis primarily relies on the 3GPP published rail wireless deployment standards [29,33] and the WRC-19 standard settings [4]. The parameters for the RIS to MR and the path loss (PL) calculation model are mainly based on the 3GPP RMa-LOS scenarios. To enhance signal propagation energy, both the TR and MR utilize millimeter-wave directional coverage antenna models [34,35], with the specific parameters detailed in Table 2.

Table 2. Simulation parameters.

Notation	Definition
Center frequency f_c/GHz	30
Bandwidth B/MHz	500
Figure 1. Parameters : D_1, D_2, D_3, D_p, h_m, D_r	580, 580, 572, 6, 3.38, 5
Tx array elevation and azimuth angles	$I(0^o, 12^o)$
MR velocity modulus $\|v_u\|$ /km/h	350
MR velocity vector $v_u(\Theta_u, \varphi_u)$	$v_u(0^o, 0^o)$
Number of TX array elements M_T	8
Number of MR array elements N_U	8
RIS spacing/λ	$\{0.8\lambda, 0.5\lambda, 0.25\lambda, 0.2\lambda\}$
RIS length in fixed scenarios A/λ	4

In this paper, we analyze the influence of R, the RIS element spacing, on the reception efficiency in high-speed motion in two scenarios: ① A remains unchanged (set as 4λ), allowing the number of RIS elements to vary without altering the coordinates of the RIS and TX; ② A is changed (the number of RIS array elements is set to 8), which also results in changes to the RIS and TX coordinates.

The coordinates for the center point of the RIS are $(0, A/2, A/2)$, for the MR are $\left[\left(5+\left(\frac{A}{2}\right)\times\sin(12^{o})\times\cos(12^{o}), \frac{A}{2}-580\times\cos(12^{o}), \frac{A}{2}\right]\right.$, and for the TX are $\left[\left(\left(\frac{A}{2}\right)\times\sin(12^{o}), \frac{A}{2}-580\times\cos(12^{o}), \frac{A}{2}\right]\right.$. The spacing of the antenna elements on both the TX and MR is 0.5λ. The specific position of each antenna element is determined cumulatively relative to the center coordinates.

Figure 3 illustrates the periodic characteristics in the spatial domain for different array structures. From Figure 3, it is evident that the power decreases rapidly on both sides as the value of d decreases, and the side lobes reach a lower level. Extending the period over $\pi/2$ (normalized value over 1), it is clear that $d = 0.5\lambda$ represents a critical condition, where power leakage occurs when $d > 0.5\lambda$. In rail deployment scenarios using $d \leq 0.5\lambda$, this issue is absent, confirming that power leakage need not be considered in the process of modeling the RIS virtual antenna.

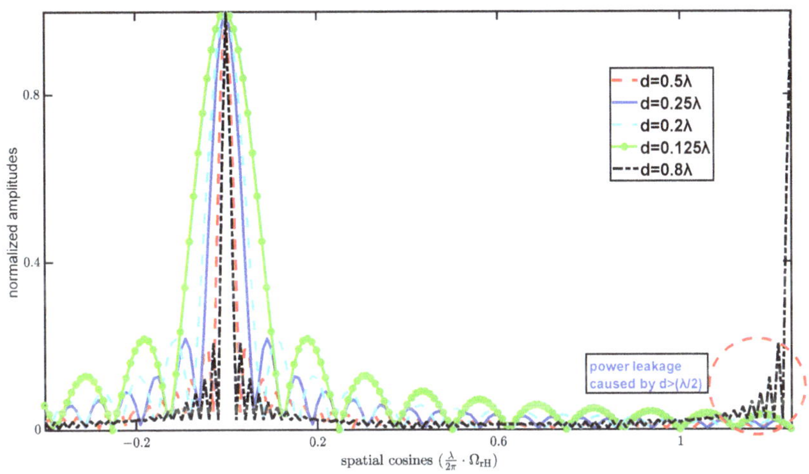

Figure 3. Spatial periodic characteristics map with different values of *d*.

Figure 4a displays the spatial beam map of an 8 × 8 RIS array with varying d values. The diagram shows that as d decreases, wavefront spreading occurs, indicating a degradation in spatial resolution. Figure 4b depicts the horizontal radiation response for different Θ cases. When the $\Theta = 0$ is smaller, the main beam is narrower, enhancing the resolution of the array. Conversely, as the Θ increases, the horizontal resolution of the array significantly diminishes. With the change in spatial angle during high-speed train operation, a discrepancy in array resolution between the RIS and MR is inevitable, leading to misalignment at the reception grid points. Consequently, the signal efficiency losses due to differences in array configuration and the time-varying spatial angle must be accounted for in the channel model.

From Equation (15), the RIS array gain response adheres to the Dirichlet function distribution, controlled by the spatial domain parameter (*d*,*k*), and peaks at the spatial cosine value of 0. Figure 5 displays the radiation direction map of the RIS, which reaches its maximum at the spatial angle of 0. As the angular distance increases, the signal energy demonstrates attenuation characteristics that align with the squared properties of the Dirichlet function, thereby validating the model's accuracy.

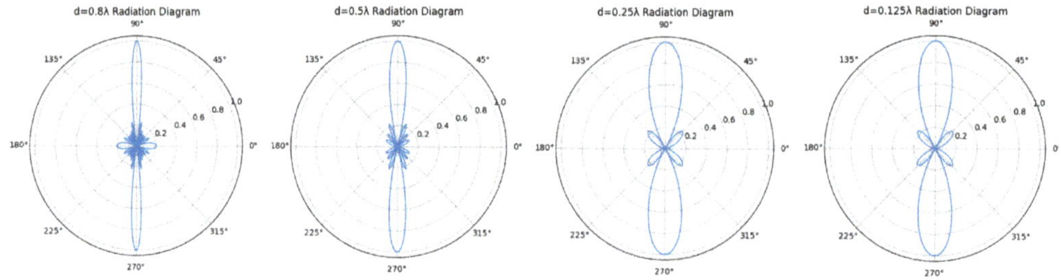

(**a**)

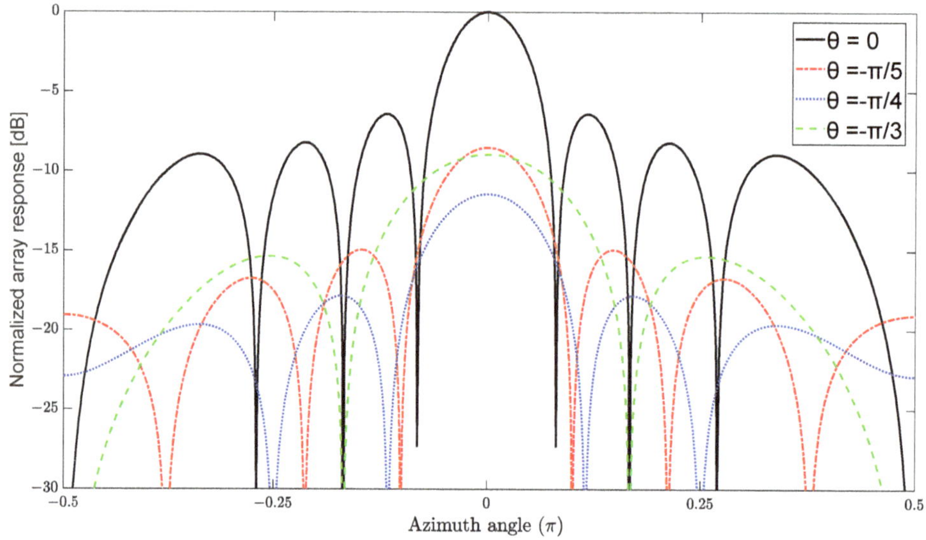

(**b**)

Figure 4. Horizontal spatial resolution for different values of d. (**a**) $\Theta = 0$ spatial array response. (**b**) $\Theta \neq 0$ spatial array response.

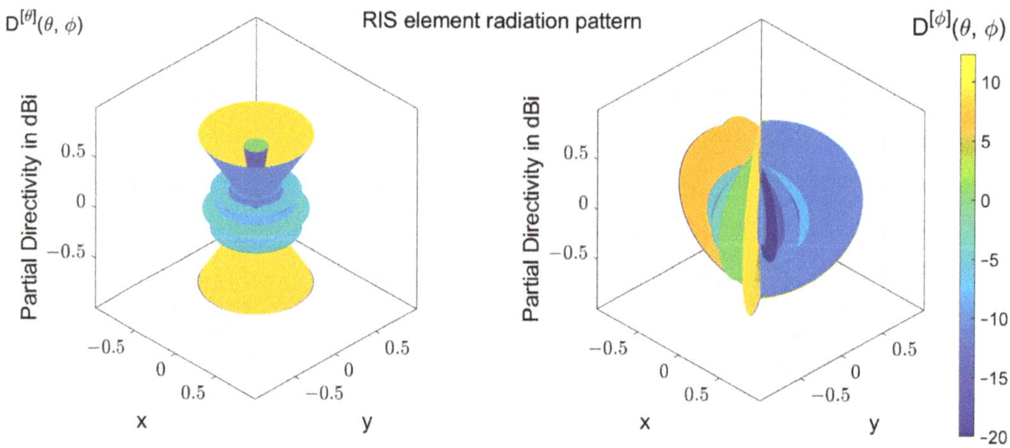

Figure 5. RIS virtual antenna direction map.

Figure 6 presents the conditions under which the RIS operates in space $d = 0.5\lambda$, alongside the MR received power. The original data are depicted in blue, while the smoothed data (using a 100 ms sliding window) are shown in red. It is evident that the high-speed movement of the train leads to sudden changes in the angle of the receiver, resulting in rapid fluctuations in reception efficiency. These variations are more pronounced than those accounted for solely by phase adjustment of the RIS. Given that the RIS is positioned between two RRUs, the train initially approaches the RIS, causing an overall rising trend in received power at the start. After reaching peak power, the train moves in the opposite direction, exhibiting a decreasing trend, consistent with findings from the literature [23].

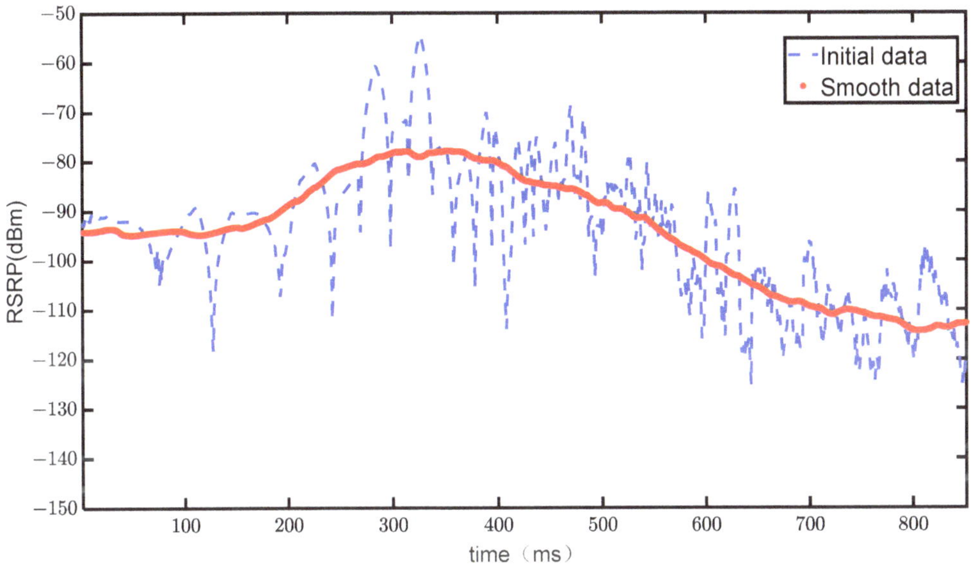

Figure 6. Received power with RIS space $d = \lambda/2$.

Figure 7 illustrates the average received power (analyzed with a 100 ms sliding window) for different RIS element spacings. The received power in Figure 7b is generally higher than in Figure 7a, attributed to the fact that, under constant RIS size conditions, reducing the RIS spacing increases the number of array elements in the same area, thereby boosting radiation efficiency—a conclusion consistent with [9]. Both Figure 7a,b indicate minimal power fluctuations under the condition of $d = 0.2\lambda$, primarily due to the reduced RIS spacing and enhanced location sampling resolution. The smoothed data further confirms that the received power of $d = 0.2\lambda$ is greater than that of $d = 0.8\lambda/0.5\lambda$, clearly demonstrating that power leakage is mitigated by reduced RIS spacing.

Figure 8 displays the time correlation function of the simulation results alongside the analytical results for various RIS element spacings. The correlation displays a decreasing trend over time intervals, suggesting that the proposed model effectively simulates position dynamics the channel in the time domain. Additionally, beyond the inherent low-frequency decreasing trend, the curve exhibits numerous burrs, attributable to differences in the reception efficiency of the RIS and MR at effective spatial grid points during train operations. From Figure 8a, it is observed that with a constant number of RIS elements, the correlation strengthens as the element spacing decreases, aligning with the conclusion that reduced element spacing leads to spatial rank deficiency and wavefront broadening. Conversely, Figure 8b shows that when the physical area of the RIS remains constant and the number of elements increases, the temporal correlation remains weak even as element space decreases. This indicates that the array gain achieved by increasing the number of elements surpasses the performance loss caused by the decreasing d-value. Particularly, a significant periodic peak occurs at $d = \lambda/2$, resulting from $d = \lambda/2$ corresponding to the Sinc function. Nonetheless, as the number of RIS elements increases, this correlation vanishes, primarily due to the increased directionality introduced by the additional RIS elements.

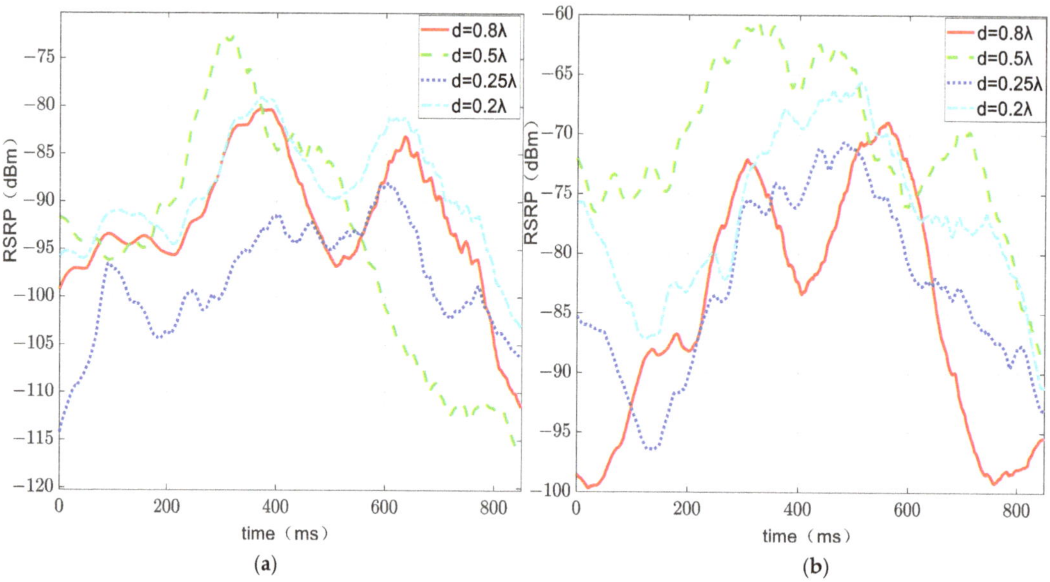

Figure 7. Received power with different RIS array spacings: (**a**) RIS size change, RIS element number without change; (**b**) RIS size without change, RIS element number change.

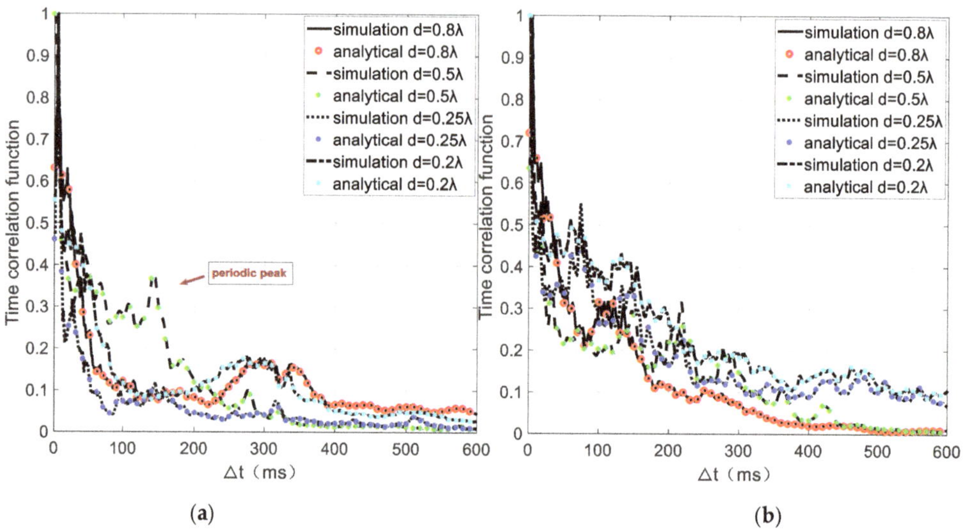

Figure 8. Time correlation function with different RIS element spaces: (**a**) RIS size change, RIS element number without change; (**b**) RIS size without change, RIS element number change.

Figure 9 illustrates the absolute values of the spatial correlation function of simulation results alongside analytical results for varying RIS element spacings. Here, the spatial correlation diminishes as the antenna index increases, supporting findings from the literature [25]. From Figure 9a, the correlation displays distinct periodicity, particularly at $d = \lambda/2$, which is largely governed by the computation method of Equation (15). In this case, the array response conforms to the Sinc function when the spacing is $\lambda/2$, and this periodicity diminishes in other array configurations. However, the correlation period shortens in Figure 9b, because the array gain, resulting from an increase in the number of RIS elements, surpasses the effect of lobe broadening due to decreased RIS spacing. Consequently, the likelihood of spatial angles emitted by the RIS failing to fall entirely within the range of the MR's corresponding receiving angles becomes smaller. The correlations at $d = 0.2\lambda$ and $d = 0.25\lambda$ decrease more gradually, mainly because the diagonal elements maintain a strong correlation as the spacing reduces. Notably, the $d = 0.2\lambda$ diagonal element exhibits the same correlation as $d = 0.5\lambda$. The discrepancy between simulation and analysis at the end of the function curve stems from the initial phase of the simulation being set to a random phase, which is particularly sensitive to the narrow beam formed by $d = 0.2\lambda$; thus, the more RIS elements, the greater the array gain.

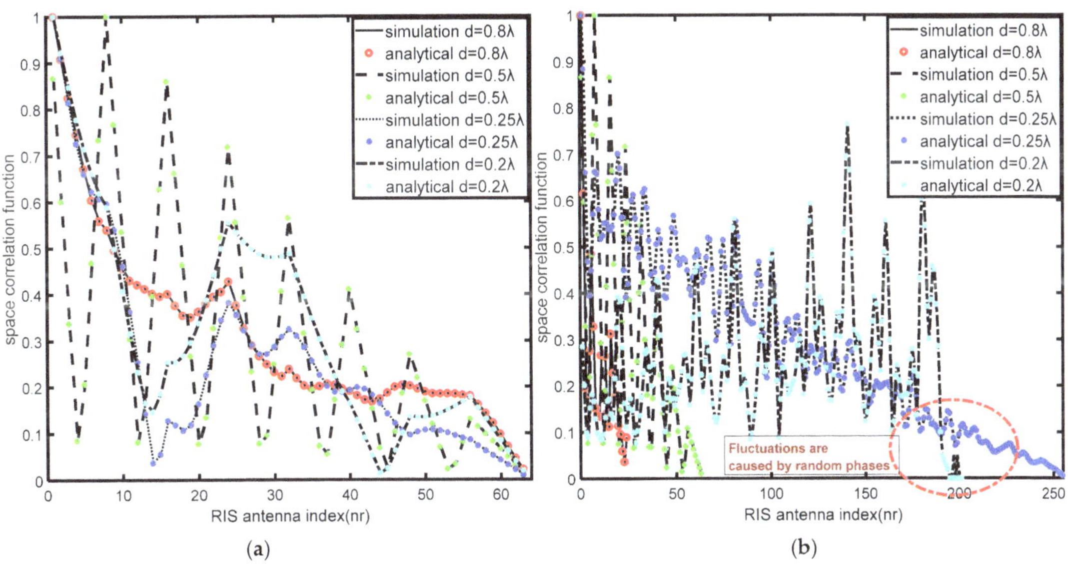

Figure 9. Spatial correlation function with different RIS array element spacing (**a**) RIS size change, RIS element number without change; (**b**) RIS size without change, RIS element number change.

5. Conclusions

In this paper, we propose an innovative GBSM model of an HSR MIMO communication channel that leverages the structure of a RIS array. The model analyzes polarized signal transmission and addresses power leakage and spatial rank deficiency resulting from varying RIS configuration features, culminating in the formation of a virtual antenna array. Using instantaneous train movement parameters derived from calculations, the model determines the expression for channel signal efficiency by integrating the virtual antenna beam pattern. The simulation results affirm that the proposed model accurately simulates the array response of RIS for a variety of array element spacings, and effectively captures the periodic spectral characteristics of the RIS and the issues of energy leakage in the time-dependent spatial domain of the channel. Additionally, the simulations highlight the significant influence of the physical characteristics of RIS on the shifting positional dynamics of the channel response. These findings are instrumental for evaluating and optimizing the performance of RIS-assisted HSR MIMO communication systems.

Author Contributions: Conceptualization, Q.X. and J.X.; methodology, Q.X. and J.X; software, Q.X. and Z.Z.; validation, Q.X. and Z.Z.; formal analysis, J.X.; writing—original draft preparation, Q.X.; writing—review and editing, J.X. All authors have read and agreed to the published version of the manuscript.

Funding: This work was partially supported by the National Natural Science Foundation of China (62161016), and the Beijing Engineering Research Center of High speed Railway Broadband Mobile Communications (BHRC-2022-1), Beijing Jiaotong University.

Data Availability Statement: The datasets generated and analyzed during the current study are available from the corresponding author upon request.

Acknowledgments: We are truly grateful to the anonymous reviewers for their constructive comments on our paper.

Conflicts of Interest: The authors declare no conflicts of interest.

Appendix A

Proof on Equation (11):

$$
\begin{aligned}
g_r(\Omega_r) &= \boldsymbol{a}_r(\boldsymbol{\theta}_{r,out}, \boldsymbol{\varphi}_{r,out}) \oslash \boldsymbol{a}_r(\boldsymbol{\theta}_{r,in}, \boldsymbol{\varphi}_{r,in}) \\
&= \left[e^{-jk(\theta_{r,in},\varphi_{r,in})^T u_1}, \ldots, e^{-jk(\theta_{r,in},\varphi_{r,in})^T u_N}\right]^* \Phi(pi,qi)\left[e^{-jk(\theta_{r,out},\varphi_{r,out})^T u_1}, \ldots, e^{-jk(\theta_{r,out},\varphi_{r,out})^T u_N}\right]^T \\
&= \sum_{q_{i=1}}^{N_V} \sum_{p_i=1}^{N_H} \left[e^{-jk(\theta_{r,out},\varphi_{r,out})^T u_n + e^{jk(\theta_{r,in},\varphi_{r,in})^T u_n}}\right]\Phi(pi,qi) \\
&= \sum_{q_{i=1}}^{N_V} \sum_{p_i=1}^{N_H} \left[e^{-j(q_i\Omega_{rH})}\right]\left[e^{-j(p_i\Omega_{rV})}\right]\Phi(pi,qi) \\
&= \left(\sum_{p_i=1}^{N_H}\left[e^{-j(q_i\Omega_{rH})}\right]\right)\left(\sum_{q_{i=1}}^{N_V}\left[e^{-j(p_i\Omega_{rV})}\right]\right)\Phi(pi,qi) \\
&= \exp\left[-j\left(\frac{N_H\Omega_{rH}}{2} + \frac{N_V\Omega_{rV}}{2} - \frac{\Omega_{rH}}{2} - \frac{\Omega_{rH}}{2}\right)\right] \times \frac{\sin(N_H\Omega_{rH}/2)}{N_H\sin(\Omega_{rH}/2)} \times \frac{\sin(N_V\Omega_{rV}/2)}{N_V\sin(\Omega_{rV}/2)}
\end{aligned}
\tag{A1}
$$

where $\Omega_{rV} = \frac{2\pi d\sin(\theta_{out})}{\lambda} - \frac{2\pi d\sin(\theta_{in})}{\lambda}$, $\Omega_{rH} = \frac{2\pi d\cos(\theta_{out})\sin(\varphi_{out})}{\lambda} - \frac{2\pi d\cos(\theta_{in})\sin(\varphi_{in})}{\lambda}$.

Appendix B

Proof on Equation (14):

$$
\begin{aligned}
g_r(\Omega_r) &= \left(\frac{d}{2\pi}\right)^2 \sum_{q_{i=1}}^{N_V} \sum_{p_i=1}^{N_H} \left[e^{-j(q_i\Omega_{rH})}\right]\left[e^{-j(p_i\Omega_{rV})}\right] \int_0^{2\pi/d}\int_0^{2\pi/d} \Omega(k,l)e^{j(dpik+dqil)}dkdl \\
&= \left(\frac{d}{2\pi}\right)^2 \int_0^{2\pi/d}\int_0^{2\pi/d} \Omega(k,l) \sum_{q_{i=1}}^{N_V} \sum_{p_i=1}^{N_H} \int_0^{2\pi/d}\int_0^{2\pi/d} \Omega(k,l)e^{j(dpik+dqil)}\left[e^{-j(\Omega_{rH}-dpik)}\right]\left[e^{-j(\Omega_{rV}-dqil)}\right] \\
&= \frac{N_H N_V d^2}{4\pi^2} e^{j\left[\frac{d(1-N_H d)\Omega_{rH}}{2} + \frac{d(1-N_V d)\Omega_{rV}}{2}\right]} \int_0^{2\pi/d}\int_0^{2\pi/d} \Omega(k,l)e^{j\left(\frac{N_H+1}{2} + \frac{N_V+1}{2}\right)} \frac{\sin[N_H(\Omega_{rH}-dk)/2]}{N_H\sin[(\Omega_{rH}-dk)/2]} \times \frac{\sin[N_V(\Omega_{rV}-dl)/2]}{N_V\sin[(\Omega_{rV}-dl)/2]}
\end{aligned}
\tag{A2}
$$

References

1. Ai, B.; Molisch, A.F.; Rupp, M.; Zhong, Z.D. key technologies for smart railways. *Proc. IEEE* **2020**, *108*, 856–893. [CrossRef]
2. Zheng, J.; Zhang, J.; Björnson, E.; Li, Z.; Ai, B. Cell-Free Massive MIMO-OFDM for High-Speed Train Communications. *IEEE J. Sel. Areas Commun.* **2022**, *40*, 2823–2839. [CrossRef]
3. European Commission. 2014. SHIFT2RAILInnovation Program 2. Available online: https://shift2rail.org/research-development/ip (accessed on 10 June 2014).
4. World Radiocommunication Conference 2019 (WRC-19). In Proceedings of the World Radiocommunication Conference 2019 (WRC-19), Sharm el-Sheikh, Egypt, 28 October–22 November 2019.
5. Björnson, E.; Luca, S.; Henk, W.; Jakob, H.; Thomas, L.M. Massive MIMO is a Reality—What is Next? Five Promising Research Directions for Antenna Arrays. *Digital Signal Proc.* **2019**, *94*, 3–20. [CrossRef]

6. Wu, Q.; Zhang, R. Towards smart and reconfigurable environment:Intelligent reflecting surface aided wireless network. *IEEE Commun. Mag.* **2020**, *58*, 106–112. [CrossRef]
7. Basar, E.; Di Renzo, M.; De Rosny, J.; Debbah, M.; Alouini, M.S.; Zhang, R. Wireless communications through reconfigurable intelligent surfaces. *IEEE Access* **2019**, *7*, 116753–116773. [CrossRef]
8. Björnson, E.; Özdogan, Ö.; Larsson, E.G. Intelligent reflectin surface vs. decode-and-forward: How large surfaces are needed to beat relaying. *IEEE Wirel. Commun. Lett.* **2019**, *9*, 244–248. [CrossRef]
9. Wu, Q.; Zhang, S.; Zheng, B.; You, C.; Zhang, R. Intelligent reflecting surface aided wireless communications: A tutorial. *IEEE Trans. Commun.* **2021**, *69*, 3313–3351. [CrossRef]
10. Huang, C.; Zappone, A.; Alexandropoulos, G.C.; Debbah, M.; Yuen, C. Reconfigurable intelligent surfaces for energy efficiency in wireless communication. *IEEE Trans. Wireless Commun.* **2019**, *18*, 4157–4170. [CrossRef]
11. Pan, C.; Ren, H.; Wang, K.; Xu, W.; Elkashlan, M.; Nallanathan, A.; Hanzo, L. Multicell MIMO communications relying on intelligent reflecting surfaces. *IEEE Trans. Wireless Commun.* **2020**, *19*, 5218–5233. [CrossRef]
12. Wu, Q.; Zhang, R. Weighted sum power maximization for intelligent reflecting surface aided SWIPT. *IEEE Wirel. Commun. Lett.* **2020**, *9*, 586–590. [CrossRef]
13. Du, H.; Zhang, J.; Cheng, J.; Ai, B. Millimeter Wave Communications With Reconfigurable Intelligent Surfaces: Performance Analysis and Optimization. *IEEE Trans. Wireless Commun.* **2021**, *69*, 2752–2768. [CrossRef]
14. Zhang, Z.; He, R.; Ai, B.; Yang, M.; Li, C.; Mi, H.; Zhang, Z. A General Channel Model for Integrated Sensing and Communication Scenarios. *IEEE Commun. Mag.* **2023**, *61*, 68–74. [CrossRef]
15. Berbineau, M.; Behaegel, R.; Garcia-Loygorri, J.M.; Torrego, R.; D'Errico, R.; Sabra, A.; Soler, J. Channel Models for Performance Evaluation of Wireless Systems in Railway Environments. *IEEE Access* **2021**, *9*, 45903–45918. [CrossRef]
16. Özdogan, Ö.; Björnson, E.; Larsson, E.G. Intelligent Reflecting Surfaces: Physics, Propagation, and Pathloss Modeling. *IEEE Wireless Commun. Lett.* **2020**, *9*, 581–585. [CrossRef]
17. Garcia, J.C.B.; Sibille, A.; Kamoun, M. Reconfigurable intelligent surfaces: Bridging the gap between scattering and reflection. *IEEE J. Sel. Areas Commun.* **2020**, *38*, 2538–2547. [CrossRef]
18. Tang, W.; Chen, M.Z.; Chen, X.; Dai, J.Y.; Han, Y.; Di Renzo, M.; Zeng, Y.; Jin, S.; Cheng, Q.; Cui, T.J. Wireless communications with reconfigurable intelligentsurface: Path loss modeling and experimental measurement. *IEEE Trans. Wirel. Commun.* **2021**, *20*, 421–439. [CrossRef]
19. Danufane, F.H.; Di Renzo, M.; De Rosny, J.; Tretyakov, S. On the path-loss of reconfigurable intelligent surfaces: An approach basedon Green's theorem applied to vector fields. *IEEE Trans. Commun.* **2021**, *69*, 5573–5592. [CrossRef]
20. Björnson, E.; Luca, S. Rayleigh fading modeling and channe hardening for reconfigurable intelligent surfaces. *IEEE Wirel. Commun. Lett.* **2021**, *10*, 830–834. [CrossRef]
21. Basar, E.; Yildirim, I.; Kilinc, F. Indoor and Outdoor Physical Channel Modeling and Efficient Positioning for Reconfigurable Intelligent Surfaces in mm Wave Bands. *IEEE Trans. Commun.* **2021**, *69*, 8600–8611. [CrossRef]
22. Najafi, M.; Jamali, V.; Schober, R.; Poor, H.V. Physics-based modeling and scalable optimization of large intelligent reflecting surfaces. *IEEE Trans. Commun.* **2021**, *69*, 2673–2691. [CrossRef]
23. Jiang, H.; Ruan, C.; Zhang, Z.; Dang, J.; Wu, L.; Mukherjee, M.; da Costa, D.B. A general wideband non-stationary stochastic channel model for intelligent reflecting surface-assisted MIMO communications. *IEEE Trans. Wireless Commun.* **2021**, *20*, 5314–5328. [CrossRef]
24. Sun, Y.; Wang, C.X.; Huang, J.; Wang, J. A 3D Non-Stationary Channel Model for 6G Wireless Systems Employing Intelligent Reflecting Surfaces With Practical Phase Shifts. *IEEE Trans. Cognitive Commun. Netw.* **2021**, *7*, 496–510. [CrossRef]
25. Sun, G.; He, R.; Ai, B.; Ma, Z.; Li, P.; Niu, Y.; Ding, J.; Fei, D.; Zhong, Z. A 3D Wideband Channel Model for RIS-Assisted MIMO Communications. *IEEE Trans. Veh. Technol.* **2022**, *71*, 8016–8029. [CrossRef]
26. Zhou, T.; Yang, Y.; Liu, L.; Tao, C.; Liang, Y. A Dynamic 3-D Wideband GBSM for Cooperative Massive MIMO Channels in Intelligent High-Speed Railway Communication Systems. *IEEE Trans. Wirel. Commun.* **2021**, *20*, 2237–2250. [CrossRef]
27. Tse, D.; Pramod, V. Spatial multiplexing and channel modeling. In *Fundamentals of Wireless Communication*, 2nd ed.; Cambridge University Press: Cambridge, UK, 2005; pp. 290–330.
28. Marzetta, T.L.; Larsson, E.G.; Yang, H.; Ngo, H.Q. The Massive Mimo Propagation Channel. In *Fundamentals of Massive MIMO*, 2nd ed.; Cambridge University Press: Cambridge, UK, 2016; pp. 163–182.
29. *3GPP TR 38.913 V17.0.0*; Study on Scenarios and Requirements for Next Generation Access Technologies. Technical Specification Group Radio Access Network: Sophia Antipolis, France, 2022; pp. 14–15.
30. Björnson, E.; Wymeersch, H.; Matthiesen, B.; Popovski, P.; Sanguinetti, L.; De Carvalho, E. Reconfigurable Intelligent Surfaces: A signal processing perspective with wireless applications. *IEEE Signal Proc. Mag.* **2022**, *39*, 135–158. [CrossRef]
31. Wan, Z.; Gao, Z.; Gao, F.; Di Renzo, M.; Alouini, M.S. Terahertz Massive MIMO with Holographic Reconfigurable Intelligent Surfaces. *IEEE Trans. Commun.* **2021**, *69*, 4732–4750. [CrossRef]
32. Gu, Q.; Wu, D.; Su, X.; Wang, H.; Cui, J.; Yuan, Y. System-level Simulation of Reconfigurable Intelligent Surface assisted Wireless Communications System. In Proceedings of the GLOBECOM 2022 - 2022 IEEE Global Communications Conference, Rio de Janeiro, Brazil, 4–8 December 2022.
33. Zhou, T.; Li, H.; Wang, Y.; Liu, L.; Tao, C. Channel modeling for future high-speed railway communication systems: A survey. *IEEE Access* **2019**, *7*, 52818–52826. [CrossRef]

34. *3GPP TR 38.901 V16.1.0*; Study on Channel Model for Frequencies from 0.5 to 100 GHz. European Telecommunications Standards Institute: Sophia Antipolis, France, 2020; pp. 27–88.
35. *3GPP TR 25.996 V17.0.0*; Spatial Channel Model for Multiple Input Multiple Output (MIMO) Simulations. European Telecommunications Standards Institute: Sophia Antipolis, France, 2022; pp. 7–37.

electronics

MDPI

Article

Inkjet-Printed Reflectarray Antenna Integrating Feed and Aperture on a Flexible Substrate Using Origami Techniques

Yi-Xin Lin, Kuan-Yu Ko, Fei-Peng Lai and Yen-Sheng Chen *

Department of Electronic Engineering, National Taipei University of Technology, 1, Sec. 3, Zhongxiao E. Rd., Taipei 10608, Taiwan; zxc9773245@gmail.com (Y.-X.L.); t107369009@ntut.org.tw (F.-P.L.)
* Correspondence: yschen@ntut.edu.tw; Tel.: +886-2-27712171-2281; Fax: +886-2-27317120

Abstract: This paper presents an innovative method for fabricating reflectarray antennas using inkjet printing technology on flexible substrates, markedly enhancing integration and manufacturability compared to traditional PCB methods. The technique employs inkjet printing to deposit conductive inks directly onto a flexible polyethylene naphthalate (PEN) substrate, seamlessly integrating feed and reflectarray components without complex assembly processes. This streamlined approach not only reduces manufacturing complexity and costs but also improves mechanical flexibility, making it ideal for applications requiring deployable antennas. The design process includes an origami-inspired folding of the substrate to achieve the desired three-dimensional antenna structures, optimizing the focal length to dimension ratio (F/D) to ensure maximum efficiency and performance. The feed and the reflectarray geometry are optimized for an F/D of 0.6, which achieves high gain and aperture efficiency, demonstrated through detailed simulations and measurements. For normal incidence, the configuration achieves a peak gain of 9.3 dBi and 48% radiation efficiency at 10 GHz; for oblique incidence, it achieves 7.3 dBi and 40% efficiency. The study underscores the significant potential of inkjet-printed antennas in terms of cost-efficiency, precision, and versatility, paving the way for new advancements in antenna technology with a substantial impact on future communication systems.

Keywords: additive manufacturing; directive antennas; flexible electronics; inkjet printing; reflectarray antennas

1. Introduction

Reflectarray antennas are increasingly pivotal in modern communication systems, offering a sophisticated alternative to traditional parabolic reflectors. The innovation of reflectarrays lies in their ability to use printed circuit board (PCB) technology to achieve high antenna gain, leveraging flat, easily manufactured unit cells in place of complex, curved surfaces. This approach significantly simplifies fabrication processes while achieving or surpassing the performance of conventional designs. Reflectarrays are distinguished by their efficient manufacturability and their capacity for high precision in directing electromagnetic waves, which has led to diverse fabrication methodologies aimed at enhancing performance, reducing costs, and facilitating easier integration with other components.

However, the integration of the feed with the reflectarray surface still represents a substantial challenge in antenna design. Conventional methods typically require auxiliary fixtures to align the feed at the focal point, complicating the manufacturing process and detracting from the efficiency advantages of PCB technology. This additional complexity highlights a crucial hurdle: achieving a seamless integration of feed and reflectarray in a manner that preserves the manufacturing simplicity and cost-effectiveness intrinsic to reflectarray technology. The necessity for external supports or complex assembly processes underscores the need for innovative approaches that can simplify antenna construction without sacrificing performance.

This paper proposes a distinctive method that integrates the feed and reflectarray aperture onto a single flexible substrate, utilizing inkjet-printed technology for fabrication. This approach eliminates the need for separate assembly processes or additional structural components, significantly simplifying the alignment of the feed and reflectarray. It also reduces the overall manufacturing complexity and cost. By fabricating both the feed and reflectarray on the same substrate, this method also enhances mechanical robustness and facilitates easier deployment, particularly in applications requiring flexible or conformable antenna solutions. The integration of the feed and reflectarray into a coherent unit manufactured via inkjet printing on a flexible substrate represents a significant advancement in antenna design, offering a streamlined, cost-effective pathway to high-performance antenna systems.

The literature reveals incremental advancements in reflectarray technology, including efforts to utilize inkjet printing for antenna fabrication, explore flexible substrates for enhanced mechanical versatility, and apply origami principles for deployable structures [1–7]. However, existing studies typically address these aspects in isolation, focusing on either the manufacturing technique, the substrate material, or the deployment mechanism without fully integrating these components into a singular, cohesive design. While inkjet printing has been explored for its potential in patterning conductive materials on substrates, its application in creating comprehensive antenna systems, including both the reflectarray and feed mechanism, remains underexplored. Similarly, the use of flexible substrates has primarily been investigated in the context of conformable or wearable technologies [8–13], with less emphasis on the integration of flexible manufacturing processes with high-performance antenna designs. Origami-inspired antennas have demonstrated potential in deployability [14–18] and mechanical reconfigurability [19–21], yet the seamless integration of origami principles with advanced manufacturing techniques and materials science to produce fully integrated, high-efficiency reflectarray systems has not been comprehensively achieved. Unlike previous work that integrated aperture and feed elements on flexible substrates using PCB techniques [22], this paper introduces the novel application of inkjet printing to achieve the same integration, which enhances design versatility and significantly reduces manufacturing complexities and costs. More specifically, inkjet printing reduces material and labor costs as it is an additive manufacturing method. In terms of time efficiency, the printing process is faster and more streamlined compared to traditional PCB fabrication and assembly. Finally, considering design flexibility, inkjet printing allows for more versatile and precise deposition of conductive inks, enabling complex and customized designs on flexible substrates. Thus, this study builds upon yet significantly advances the concepts introduced earlier.

In particular, inkjet printing shows distinct features as compared to other electronic printing methods such as aerosol jet printing (AJP), direct ink writing (DIW), and electrohydrodynamic printing (EHD). AJP can print finer features with higher resolution than inkjet printing. However, it requires more complex equipment and higher operational costs. Inkjet printing is more accessible and cost-effective, making it suitable for rapid prototyping and low-to-medium volume production. DIW offers the ability to print highly viscous materials and create three-dimensional (3D) structures. However, it lacks the high resolution and speed of inkjet printing. For applications requiring detailed patterns on flexible substrates, inkjet printing provides better performance. EHD can achieve extremely high resolution, but it involves complex setups and is limited by slower printing speeds. Inkjet printing strikes a balance between resolution, speed, and cost, making it ideal for producing flexible electronic components like antennas. Thus, inkjet printing is pursued in this research due to its balance of precision, cost-effectiveness, and suitability for flexible substrates, offering significant advancements in the design and fabrication of high-performance reflectarray antennas.

In demonstrating the proposed concept, this study focuses on designing and fabricating specific structures that validate the integrated feed and reflectarray approach. By employing a flexible polyethylene naphthalate (PEN) substrate, the paper delineates the process of inkjet printing conductive patterns to form the reflectarray and feed elements, followed by the origami-inspired folding of the substrate to assemble the 3D antenna structure. The examination includes detailed performance metrics, such as gain, efficiency, and beam pattern, comparing these results with traditional antenna designs to underscore the advantages of the proposed method. This empirical validation not only proves the feasibility of the concept but also establishes a foundation for future research and development in integrated antenna systems, offering a clear, tangible pathway toward more efficient, cost-effective, and versatile antenna technologies.

2. Fabrication Process

2.1. Inkjet Printing

Before demonstrating the design and manufacturing of the proposed reflectarrays, we first explain the inkjet-printed process used in this study [23]. The inkjet-printed process is performed through the FUJIFILM Dimatix 2850 nanoparticle inkjet printer (DMP-2850). Its components include an ink cartridge, a cleaning pad for nozzle clogging, a printing platform, and cameras for droplet observation and positioning.

The printer has two nozzle systems: a single-nozzle system and a multi-nozzle system. The single-nozzle system operates by moving the printing platform in the x–y direction to inkjet the required pattern, where the droplet spacing and position are also determined by the movement of the platform. The multi-nozzle system arranges all nozzles in a row, operating at an angled cartridge, where the printhead movement along the substrate determines the x–spacing and the angle determines the y–spacing of the droplets. The DMP-2850 has the advantage of being compatible with ink cartridges that can use multiple nozzles at once, and current models on the market have up to 16 nozzles, effectively reducing printing time and avoiding nozzle clogging issues.

The inkjet printing process begins with exporting the finalized antenna design file to the pattern editing software, such as ACE-3000 used in this study, where it undergoes file conversion. Before initiating the printing, the conductive ink is loaded into the ink cartridge in a single-fill operation, with a maximum ink volume of 1.5 ml. The cartridge and a cleaning pad are then installed into the printer. The DMP-2850 operation primarily involves two components: the Droplet Watcher and the Fiducial Camera. The Droplet Watcher is tasked with calibrating and cleaning nozzles 1 to 16, verifying nozzle conditions within the cartridge, and selecting the nozzle number for printing.

The waveform used for ink droplet ejection in the FUJIFILM Dimatix DMP-2850 inkjet printer is critical for controlling droplet formation and ensuring consistent ejection. The optimization of these waveform parameters was undertaken to achieve control over droplet size and ejection consistency, which is essential for high-quality printing. Initially, standard settings provided by the manufacturer were used to establish a baseline. The jetting voltage was then varied to control droplet velocity and size. Higher voltages increase droplet speed, but if set too high, they can cause satellite droplets, leading to print defects. The optimal jetting voltage was determined to be 30.5 V, which balanced droplet formation and minimized defects. The pulse frequency and shape were also adjusted to optimize droplet breakup and ensure uniformity. The pulse frequency was set to 5 kHz, which was found to provide a balance between droplet formation and ejection speed. Using the Droplet Watcher system, the droplet formation was observed in real-time, allowing for adjustments to the waveform shape and duration. This real-time observation ensured consistent droplet size and minimized any satellite droplets. Finally, the optimized settings for the waveform included a specific pulse shape tailored to the viscosity and surface tension of the ink. These settings ensured stable and consistent droplet formation, which was critical for achieving high-quality and high-resolution prints.

Prior to placing the substrate on the printing platform, it is essential to test the surface tension of the substrate to ensure compatibility between the conductor ink and the substrate. Once the substrate is positioned on the platform, a vacuum is activated to secure it in place. For substrates less than 0.5 mm thick, the platform can accommodate a maximum printing area of 210×315 mm^2; substrates between 0.5 and 25 mm thick are limited to a maximum area of 210×260 mm^2. The Fiducial Camera then performs a calibration of the substrate and printing points, covering an observation range of 1.62×1.22 mm^2 and a resolution of 2.54 µm/pixel.

After the printing is complete, the substrate with the printed conductor ink patterns is immediately transferred to a hot air circulating oven for curing. This study employs the SIGMA BS400 oven, which can be temperature-adjusted up to 200 °C. The oven is equipped with an electronic temperature controller, an over-temperature controller, an ammeter, and an electric heating switch. Prior to usage, the external power main and the exhaust system are activated. The over-temperature controller is then adjusted to the rated temperature to prevent overheating inside the oven. Subsequently, the power and startup switches are turned on, and the required baking temperature is set using the displacement and up/down keys on the electronic temperature controller. The SV display indicates the currently set temperature, while the PV display shows the internal oven temperature. More specifically, the post-heat treatment process was performed to ensure the reliability and performance of the inkjet-printed reflectarray. The printed substrates were initially baked at a temperature of 200 °C for 30 minutes. However, to address issues such as the coffee ring effect caused by rapid evaporation of the ink at the edges, a sequential temperature increase method was adopted. This method involved baking the substrates at progressively higher temperatures to improve the uniformity and conductivity of the printed patterns. The curing process was conducted in seven stages, with each stage involving a different temperature and duration to achieve optimal results. The stages started with baking at 30 °C and were terminated by baking at 200 °C. This gradual temperature increase ensured that the ink particles properly coalesced, significantly improving the surface uniformity and conductivity of the printed patterns. Finally, the baking process completed the inkjet printing process.

2.2. Material Properties

After describing the inkjet printing process, this study analyzes the material characteristics of the substrate and conductors before proceeding with the antenna design. Furthermore, these material characteristics complement the inkjet printing process, necessitating corresponding baking, coating, and substrate selection to complete the set material characteristics.

Initially, the study analyzes the properties of conductors. Conductive ink is an indispensable part of the inkjet printing process as the ink distributed from the nozzles forms droplet characteristics on the substrate such as viscosity, surface tension, thermal performance, and substrate compatibility. These factors are crucial for the designed printing process. Common conductive inks used in inkjet printing include polymer-based and metal nanoparticle inks, with nanoparticle inks recently being widely used to achieve high-frequency conductive structures on organic or composite substrates.

This research utilizes silver-based nanoparticle ink. More specifically, the ink employed is the DGP 40LT-15C, sourced from FUJIFILM Dimatix. This silver nanoparticle ink is chosen due to its high conductivity of 9.1×10^6 S/m and low curing temperature of 120 °C. These properties make it ideal for low-temperature substrates such as paper, polyethylene terephthalate (PET), and polymers. Specifically, the use of dyne pens to measure surface tension, which directly correlates with the adhesion properties of the substrate, has been conducted. The test method involves drawing a line on the surface of the substrate with the ink from the dyne pen, which quickly reveals the surface tension and wettability of the substrate. Surface tension is related to the wetting angle, which is the angle at which the ink droplet contacts the substrate. This angle can indicate whether the surface is hydrophilic or hydrophobic. If the wetting angle is too large, the ink droplet will form a spherical shape on the substrate; if the wetting angle is too small, the ink droplet will spread out, affecting conductivity. By changing the dyne level of the test pens until the ink line remains unshrunk, we determine the adhesion strength of the substrate. If the ink line on the substrate does not form any droplets within two seconds, it indicates that the surface tension of the substrate is greater than or equal to the dyne level of the pen, which is a hydrophilic phenomenon, showing that the ink adheres well to the substrate. If the ink line slowly shrinks, it indicates that the surface tension of the substrate is less than the dyne level of the pen. If the ink line immediately shrinks and forms droplets, it indicates that the surface tension of the substrate is much lower than the dyne level of the pen, which is a hydrophobic phenomenon, showing that the ink does not adhere well to the substrate. Our results showed that the surface tension of the PEN substrate is approximately 30–32 dyn/cm. The surface tension of the conductive ink used in this study is about 37 dyn/cm, indicating that the surface tension of the substrate is not significantly lower than that of the conductive ink. This result demonstrates good compatibility between the substrate and the ink, confirming the adhesion strength of the selected materials. Since the ink deposited on the substrate is not a continuous conductor but rather closely spaced silver particles, this indicates its low conductivity. Thus, it is necessary to heat the conductor ink and melt the nanoparticles together to achieve high conductivity. Some studies have used methods like thermal curing, laser curing, and ultraviolet irradiation to cure the ink. However, variables such as ink viscosity, surface tension, number of printing layers, oven curing temperature, and time all affect its conductivity, meaning that the ink has different conductivities in its liquid and solid states. Therefore, this paper uses thermal curing to solidify the conductor ink and achieve the optimal settings for a higher conductivity structure.

To characterize the conductivity of the printed conductor layers, this study first predicts the impact of conductivity on the impedance matching of a microstrip antenna. While the detailed structure of the microstrip antenna will be detailed in the next section, this section analyzes the effect of the conductivity on the antenna performance. The conductivity, denoted by σ, is adjusted from 1×10^5 to 1×10^9 S/m, which affects the impedance matching as shown in Figure 1a. When σ is lower, it can be observed that the resonance frequency shifts to a lower frequency, causing poor antenna matching. Figure 1b shows the radiation efficiency of the antenna at 10 GHz, illustrating that the conductivity significantly affects its efficiency. With a conductor conductivity of 1×10^5 S/m, the radiation efficiency is only 16%, indicating poor conductivity of the antenna, which also affects the radiation efficiency and gain. However, based on the above observations, it is known that by increasing the number of printed ink layers, curing temperature, and time, the conductivity can be increased. As a result, this study designs an antenna printed with three layers of conductor. This indicates that the conductivity of the printed three layers of conductor is approximately in the range of 1×10^5 to 1×10^7 S/m, and the optimum radiation efficiency is about 55%.

The sintering process for the DGP 40LT-15C silver nanoparticle ink used in our study involves a thermal curing method. Given that PEN has a glass transition temperature (Tg) of 120 °C, the sintering temperature for the silver nanoparticle ink was selected to align with this constraint. The sintering process was conducted at 110 °C for 1 h to ensure effective curing of the ink while preventing any deformation or damage to the PEN substrate. This temperature is slightly below the Tg of PEN, ensuring the structural integrity of the substrate while achieving optimal electrical conductivity of the printed film. Additionally, maintaining the sintering temperature just below Tg prevents issues like warping or changes in the physical properties of the PEN substrate, which could affect the overall performance and reliability of the antenna. This consideration ensures the final product maintains both structural integrity and high functionality. Additionally, to address the coffee ring effect and ensure uniform ink distribution, the curing process was performed in a hot air circulating oven with a controlled temperature ramp. The gradual increase in temperature helps to avoid rapid solvent evaporation, which can cause uneven ink distribution and affect the quality of the printed patterns. The resultant film thickness of the printed conductor on the PEN substrate, after three layers of printing, was measured to be 25 μm. This was determined using a scanning electron microscope (SEM). An example of the SEM image is shown in Figure 2a, which indicates the 3-layer cross-sectional profile for the depositions. By measuring the thickness at the 10 sample points, the resultant film thickness of the printed conductor on the PEN substrate, after three layers of printing, can be determined. Similarly, another example is the 1-layer of printing, as shown in Figure 2b. The final thickness was determined by the average, which was found to be 10 μm on different substrates.

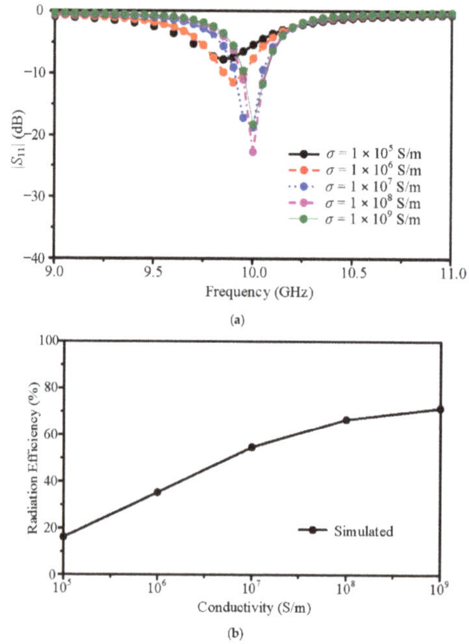

Figure 1. Impact of different conductivities on the (**a**) impedance matching and (**b**) radiation efficiency of the antenna.

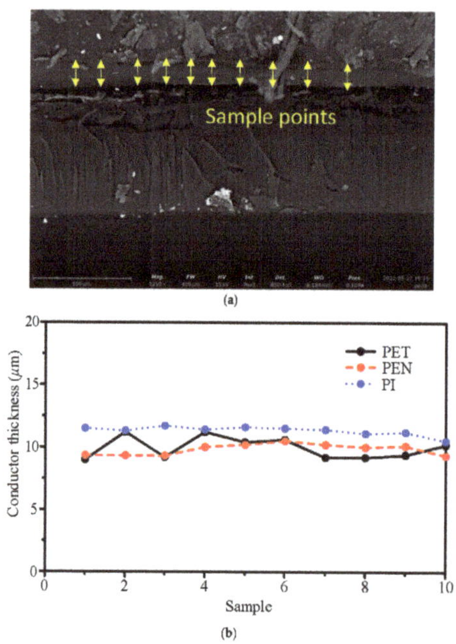

Figure 2. (**a**) Three-layer cross-sectional image and (**b**) the thickness obtained by the measurement using SEM.

Next, we examined the material characteristics of dielectric substrates utilized in the design of a reflectarray. Specifically, we analyzed flexible substrates such as PET, PEN, and polyimide (PI), focusing on how their thickness impacts the adjustable phase range of reflectarray unit cells. Generally, thicker substrates offer a larger adjustable phase range, though this comes with increased loss. A key feature of the proposed technique is the ability to conveniently increase the board thickness by folding flexible substrates. We varied the thicknesses of PET, PEN, and PI to simulate their adjustable phases, as presented in Figure 3. The results demonstrate that PEN, in particular, provides superior phase adjustability with increasing thickness compared to PET and PI, making it the optimal choice for subsequent antenna designs. This is illustrated in Figure 3, where the simulated results show a steep increase in the adjustable phase range for PEN, compared to the more moderate improvements seen with PET and PI. This capability to adjust phase effectively by altering substrate thickness is crucial for optimizing antenna performance while minimizing losses, solidifying PEN as the substrate of choice in the development of advanced antenna structures.

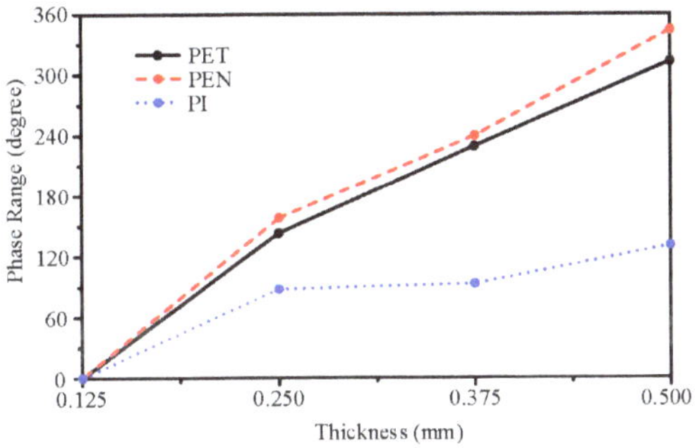

Figure 3. Variation in the adjustable phase range corresponding to changes in the thickness of flexible substrates.

More specifically, the PEN utilized is sourced from Goodfellow Corporation. The surface free energy of PEN is measured to be approximately 42.8 dyn/cm, which ensures good wettability and adhesion with the conductive ink used in the inkjet printing process. Additionally, the surface roughness of the PEN substrate was characterized using atomic force microscopy (AFM), revealing an average roughness of 2.3 nm. In particular, the initial layer of ink is deposited on the PEN substrate, and its interaction is primarily governed by the surface energy of the substrate and the surface tension of the ink. The ink had a lower surface tension than the critical surface tension of the PEN substrate to achieve proper wetting and adhesion. The contact angle measurement indicated a favorable wetting behavior, crucial for forming a uniform initial layer. For subsequent layers, the interaction changes as the ink is deposited on a layer of previously deposited ink. The critical factors here include ink penetration and merging, layer homogeneity, and sintering effects. The process starts by ensuring that new ink droplets merge seamlessly with the underlying layer to avoid delamination. The controlled print speed and droplet spacing facilitate uniform droplet merging. Afterwards, the uniformity of the underlying layer was maintained by optimizing the droplet spacing (30 µm), providing consistent overlap and coverage. Sintered layers can have different surface roughness, impacting wetting behavior. The sintering process aims to achieve a smooth and uniform surface, enhancing the adhesion of subsequent layers.

2.3. Parametric Effects

To clarify the parameters of inkjet printing, the ink's rheological properties and kinematic settings were further analyzed. The kinematic settings of the FUJIFILM Dimatix DMP-2850 printer were carefully optimized to achieve high-quality prints. The printer's multi-nozzle system allowed for precise control over droplet spacing, set at 30 µm, which balanced the need for high resolution and uniform deposition. This careful control helped mitigate issues such as the coffee ring effect, where ink accumulates at the edges of printed patterns, leading to non-uniform conductivity. By maintaining these precise settings, we ensured the production of a consistent and high-performance conductive layer necessary for the functionality of the reflectarray.

Moreover, the printing speed was set at 10 mm/s, providing adequate control over the deposition process. This speed was selected to prevent defects such as splattering or insufficient coverage while maintaining high throughput. The layer thickness achieved per pass was approximately 1.2 μm, optimized to ensure electrical performance while maintaining the flexibility of the substrate. Multiple layers were printed to achieve the desired thickness. The curing temperature and time are crucial for achieving the desired electrical properties of the printed ink. For PI substrates, the printed patterns were cured at 150 °C for 30 minutes; nevertheless, for substrates with lower Tg, such as PEN and PET, the curing temperatures were set below their respective Tg values to ensure mechanical integrity. Specifically, curing was conducted below 120 °C for PEN and below 75 °C for PET substrates. In practical terms, the antenna is suitable for use in a variety of applications, including wearable electronics, flexible sensors, and other devices operating within the temperature range of –40 °C to 115 °C. For applications requiring operation above 120 °C, alternative substrate materials with higher Tg values may be considered to ensure durability and performance.

In addition, to ensure the reliability and durability of the printed antenna, we conducted bending tests to assess the relative resistance change under mechanical stress. In our experiments, the relative resistance change of the inkjet-printed silver nanoparticle traces on the PEN substrate was measured after multiple bending cycles. The specimens, sized 60 mm × 60 mm, were subjected to repeated bending with radii of 20 mm, 30 mm, 40 mm, and 50 mm using polystyrene fixtures. The initial resistance was recorded using a multimeter, followed by resistance measurements after every 1000 bending cycles. The measured results are shown in Figure 4. Our results showed that after 5000 bending cycles, the relative resistance change was less than 5%, indicating excellent mechanical durability and reliability of the printed antenna.

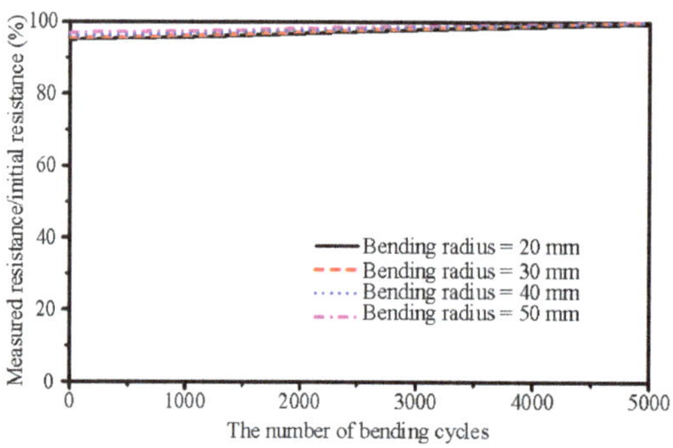

Figure 4. Mechanical durability of the materials.

In addition to using dyne pens to assess the surface tension of the PEN substrate, we conducted a Scotch tape test to evaluate the adhesion strength of the inkjet-printed electrodes [24]. The test involved applying Scotch tape to the printed electrodes and subjecting the samples to multiple peeling cycles (5, 10, 30, and 50 cycles). Visual inspection after each set of cycles revealed no delamination or damage. Furthermore, the relative resistance change was measured to be less than 2% after 50 peeling cycles, indicating strong adhesion and minimal impact on electrical performance.

These detailed considerations of ink rheological properties, kinematic settings, and bending durability underscore the robustness of our inkjet printing methodology and highlight the innovative aspects of our approach in integrating both feed and aperture elements onto a single flexible substrate.

3. Inkjet-Printed Reflectarray with Normal Incidence

Sections 3 and 4 of this paper are dedicated to demonstrating the implementation of reflectarray antennas with normal and oblique incidence feeds, respectively, building upon the foundational methodologies established in Section 2. The logical progression from the fabrication techniques detailed previously into practical antenna applications underscores the versatility and potential of inkjet-printed, flexible substrates in advanced antenna designs. Section 3, in particular, focuses on detailing the geometric structuring and integration of feed elements and reflectarrays into a single flexible substrate, advancing from mere component fabrication to functional antenna assembly. This approach aims to overcome the traditional challenges associated with the manufacturing and alignment sensitivities of conventional reflectarray antennas by proposing and demonstrating a novel technique of folding the flexible substrate into predetermined 3D shapes.

Illustrating these concepts, Figure 5 showcases the unfold and 3D structure of the reflectarray antenna within a hexahedral framework. This visualization provides the antenna geometry and highlights the practical application of the folding and unfolding mechanisms in tailoring antenna characteristics to specific requirements. The presented designs of polyhedral structures, specifically a hexahedron and a tetrahedron, exemplify this integrated approach. These configurations are optimized based on theoretical designs that dictate the focal length, denoted by F, and dimensions, denoted by D, of the reflectarray (with an optimal F/D of 0.6 for maximum efficiency), thus ensuring the practical efficacy of the antenna in real-world applications.

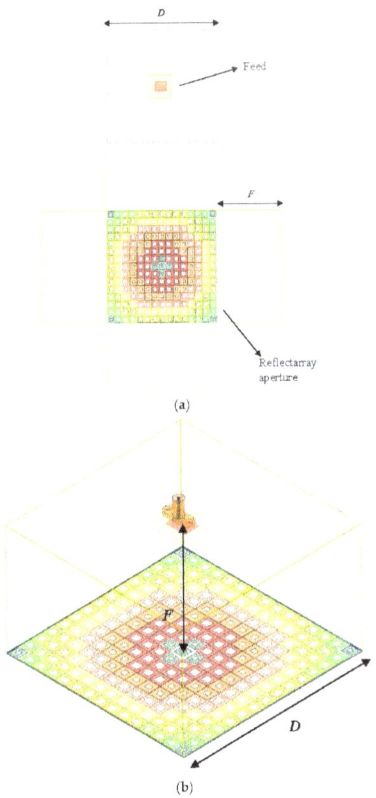

Figure 5. Reflectarray antenna with normal incidence on a hexahedral structure, showing (**a**) an unfolded diagram and (**b**) a three-dimensional diagram.

3.1. Feed Antenna

As a proof of concept, the proposed inkjet-printed origami reflectarray is designed at 10 GHz. The substrate material is polyethylene naphthalate (PEN) with a dielectric constant of 3.43 and a loss tangent of 0.005, with a thickness of 0.25 mm. Figure 6 shows the detailed feed antenna structure. This patch antenna is designed and optimized with parameters such as the patch width (W_{pa} = 10.8 mm), length (L_{pa} = 7.85 mm), and ground dimensions (W_{gnd} = L_{gnd} = 20 mm). The patch antenna is fed at the corner, which typically can be associated with attempts to achieve circular polarization. However, in this particular design, the corner feeding is employed primarily for impedance matching purposes. Based on the radiation patterns, which will be shown as follows, the antenna remains linearly polarized.

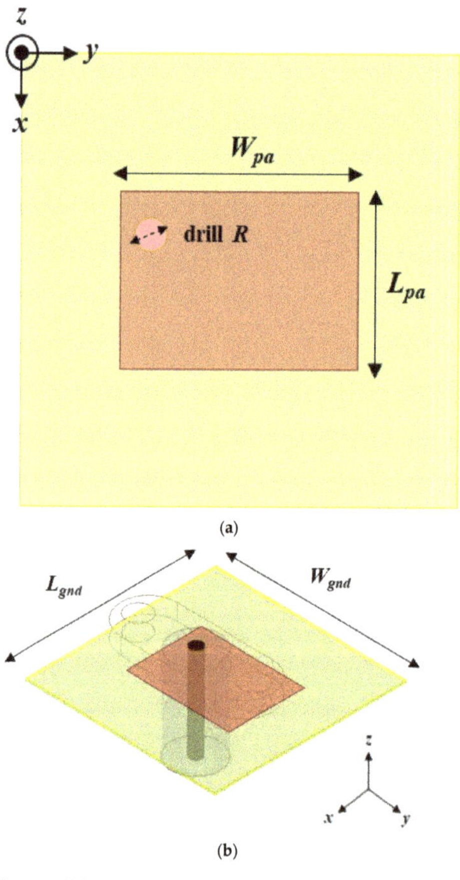

Figure 6. Microstrip antenna structure (**a**) top view and (**b**) side view.

The feed antenna is simulated and measured. Figure 7 illustrates the reflection coefficient across 8–12 GHz, while Figure 8 displays the peak gain. The simulated impedance bandwidth, characterized by a reflection coefficient less than −10 dB, is observed to be 1.6%. This narrow bandwidth is typical for microstrip patch antennas, reflecting their inherent limitation in bandwidth capacity due to the high Q factor of the resonant structure. The measurements closely aligned with the simulated results, showcasing a reflection coefficient of −18 dB at 10 GHz and revealing a slightly broader actual bandwidth percentage of 2.6%. The peak gain of 4.9 dBi is observed at 9.98 GHz—closely matching the targeted 10 GHz operation. Additionally, Figure 9 shows the radiation patterns in the E-plane and

H-plane, demonstrating a front-to-back ratio (FBR) of 17.1 dB and a cross-polarization discrimination (XPD) of 24.8 dB. The half-power beamwidths (HPBW) are 80.6° and 74.7° in the E-plane and H-plane, respectively. The measured cross-polarization level is higher than the simulated one in Figure 9 due to several factors related to the experimental setup and the inherent limitations of practical measurements; nevertheless, the radiation performance indicates a directed beam, suitable for serving the reflectarray as a feed.

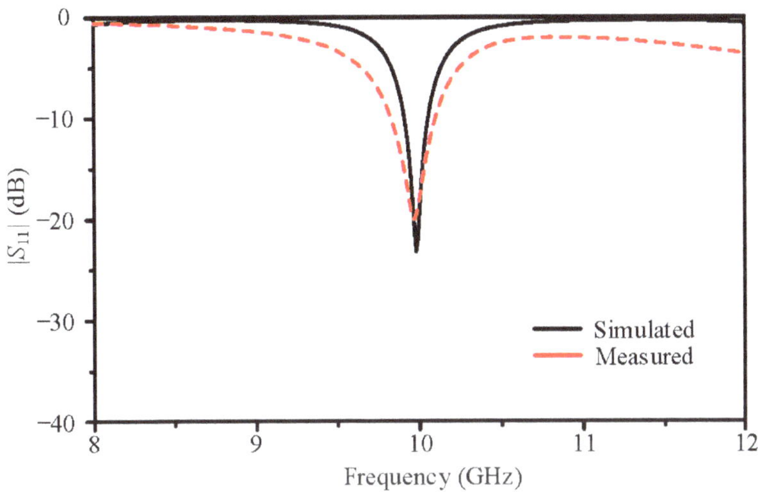

Figure 7. Simulation and measurement of the reflection coefficient for the feeding microstrip antenna.

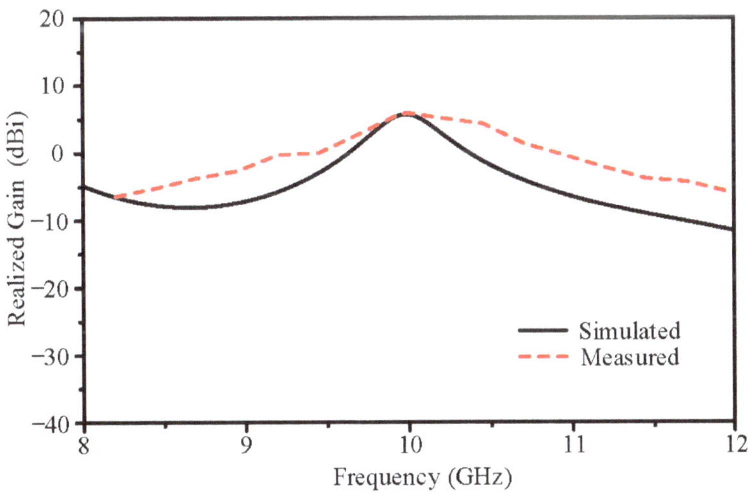

Figure 8. Simulation and measurement of the realized gain for the feeding microstrip antenna.

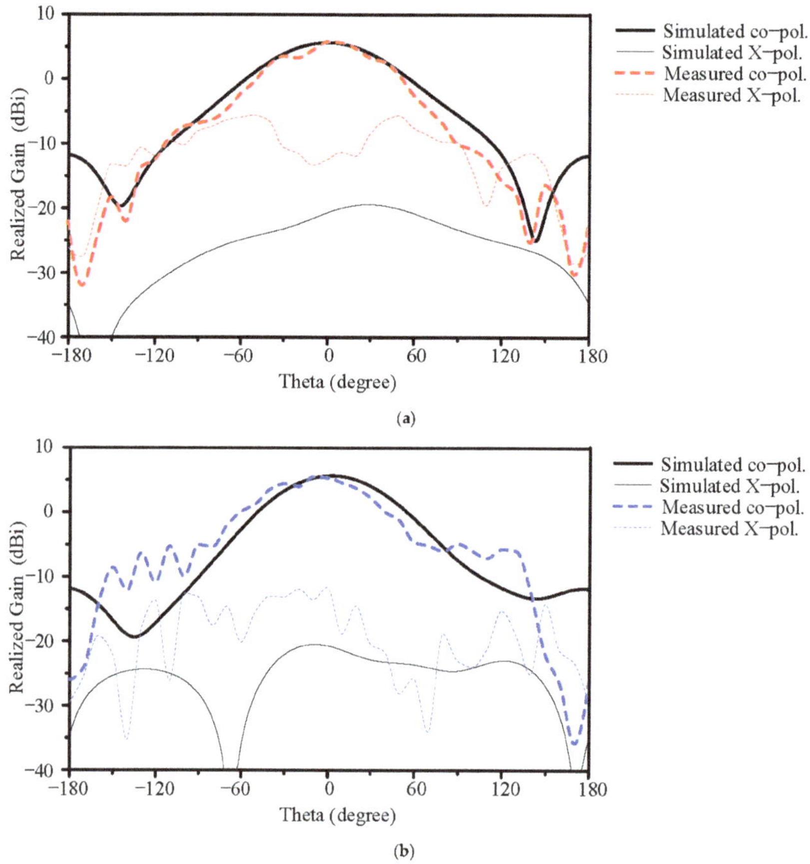

Figure 9. Simulation and measurement of the (**a**) E-plane and (**b**) H-plane patterns for the feeding microstrip antenna.

In the design of the reflectarray antenna, the radiation pattern of the feed antenna plays a crucial role in determining the optimal F/D. For effective design integration, the feed pattern was approximated by the function $\cos^q\theta$, where q is an exponent that aligns with the characteristics of the pattern. It was identified that when q = 2.8 the $\cos^q\theta$ pattern closely aligns with the simulated radiation patterns of the E-plane and H-plane. Integrating this value of q into the analyses for illumination efficiency and spillover efficiency, the overall antenna efficiency was computed [25]. The findings indicated that an F/D of 0.6 yields the highest overall efficiency, evaluated as 59%. This efficiency benchmark guided the structural design of the antenna, culminating in the 3D architecture shown in Figure 5.

3.2. *Reflectarray Antenna*

Next, the unit cells of the reflectarray were designed to achieve a full phase adjustable range, essential for beam steering. Five distinct unit cell shapes were considered: rectangular, circular, square-ring, circular-ring, and cross-shaped, optimized for a target frequency of 10 GHz on the PEN substrate. The substrate thickness was set at 0.5 mm to balance phase resolution and minimize physical size. Among the configurations, the square-ring unit cell exhibited superior performance, achieving a phase adjustable range up to 343° with a loss of 2.91 dB, demonstrating its effectiveness in providing extensive phase control with minimal loss. The simulated insertion loss of 2.91 dB in the reflectarray element is primarily caused by a combination of factors, including the properties of the substrate and the conductivity of the printed conductor layer. The flexible PEN substrate, while chosen for its mechanical and thermal properties, does introduce some dielectric loss. However, the more significant contributor to the insertion loss is the lower conductivity of the printed silver nanoparticle ink compared to bulk metal conductors. Additionally, the multi-layer printing process, while improving the thickness and overall conductivity of the printed traces, cannot entirely mitigate the inherent losses associated with the lower conductivity of the printed ink. The unit cell spacing was maintained at a 0.22 wavelength to reduce mutual coupling while conserving array compactness.

Figure 10 demonstrates the profile of the unit cell. This configuration leveraged a multi-layer stacking approach, using a minimal gap of 0.01 mm between layers. Such a multi-layer structure can be efficiently created, as the proposed technique can fold a planar substrate into a 3D and multi-layer configuration. This method effectively utilized the substrate characteristics and the geometrical flexibility of the unit cells to meet the design requirements for high-performance beam steering in reflectarray applications.

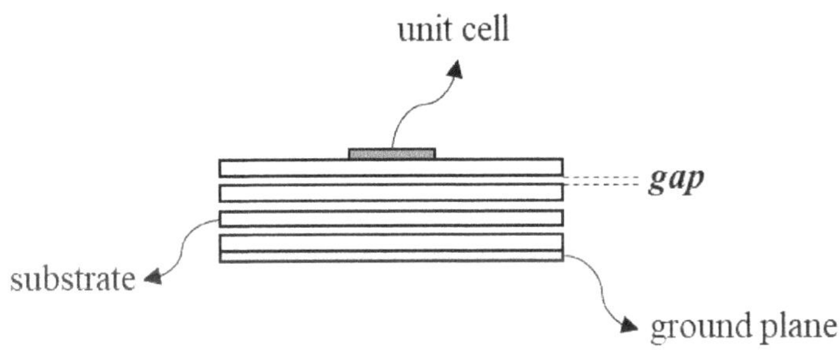

Figure 10. Unit cell structure (side view).

In the design of the reflectarray, the phase distribution across the aperture is critical for achieving in-phase radiation from all unit cells. By employing the square-ring unit cell sizes, the reflectarray is configured to cover a full 360° adjustable phase range, which allows for fine-tuning the phase of each unit cell to achieve a coherent wavefront.

Specifically, the reflectarray utilizes eight different square-ring unit cell structures, arranged to incrementally cover the required phase range with each unit providing a distinct phase shift. This setup ensures that the phase shift between adjacent cells increments by 45. This arrangement is chosen to ensure a phase progression across the reflectarray, facilitating uniform wavefront shaping and maintaining high directivity.

Each unit cell is positioned within the reflectarray aperture, measuring 99×99 mm^2, composed of 15×15 unit cells, to match the design criterion of $F/D = 0.6$. The layout of the unit cells is such that the transitions between the phase states are smooth, minimizing discontinuities and ensuring that the reflectarray can effectively focus the radiated energy. Figure 11 shows the arrangement of the unit cells, which allows the reflectarray to perform with enhanced gain and efficiency.

Finally, the reflectarray aperture is created using inkjet printing techniques, which facilitates the placement of the conductive ink on a flexible substrate. To ensure alignment during printing, structural patterns are placed beneath the printing substrate. Following the inkjet printing process, the metallic ground plane, originally intended to be inkjet-printed, is substituted with a copper foil tape to enhance practicality and cost-effectiveness and reduce printing time. The inkjet-printed reflectarray is then folded into a hexahedral structure, as depicted in Figure 12. This transformation from a flat aperture to a 3D form is critical for achieving the desired F/D, allowing for focus and directionality of the beam.

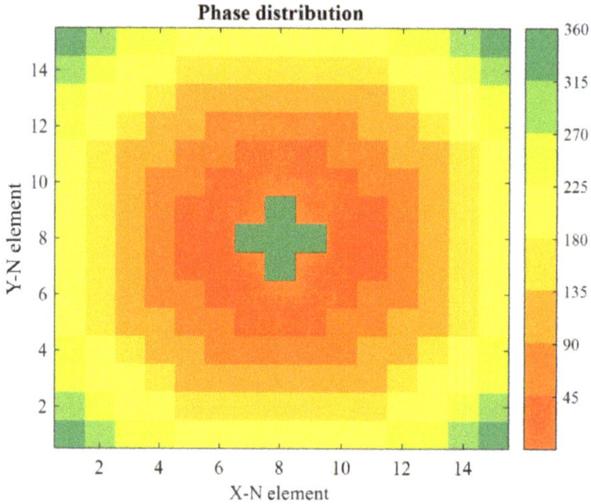

Figure 11. Discrete phase distribution of the reflectarray with normal incidence.

Figure 12. Photograph of the reflectarray antenna with normal incidence.

Performance evaluations of the integrated reflectarray were conducted, comparing simulation with measurements. The peak antenna gain, shown in Figure 13, reached 9.3 dBi at 10 GHz, which differs by 3.7 dB from the simulated gain. Additionally, the radiation efficiency was 48%, with a marginal 2% difference between the measured and simulated values. Finally, the antenna patterns in the E-plane and H-plane are detailed in Figure 14, where measured peak gains of 8.7 and 9.3 dBi, respectively, and FBR of 19.7 and 25.3 dB are observed. Thus, the proposed approach successfully demonstrates the creation of a high-performance reflectarray antenna by inkjet printing all necessary components onto a single flexible substrate, which is subsequently folded into a 3D structure to meet specific design parameters. This method streamlines the fabrication process, ensures precise component alignment, and effectively integrates the feed and reflectarray elements, exemplifying a significant advancement in antenna technology.

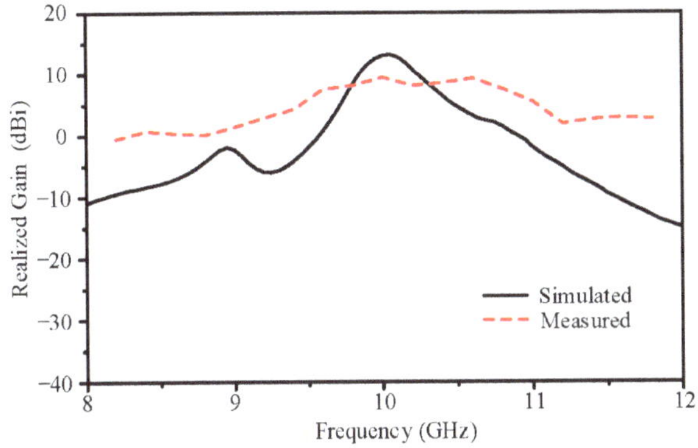

Figure 13. Simulation and measurement of the realized gain for the reflectarray antenna with normal incidence.

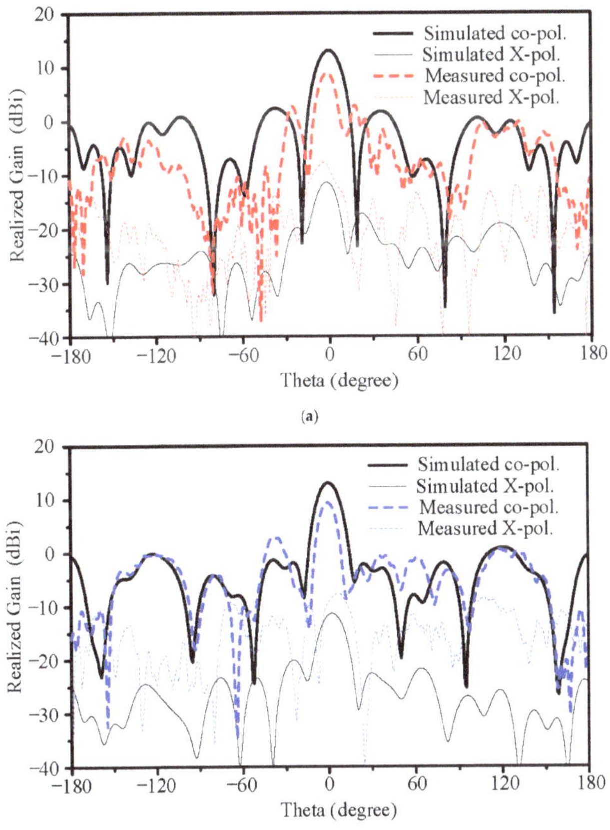

Figure 14. Simulation and measurement of the (**a**) E-plane and (**b**) H-plane patterns for the reflectarray antenna with normal incidence.

4. Inkjet-Printed Reflectarray with Oblique Incidence

Following the evaluation under normal incidence conditions, Section 4 extends the analysis to oblique incidence feeds. This section explores the adaptability and versatility of the fabricated reflectarrays in handling incoming signals from various angles, which is crucial for practical applications where signal incidence can vary significantly. The goal is to illustrate the robustness of the design and fabrication techniques in maintaining performance metrics across a range of incidence angles, highlighting the antenna capability in more dynamic operational environments.

Figure 15 illustrates the unfolded and folded tetrahedral structures for the oblique incidence reflectarray antenna. This geometry is chosen to optimize the F/D to match the previously optimized value of 0.6, ensuring effective antenna performance. The tetrahedral structure allows for the integration of the reflectarray on a single facet, aligning the feed at the optimal angle for maximum aperture efficiency.

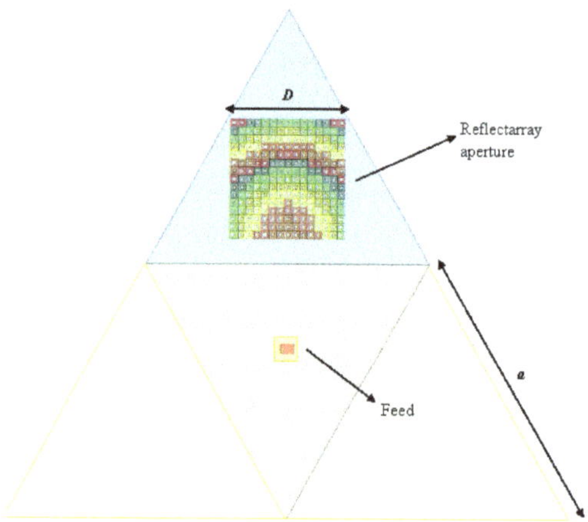

(a)

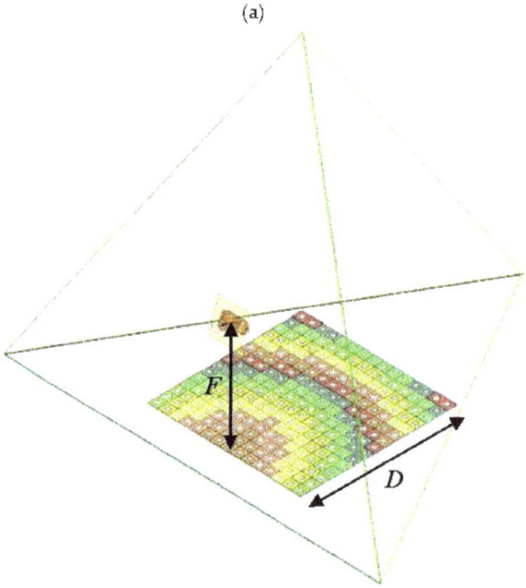

(b)

Figure 15. Reflectarray antenna with oblique incidence on a hexahedral structure, showing (**a**) the unfolded diagram and (**b**) the three-dimensional diagram.

The reflectarray is designed to fit within the tetrahedral envelope, with each side of the tetrahedron accommodating the printed antenna array. Given the antenna aperture size of 99 mm, the edge length, denoted by a, of the tetrahedron is calculated to be 242.5 mm to ensure the reflectarray fits precisely without altering the F/D. The theoretical framework for the tetrahedron involves calculating dihedral angles and other geometric parameters to maintain the F/D at 0.6. This is confirmed by deriving the relationship between the edge length a and the focal length F through geometric calculations. The results indicate that an edge length a of 242.5 mm yields a focal length F that maintains the F/D close to the ideal value, confirming the practicality of the proposed structure for oblique feed incidence.

Continuing from the structural design outlined, the feed mechanism for the oblique incidence reflectarray remains consistent with the approach used in Section 3 for the normal incidence configuration. Located at an inclined plane of the tetrahedron, the placement of the feed antenna minimizes interference and optimizes energy distribution across the reflectarray aperture. With the reflectarray now on the base within the tetrahedral structure, the unit cell configurations initially designed for normal incidence require adjustments. These changes are necessary to address the different path lengths and angles due to the new incidence angle.

Figure 16 shows the phase distribution for the arrangement of unit cells, each adjusted to specific phase delays tailored for the geometric constraints of the oblique feed. This setup ensures that reflected waves from different array sections constructively interfere in the desired direction, enhancing the directional property of the antenna. In this configuration, the phase adjustments across the reflectarray aim at the maintenance of a uniform phase front.

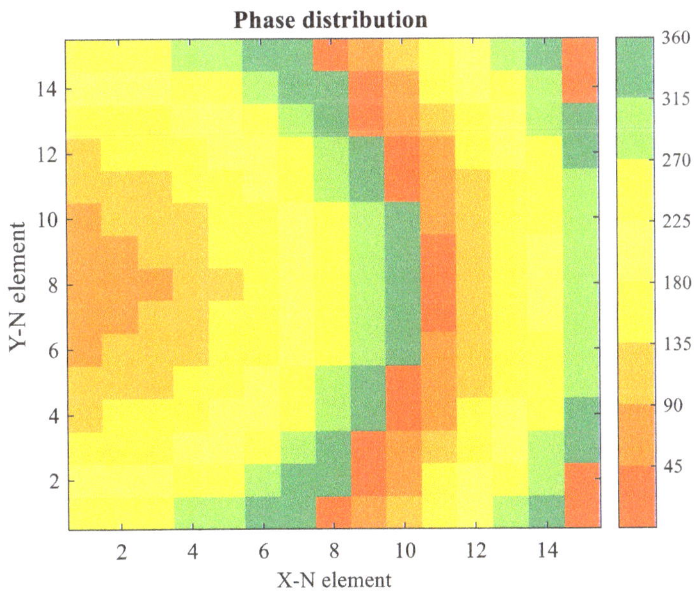

Figure 16. Discrete phase distribution of the reflectarray with oblique incidence.

The reflectarray antenna with oblique incidence was fabricated and tested. The reflectarray, with an aperture size of 99 mm, was integrated onto a tetrahedral structure, which was segmented and bonded to form a complete 3D configuration, as shown in Figure 17. To circumvent the limitations of inkjet printing, the metal ground plane was substituted with copper foil tape, enhancing the structural integrity and cost-effectiveness of the assembly.

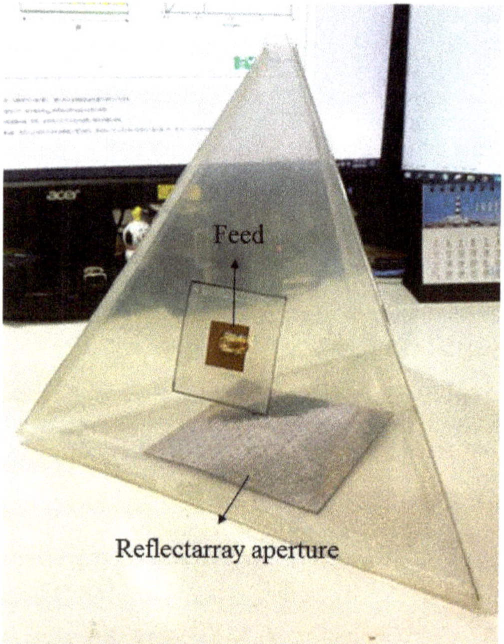

Figure 17. Photograph of the reflectarray antenna with oblique incidence.

Subsequently, the antenna performance was evaluated through simulations and measurements. Figure 18 details the peak gain response, achieving a peak gain of 7.3 dBi at 10 GHz, differing by 2.8 dB from the simulated gain. The measured radiation efficiency is 40%, with a 10% discrepancy compared to the simulated values. The E-plane and H-plane radiation patterns, depicted in Figure 19a,b, reveal peak gains of 7.3 and 7.1 dBi, respectively, and FBR of 22.7 and 30.1 dB, with a HPBW of 12° and 10°.

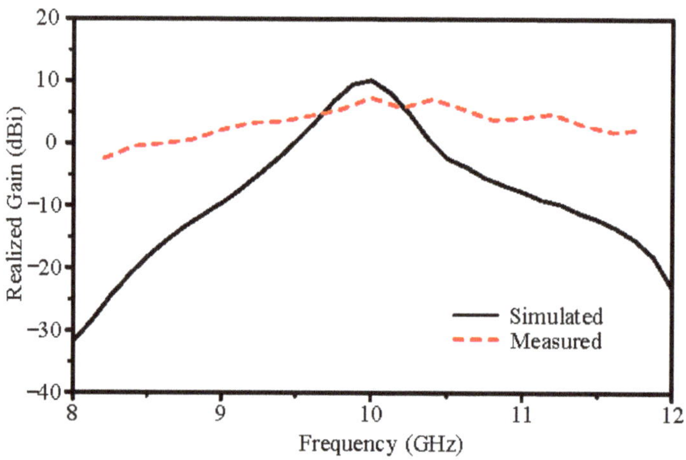

Figure 18. Simulation and measurement of the realized gain for the reflectarray antenna with oblique incidence.

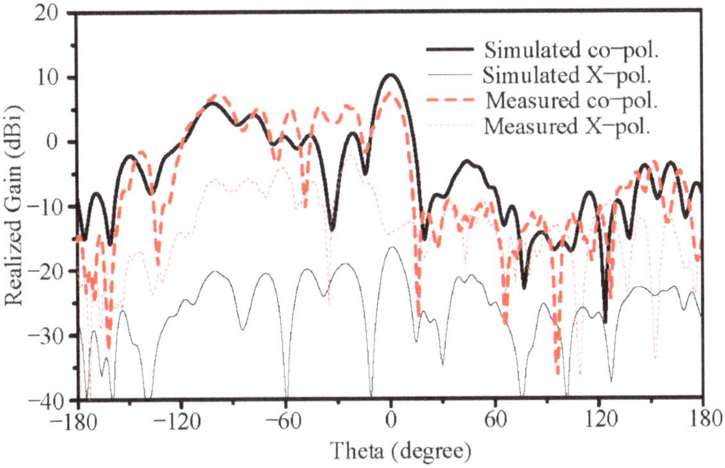

(a)

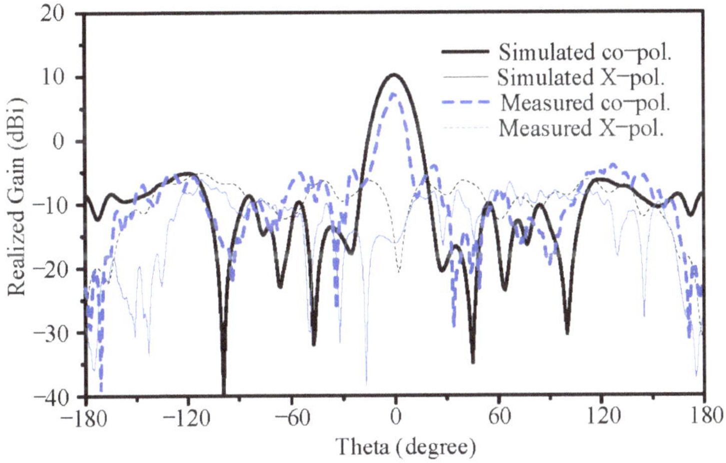

(b)

Figure 19. Simulation and measurement of the (**a**) E-plane and (**b**) H-plane patterns for the reflectarray antenna with oblique incidence.

For the two reflectarray antennas, the corresponding effective apertures are 609.6 mm^2 and 384.6 mm^2, which result in an aperture efficiency of 6.2% and 3.9%. The primary reason for the lower aperture efficiency is the use of inkjet printing and flexible substrates. The inkjet printing method results in lower radiation efficiency because the conductivity of the printed silver nanoparticle ink is not as high as that of copper. Additionally, the feed antenna used in this study is a patch antenna, which has a relatively significant backlobe compared to the horn antennas widely used in the literature. This backlobe further reduced the aperture efficiency. Improving the gain and aperture efficiency can be achieved by employing multiple reflectarray surfaces. In addition, the differences between the simulated and measured gains are likely due to several factors. First of all, the inkjet printing resolution affected the precision of the unit cell boundaries, observed through electron microscopy. Secondly, the file conversion for printing patterns led to a dimensional increase of 0.2 mm on the x-axis, impacting the continuous design of the reflectarray aperture. Thirdly, the alignment of the feed antenna with the coaxial cable influenced the radiation pattern. In addition, manufacturing tolerances during the inkjet printing process introduce small deviations from the ideal design, resulting in structural asymmetries that affect radiation characteristics. These inaccuracies stem from variations in droplet size, placement, and layer thickness, which simulations cannot fully account for. Despite these challenges, the trend in simulated and measured results remains consistent, confirming the efficacy of the proposed structural adjustments.

5. Conclusions

This study introduces a novel inkjet-printed reflectarray antenna integrated onto a flexible substrate, presenting a significant advancement over traditional PCB-based techniques by simplifying the manufacturing process and enhancing design flexibility. The proposed method combines inkjet printing and origami-inspired folding techniques to fabricate both the feed and reflectarray elements on a single substrate, which is then configured into precise 3D structures. Distinct from the existing literature, this approach eliminates the need for multiple assembly processes and structural supports, which are typically required in conventional antenna designs. The proposed technique not only reduces manufacturing complexity but also substantially lowers production costs. Furthermore, it provides an unprecedented level of precision in component alignment and integration, facilitating superior antenna performance. The effectiveness of the proposed method is validated through a series of simulations and measurements which underscore the capability of the proposed technique to maintain high performance across different configurations and operational conditions. The integration of inkjet printing with flexible substrate technologies not only adheres to the demands of modern antenna systems but also opens up new avenues for deployable and wearable antenna applications. The success of this research in achieving a streamlined fabrication process, coupled with the high performance of the resultant antenna systems, positions this study to significantly impact future developments in antenna technology, potentially influencing both academic research and industry practices in communication systems.

Author Contributions: Implementation, Y.-X.L., K.-Y.K., and F.-P.L.; conceptualization: Y.-S.C. All authors have read and agreed to the published version of the manuscript.

Funding: This work was supported by the Ministry of Science and Technology, Taiwan, under Contract MOST 111-2221-E-027-047-MY2.

Data Availability Statement: Data is contained within the article.

Conflicts of Interest: The authors declare no conflict of interest.

References

1. Kaddour, A.S.; Velez, C.A.; Hamza, M.; Brown, N.C.; Ynchausti, C.; Magleby, S.P.; Howell, L.L.; Georgakopoulos, S.V. A foldable and reconfigurable monolithic reflectarray for space applications. *IEEE Access* **2020**, *8*, 219355–219366. [CrossRef]
2. Rubio, A.J.; Kaddour, A.S.; Georgakopoulos, S.V.; Brown, N.; Ynchausti, C.; Howell, L.; Magleby, S. An origami-inspired foldable reflectarray on a straight-major square-twist pattern. In Proceedings of the 2021 IEEE 21st Annual Wireless and Microwave Technology Conference, Sand Key, FL, USA, 28–29 April 2021; pp. 1–4.
3. Velez, C.A.; Kaddour, A.S.; Georgakopoulos, S.V.; Bolanos, D.S.; Ynchausti, C.; Magleby, S.; Howell, L.L. Reconfigurable and deployable Miura-Ori RA analysis for satellite applications. In Proceedings of the 2021 IEEE 21st Annual Wireless and Microwave Technology Conference, Sand Key, FL, USA, 28–29 April 2021; pp. 1–3.
4. Velez, C.A.; Kaddour, A.S.; Georgakopoulos, S.V.; Ynchausti, C.; Magleby, S.; Howell, L.L. A novel Miura-Ori origami reflectarray antenna for CubeSat applications. In Proceedings of the 2021 IEEE International Symposium on Antennas and Propagation & USNC/URSI Radio Science Meeting, Singapore, 4–10 December 2021; pp. 1759–1760.
5. Kaddour, A.S.; Velez, C.A.; Georgakopoulos, S.V. A deployable and reconfigurable origami reflectarray based on the Miura-Ori pattern. In Proceedings of the 2020 IEEE International Symposium on Antennas and Propagation & North American Radio Science Meeting, Montreal, QC, Canada, 5–10 July 2020; pp. 91–92.
6. Rubio, A.J.; Kaddour, A.S.; Ynchausti, C.; Magleby, S.; Howell, L.L.; Georgakopoulos, S.V. A foldable reflectarray on a hexagonal twist origami structure. *IEEE Open J. Antennas Propag.* **2021**, *2*, 1108–1119. [CrossRef]

7. Miguélez-Gómez, N.; Parkhurst, J.M.; Pepin, K.; Moline, N.; Leblanc, S.; Udrea, B.; Rojas-Nastrucci, E.A. Thickness-accommodation in X-band origami-based reflectarray antenna for small satellites applications. In Proceedings of the 2020 IEEE International Conference on Wireless for Space and Extreme Environments, Vicenza, Italy, 12–14 October 2020; pp. 54–59.
8. Elmobarak, H.A.; Rahim, S.A.; Himdi, M.; Castel, X.; Rahman, T.A. Low cost instantly printed silver nano ink flexible dual-band antenna onto paper substrate. In Proceedings of the 2017 11th European Conference on Antennas and Propagation, Paris, France, 19–24 March 2017; pp. 3061–3063.
9. Yang, L.; Rida, A.; Vyas, R.; Tentzeris, M.M. RFID tag and RF structures on a paper substrate using inkjet-printing technology. *IEEE Trans. Microw. Theory Tech.* **2007**, *55*, 2894–2901. [CrossRef]
10. Cook, B.S.; Shamim, A. Inkjet printing of novel wideband and high gain antennas on low-cost paper substrate. *IEEE Trans. Antennas Propag.* **2012**, *60*, 4148–4156. [CrossRef]
11. Jilani, S.F.; Alomainy, A. An inkjet-printed MMW frequency-reconfigurable antenna on a flexible PET substrate for 5G wireless systems. In Proceedings of the Loughborough Antennas & Propagation Conference, Loughborough, UK, 13–14 November 2017; pp. 1–3.
12. Khinda, G.S.; Umar, A.; Cadwell, R.J.; Alhendi, M.; Stoffel, N.C.; Borgesen, P.; Poliks, M.D. Flexible inkjet-printed patch antenna array on mesoporous PET substrate for 5G applications with stable RF performance after mechanical stress cycling. In Proceedings of the 2020 IEEE 70th Electronic Components and Technology Conference, Orlando, FL, USA, 3–30 June 2020; pp. 1824–1831.
13. Jun, S.; Sanz-Izquierdo, B.; Summerfield, M. UWB antenna on 3D printed flexible substrate and foot phantom. In Proceedings of the 2015 Loughborough Antennas & Propagation Conference, Loughborough, UK, 2–3 November 2015; pp. 1–5.
14. Hamza, M.; Zekios, C.L.; Georgakopoulos, S.V. A thick origami reconfigurable and packable patch array with enhanced beam steering. *IEEE Trans. Antennas Propag.* **2020**, *68*, 3653–3663. [CrossRef]
15. Zhong, Z.; Xu, L.; Zhang, H.; Zhang, P. A foldable circular polarized microstrip antenna array for satellite communication. In Proceedings of the 2018 12th International Symposium on Antennas, Propagation and EM Theory, Hangzhou, China, 3–6 December 2018; pp. 1–4.
16. Seiler, S.R.; Bazzan, G.; Fuchi, K.; Alanyak, E.J.; Gilman, A.S.; Reich, G.W.; Cook, A.; Buskohl, P.R.; Pallampati, S.; Huff, G.H. An origami inspired circularly-polarized folding patch antenna array. In Proceedings of the 2018 IEEE International Symposium on Antennas and Propagation & USNC/URSI National Radio Science Meeting, Boston, MA, USA, 8–13 July 2018; pp. 181–182.
17. Seiler, S.R.; Bazzan, G.; Fuchi, K.; Alanyak, E.J.; Gillman, A.S.; Reich, G.W.; Buskohl, P.R.; Pallampati, S.; Sessions, D.; Huff, G.H. Physical reconfiguration of an origami-inspired deployable microstrip patch antenna array. In Proceedings of the 2017 IEEE International Symposium on Antennas and Propagation & USNC/URSI National Radio Science Meeting, San Diego, CA, USA, 9–14 July 2017; pp. 2359–2360.
18. Shah, S.I.H.; Sarkar, A.; Lim, S. Electromechanically deployable high-gain pop-up antenna using shape memory alloy and kirigami technology. *IEEE Access* **2020**, *8*, 225210–225218. [CrossRef]
19. Russo, N.E.; Zekios, C.L.; Georgakopoulos, S.V. A capacity reconfigurable multimode origami MIMO antenna. In Proceedings of the 2019 IEEE International Symposium on Antennas and Propagation & USNC/URSI Radio Science Meeting, Atlanta, GA, USA, 7–12 July 2019; pp. 411–412.
20. Russo, N.E.; Zekios, C.L.; Georgakopoulos, S.V. Origami multimode ring antenna based on characteristic mode analysis. In Proceedings of the 2019 IEEE International Symposium on Antennas and Propagation & USNC/URSI Radio Science Meeting, Atlanta, GA, USA, 7–12 July 2019; pp. 2037–2038.
21. Shah, S.I.H.; Gosh, S.; Tentzeris, M.M.; Lim, S. A Novel bio inspired pattern reconfigurable quasi-Yagi helical antenna using origami DNA. In Proceedings of the 2018 International Symposium on Antennas and Propagation, Busan, Republic of Korea, 23–26 October 2018; pp. 1–2.
22. Ko, K.-Y.; Lai, F.-P.; Chen, Y.-S. Origami-driven manufacturing of reflectarrays or transmitarrays with feeds and fixture. *IEEE Trans. Compon. Packag. Manuf. Technol.* **2024**, *14*, 669–680. [CrossRef]
23. Lin, T.-H.; Bahr, R.; Tentzeris, M.M.; Raj, P.M.; Sundaram, V.; Tummala, R. Novel 3D-/Inkjet-Printed Flexible On-package Antennas, Packaging Structures, and Modules for Broadband 5G Applications. In Proceedings of the 2018 IEEE 68th Electronic Components and Technology Conference (ECTC), San Diego, CA, USA, 29 May–1 June 2018; pp. 214–220.
24. *ASTM D3359-17*; Standard Test Methods for Measuring Adhesion by Tape Test. ASTM International: West Conshohocken, PA, USA, 2017. Available online: https://webstore.ansi.org/standards/astm/astmd335917 (accessed on 11 May 2024).
25. Chen, Y.-S.; Wu, Y.-H.; Chung, C.-C. Solar-powered active integrated antennas backed by a transparent reflectarray for CubeSat applications. *IEEE Access* **2020**, *8*, 137934–137946. [CrossRef]

electronics

MDPI

Review

The Current Progress and Future Prospects of Path Loss Model for Terrestrial Radio Propagation

Jian Wang [1,2,3,†], Yulong Hao [1,†] and Cheng Yang [1,2,*]

[1] School of Microelectronics, Tianjin University, Tianjin 300072, China; wangjian16@tju.edu.cn (J.W.); haoyulong98@tju.edu.cn (Y.H.)
[2] Qingdao Institute for Ocean Technology, Tianjin University, Qingdao 266200, China
[3] Shandong Engineering Technology Research Center of Ocean Information Awareness and Transmission, Qingdao 266200, China
* Correspondence: ych2041@tju.edu.cn
† These authors contributed equally to this work.

Abstract: The radio channel model is a major factor supporting the whole life cycle of the terrestrial radio system, including the demonstration, design, validation, operation, and so on. To improve the spectrum sharing and spectral efficiency in terrestrial radio services, we analyze three types of path loss models in detail: deterministic, empirical, and semi-empirical models, to meet the requirements of path loss modeling for supporting traditional band expansion and reuse. Then, we conduct a comparative analysis based on the characteristics of the current models. Furthermore, a preview of the future terrestrial path loss modeling methods is provided, including intelligent modeling processes and multi-model hybridization methods. Finally, we look forward to the potential technology that can be used in future wireless communication, such as terahertz communication, reconfigurable intelligent surface technology, and integrated communication and sensing technology. The above research can provide a reference for the development of terrestrial radio channel modeling, promoting the technologies of terrestrial channel modeling. We hope this paper will stimulate more interest in modeling terrestrial radio channels.

Keywords: radio propagation; path loss; model; current progress; future prospect

1. Introduction

As the elaboration of transmission effects between the transmitter and receiver of the radio system, the radio channel is subject to many factors [1]. For instance, terrain and surface features, meteorological parameters of the troposphere and ionosphere, solar activity, etc. [2–4]. The characteristic of the channel determines the performance of the electronic information system, which is also very significant in the whole life cycle, such as design, development, production, validation, operation, and maintenance [5]. The electromagnetic spectrum is the strategic resource that all radio-electronic warfare relies on and is a hub of cross-domain joint operations linking land, sea, air, and space networks. Therefore, the security of spectrum management is directly related to national security and radio system construction [6]. The rapid growth of radio systems deployment leads to a shortage of electromagnetic spectrum resources. As a result, the possibility of radio interference will increase, and the performance of radio systems will be reduced. The radio channel is strongly related to the electromagnetic spectrum [7]. To avoid the frequency utilization problem caused by the congestion of the electromagnetic spectrum, one of the most effective and frequently used methods is the utilization of channel models in frequency planning, assignment, coordination, and so on. New channel models are proposed to increase the accuracy and robustness at the frequency bands in use, reduce the mutual interference between radio systems, and supply further support for expanding and reusing the existing spectrum.

According to the modeling principle, the existing models can be categorized as deterministic, empirical, and semi-empirical. The deterministic models are derived from the electromagnetic propagation formula, and Ray Tracing (RT) [8] and Parabolic Equations (PE) [9] are the typical deterministic models. The empirical models (also called the statistical channel model) are based on statistical theory: the Okumura model [10], the Okumura–Hata model [11], the ITU-R P.1546 (below referred to as P.1546) model [12], and the COST-231 model [13] are the typical empirical models. The semi-empirical models mix the features of deterministic and empirical channel models. The COST-259 model [14] and the International Mobile Telecommunications (IMT)-Advanced model [15] are typical semi-empirical models. Traditional modeling methods used the manual statistical analysis of measured data under specific scenarios and radio wave propagation conditions. In addition, the development of computer and artificial intelligence, in which machine learning (ML) is an important branch, provides an accurate and efficient modeling method with self-learning and self-adaptive capabilities. ML is a predicted and classified method that can mine hidden rules from a large number of data [16]. Theoretically, many problems in radio channel modeling can be considered regression, clustering, and classification in ML. The efficiency of radio channel modeling is consistent with the capabilities of ML. Therefore, we can intelligently structure the channel model through ML to learn the propagation characteristics and hidden rules with measured data [17,18]. In [19], using Artificial Neural Networks (ANNs), the authors developed a new method for multiband heterogeneous radio network scenarios. To predict path loss in the global mobile communication system band, Eichie et al. developed a Multilayer Perceptron (MLP) neural model [20]. Sotiroudis et al. proposed an urban environment model, and the research results show that as long as the size of an ANN is correctly selected, the path loss model based on the ANN will operate efficiently [21]. Wang et al. proposed a new method based on Backpropagation (BP) ANN for radio wave propagation prediction. Thrane et al. proposed a comprehensive model based on deep learning, supplemented by a random and RT [22].

Therefore, this paper aims to put forward the research prospect of future radio channel methods based on summarizing the research results of terrestrial radio channel models and combining the challenges faced in developing radio communication. For the above purposes, the following arrangement is made: Section 2 overviews the category of existing models based on the modeling methods, including the deterministic, empirical, and semi-empirical models. At the same time, we compare and summarize the features of the three modeling methods. Then, the outlook of the future challenge and modeling methods with the development trend of new techniques, applications, and scenarios is given in Section 3. Finally, we summarize the development of terrestrial channel modeling.

2. Current Progress of Channel Model

Ultra-short wave, microwave, and millimeter wave are the primary communication frequency bands for terrestrial radio communication. The propagation of these frequency bands works together through various propagation mechanisms. Terrestrial radio communication modes include single-input single-output (SISO) and multi-input multi-output (MIMO); MIMO includes single-input multi-output (SIMO), multi-input single output (MISO), and multi-input multi-output (MIMO). The communication scenes cover large-scale and small-scale scenes. Among them, the large-scale scenarios include land, sea, and mixed scenes, and the small-scale scenarios include indoor and street scenes. Figure 1 shows the knowledge graph of terrestrial wireless communication.

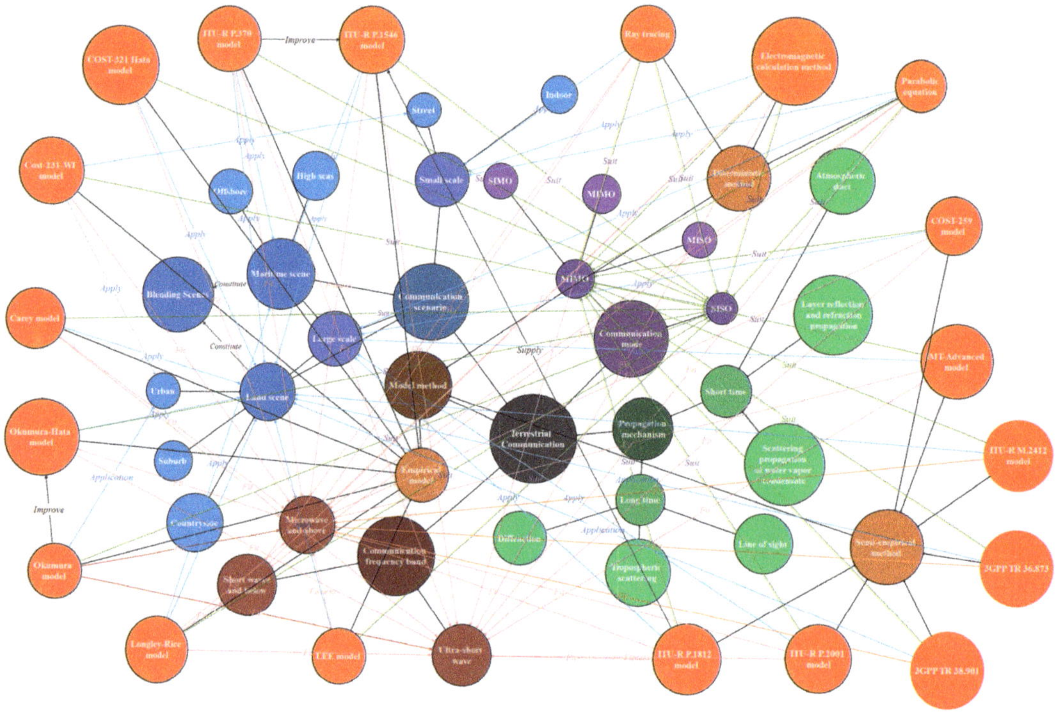

Figure 1. Knowledge graph of terrestrial wireless communication channels.

Next, we will analyze and compare the three modeling methods according to their typical channel models. In addition, the current status of channel models, which are mostly based on the references published in the IEEE, IET, and Hindawi between 2012 and 2022, are also described. To conclude, each modeling method will be discussed in terms of its advantages and disadvantages.

It should be noted that there are three types of radio channel models based on different modeling principles: the deterministic model, the empirical model, and the semi-empirical model, respectively. The development history is shown in Figure 2. From the figure, we can observe that models proposed based on empirical methods are the most abundant, while those derived from deterministic methods are the least prevalent. The majority of models proposed in the early and middle stages are based on empirical methods, with recent years witnessing a surge in models based on semi-empirical methods. Deterministic models still in use today include ray tracing and parabolic equation methods. It can be inferred that semi-empirical modeling methods will be the predominant approach in future modeling. Empirical methods, given their simplicity and acceptable accuracy, are likely to continue being widely employed in small-area modeling or localized adjustments to models in the future.

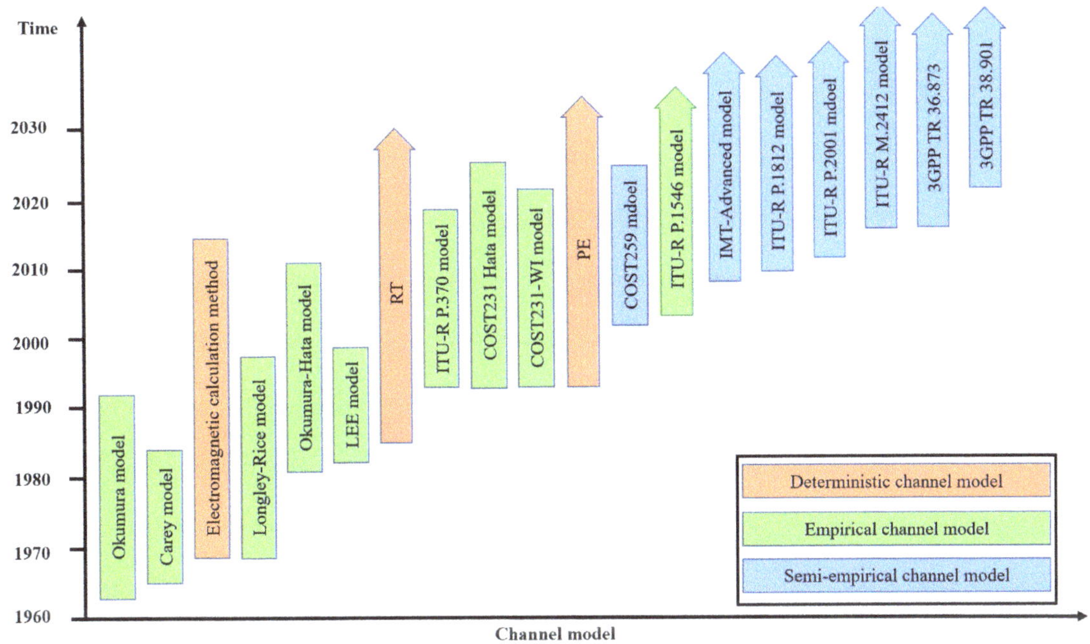

Figure 2. The history of typical channel models.

2.1. Deterministic Model

2.1.1. The Typical Model

The deterministic channel modeling method, which analyzes the radio channel propagation characteristics by solving Maxwell equations or using geometric approximation numerical methods, is derived from the radio wave propagation mechanism [23]. Figure 3 shows the typical deterministic models.

RT is the most widely used deterministic channel model method [24–26]. According to the modeling principle, RT can be classified into forward RT, backward RT, and mixed RT. Shooting and Bouncing Ray (SBR), first proposed in 1986, is the widely used forward RT [27]. Forward RT traces every ray from transmitting to receiving equipment after the initial ray is emitted. The RT process of the current ray is terminated when the ray has experienced a number of reflections and diffraction, or the ray hits the receiving model. However, the computation of forward RT will increase with time. Because of the simple principle, this method is widely used in various fields to develop RT acceleration technology and hardware computing efficiency. The core logic of backward RT and forward RT is entirely different. Backward RT is used to find the propagation path of direct, reflection, diffraction, or other combined forms by using the mirror method to traverse the basic units of triangles or other structures in the scene when the locations of transmitters and receivers are known [28]. The principle of backward RT is relatively simple, but it needs the locations of transmitters and receivers to be known. Mixed RT was first proposed to predict street radio wave propagation [29]. It is an RT method that combines the characteristics of forward RT and backward RT. The method first uses forward RT to obtain the approximate propagation path and then uses backward RT to modify and verify the obtained propagation path further.

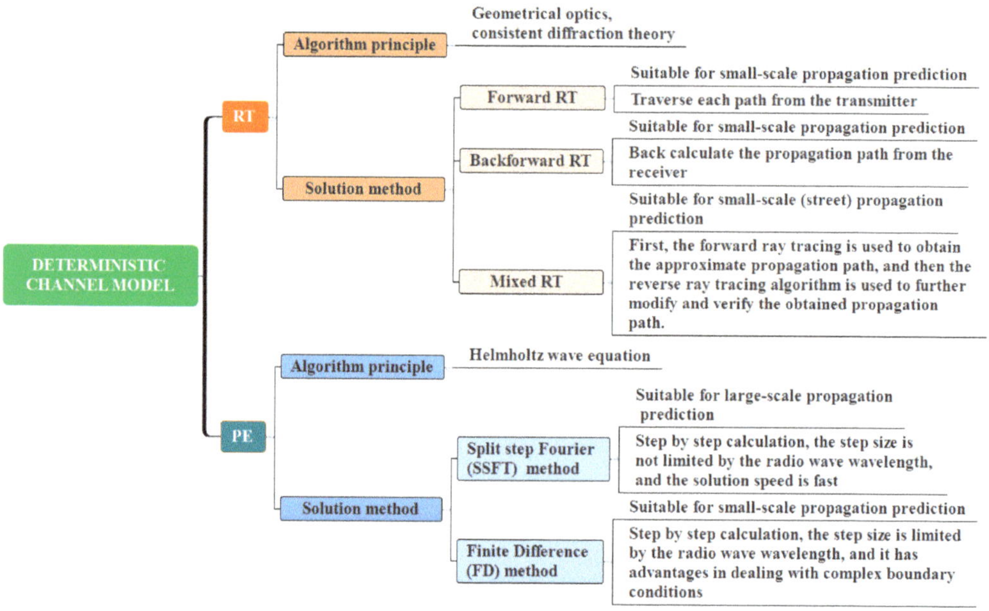

Figure 3. Typical deterministic channel mode.

In 1946, Lenontovich and Fock proposed PE to solve the problem of radio wave diffraction. In 1977, Tappert solved PE in acoustic waves by using the Split-Step Fourier transform (SSFT) method, which attracted people's attention and research [30]. In 1991, Kuttler and Dockery systematically deduced the PE of radio wave propagation in the troposphere for the first time. Currently, the leading solution methods of PE are the SSFT and Finite Difference (FD) methods, both of which are step algorithms. The step size is almost not limited by the wavelength of the radio wave when using SSFT to solve PE. Therefore, the result can be obtained quickly when solving large-scale problems. However, it is not easy to deal with complex boundary conditions. The calculation process of the FD method involves many matrix operations. It has significant advantages in dealing with complex boundary conditions. However, its step size is limited by the radio wave wavelength, so it is unsuitable for solving large-scale problems. It is mainly used to calculate the electromagnetic scattering characteristics of targets and radio propagation prediction in small-scale areas.

The electromagnetic calculation methods are mainly used to solve the radio wave propagation problem by calculating the Maxwell equations strictly [31]. However, these methods have not been used on a large scale due to the shortcomings of a large amount of computation and high requirements for hardware platforms.

2.1.2. Other Models

Figure 4 shows the principle and the scenario of the deterministic models in the references. Afsharinejad et al. proposed a log-distance path loss model based on Monte Carlo simulations. Shadow fading of air and leaves is considered in the model, and path loss variations are defined. Since the latter model only considers transmission distance, it can be interpreted as a simplified version of the theoretical model [32]. In [33], the authors proposed a theoretical model based on the Fresnel zone theory to solve the very near-ground path loss problem. In [34], a model based on Green's theorem is proposed. The model is for the blending scenario. In addition, the authors propose channel models based on the standard models and propagation theories. For example, a ground reflection model based on double rays, along with other propagation mechanisms, is proposed in [35]. In [36], they use the RT and edge peak diffraction theory to propose a model.

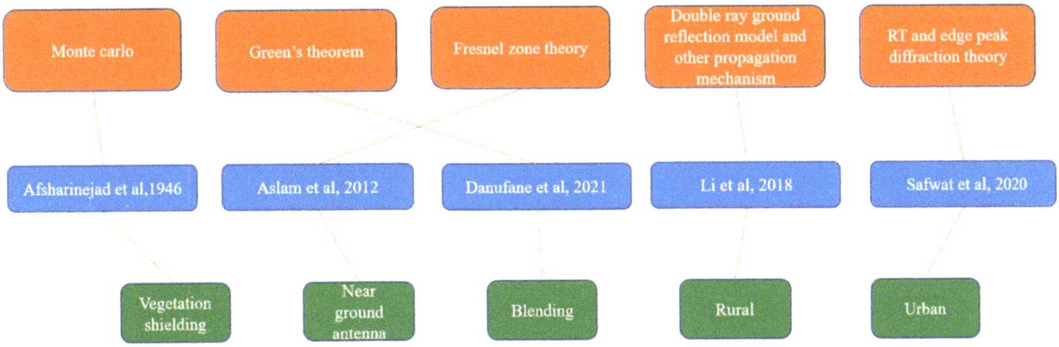

Figure 4. The method and the scenario of the deterministic models in the references [32–36].

From the above references, it is evident that deterministic models are primarily applicable to small-scale scenarios within a prediction range of 200 m. This limitation arises because achieving precise predictions with deterministic models requires a substantial amount of environmental parameters and computational resources [37–39]. The expansion of the prediction range imposes higher demands on data collection and computing power, which is a key reason why deterministic methods are not as widely employed as the other two methods.

2.2. Empirical Model

The empirical modeling method is a channel parameters modeling method that uses mathematical statistics. The empirical model uses standardized propagation characteristics, testing cumulative data or statistical curves to interpolate or extrapolate according to equipment frequency parameters, and completes prediction analysis through a series of corrections [40]. Table 1 shows the typical empirical model parameter information.

Table 1. Parameter Information of Typical Empirical Channel Model.

Model	Scenarios	Distance/km	Frequency/MHz	Transmitting Height/m	Receiving Height/m	Author	Proposed Time
Okumura	Quasi-smooth urban area	1–20	150–1500	30–200	1–10	Okumura et al.	1962
Carey	Flat ground	<130	35–460	30–1500	Average 1.8	FCC	1964
Longley-Rice	Ground, sea	1–2000	20–40,000	1–1500	1–9	Longley and Rice	1968
Lee	Urban, suburb, rural	>20	450–2000	30	3	W. C. Y. Lee	1982
Okumura-Hata	Urban, suburb, rural	1–20	150–1500	30–200	1–10	Hata et al.	1980
COST-231-Hata	Urban, suburb, rural	1–20	1500–2000	30–200	1–10	EURO-COST	1991
COST-231-WI	Urban, dense urban	0.02–5	800–2000	4–50	1–3	EURO-COST	1991
ITU-R P.370	Multi-scenario	<1000	30–1000	<1200	Average 10	ITU	1991
ITU-R P.1546	Multi-scenario	1–1000	30–4000	<3000	Average 10	ITU	2001

2.2.1. The Typical Model

The Okumura model is one of the earliest empirical models based on field intensity measurement. Okumura was committed to studying the relationship between power density and distance in Tokyo [10]. Because the model is entirely based on statistics, its result is a fitting curve rather than a specific formula. The Longley–Rice and Okumura models were proposed at the same time. The Longley–Rice model already has a general computer program to calculate path loss. For a given transmission, the computer calculates path loss according to frequency, path distance, polarization direction, antenna height, surface diffraction, ground radius, and ground conductivity [41]. In order to extend Okumura's research result to areas other than Tokyo, Hata proposed the Okumura–Hata model by formulating Okumura's fitting curve. The Okumura–Hata model is very suitable for large-scale scenes where the transmitter is higher than the receivers and surrounding buildings. The arguments of the LEE model are easily available, so it is very popular. The LEE model can be broken down into the macro and microcell sub-models. The macrocell sub-model firstly considers terrain as flat, only considers the impact of buildings, and after this, adds the effects of terrain and landform. The microcell model assumes that the attenuation of signals is highly correlated with the length of buildings on the propagation path. The LEE model also divides terrain and geomorphology influence into three situations for calculation: non-blocking situation, blocking situation, and water surface reflection situation [42]. COST-231 models were proposed by EURO-COST in 1991. Compared with the Okumura–Hata model, COST-231 models are more often used. The Okumura–Hata model is the basis of the COST-231–Hata model. The range of its frequency is extended from 1500 MHz to 2000 MHz. The COST-231–WI model combines the result of the urban environment calculated by the Walfisch–Bertoni model [43] and the corrected function of the street trend of the Ikegami model and the experimental correction. Therefore, this model is very suitable for predicting path loss of the street scene. However, this model also has some defects; when the transmitting antenna height changes slightly, the path loss jumps steeply.

Therefore, when using this model, the transmitter's antenna should be installed at a height several meters higher than the street building. The Carely model was extended from the CCIR curve. It was included as one of the standard models by the Federal Communications Commission (FCC) of the US. This model provides the field strength–distance curve when the average height of the mobile station antenna is 1.8 m, the height of the base station antenna is 30–1500 m, and the coverage is less than 130 km on flat ground. Moreover, the ITU-R P.370 (below referred to as P.370) and P.1546 models proposed by ITU are also mainstream empirical models. The P.370 proposal is intended to forecast the radio wave propagation field strength of meter wave and decimeter wave bands under various climatic conditions for broadcast service planning engineers in various countries [44]. P.1546 is proposed based on the P.370 proposal. Although the prediction accuracy and application range are significantly improved compared with the P.370 model, both recommendations are revised based on data statistics according to empirical values, and the diffraction effect of obstacles is not strictly considered. Therefore, when diffraction is the main factor affecting propagation in mountainous areas, the prediction error of these two propagation models increases significantly [12].

Most empirical models are large-scale propagation and prediction models. The Rice and the Rayleigh models are small-scale propagation models. These two models apply to indoor or other object-intensive areas. The difference is that the Rice model applies to a situation where there is a direct path, and when there is no direct path, the Rayleigh model applies.

2.2.2. Other Models

The empirical methods can be divided into three main types. The first is based on traditional linear regression, the second is based on ML, and the last is the mixed approach. To evaluate the accuracy and improvement of these models, we introduced evaluating standards of the mean absolute error (MAE), mean relative error (MRE), root-mean-square error (RMSE), and relative root-mean-square error (RRMSE) to reflect the deviation from the measurements and the stability and robustness of the prediction model. The definitions and characteristics of these indexes are listed in Table 2.

Table 2. Equation and characteristics of the performance criteria.

Index	Definition	Characteristic
MAE	$\frac{1}{n}\sum_{i=1}^{n}\left\|y_i - y_i^p\right\|$	Evaluate the absolute deviation between the predicted value (y_i^p) and measured value (y_i), where n is the number of samples.
MRE	$\frac{1}{n}\sum_{i=1}^{n}\frac{\left\|y_i - y_i^p\right\|}{y_i}$	Evaluate the relative deviation between the predicted value and measured value.
RMSE	$\sqrt{\frac{1}{n}\sum_{i=1}^{n}\left(y_i - y_i^p\right)^2}$	Evaluate the root-mean-square error between the predicted value and the measured value.
RRMSE	$\sqrt{\frac{1}{n}\sum_{i=1}^{n}\left(\frac{y_i - y_i^p}{y_i}\right)^2}$	Evaluate the relative root-mean-square error between the predicted value and measured value.

Figure 5 shows the principle and the scenario of the empirical models in the references. From Figure 5, we can see that almost half of the studies are based on traditional linear regression. On the one hand, some are directly modeled using linear regression based on measured data for street [45], bleeding [46–48], indoor [49–54], urban [55–57], vegetation shielding [58–60], vehicle-to-vehicle channels (V2V) [61], and station scenarios [62]. At the same time, some researchers have used linear regression to fit the standard models. For example, He et al. used linear fitting to assist the Hata model based on the measurement data for the GSM railway [63]. Nossire et al. used linear fitting to correct the Okumura model based on the measurements for an urban indoor scenario [64]. In [65], a standard Macrocell model was proposed with the rectification of measurements to predict the path loss of the blending scenario.

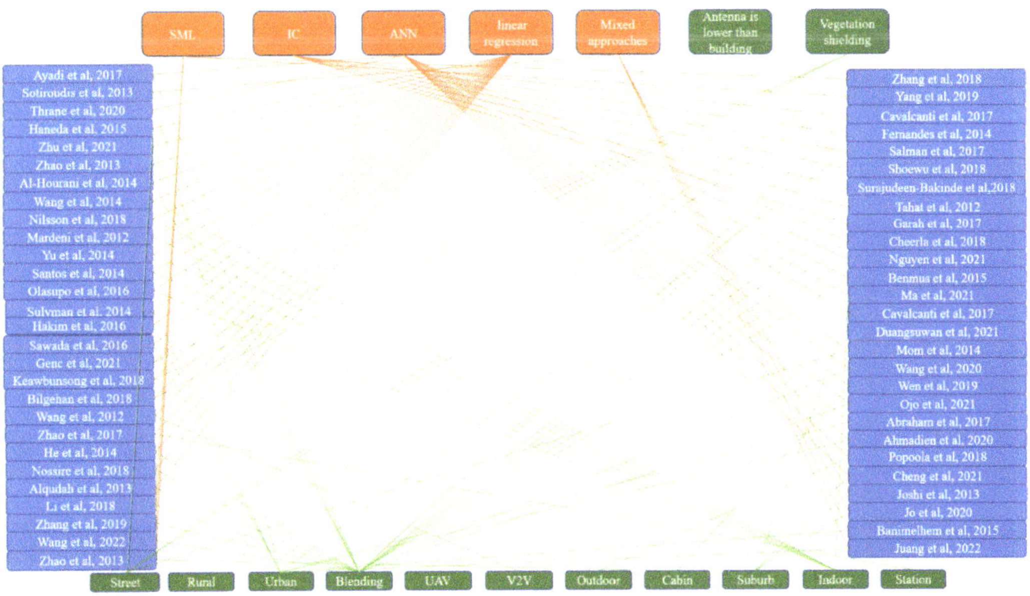

Figure 5. The principle and the scenario of the empirical models in the references [19,21,22,37,46–96].

As an artificial intelligence method, ML has been widely used in modeling. These methods include Statistic Machine Learning (SML), Intelligent Computation (IC), ANN, etc. The most commonly used algorithms are the GMM, SVM, and random decision forests of SML. In [66], The GMM solves the problem of clustering in wireless propagation channels.

In [67–69], new methods based on SVM are proposed for blending and cabin scenarios. In [70,71], the authors established channel models in UAV scenarios using the random forest and KNN algorithms, and the comparative results are presented in Table 3. From the table, it can be concluded that in UAV scenarios, the predictive accuracy of the channel model established with the random forest outperforms that of the KNN algorithm.

Table 3. The result of Refs. [70,71].

Reference	Scenario	Evaluation Parameters	Conclusion
[70]	UAV	MAE, RMSE	Random forest algorithm is better than KNN in channel modeling
[71]	UAV	RMSE	

Moreover, the IC is also widely used in channel modeling. Cavalcanti et al. proposed a GA-based optimization method to tune constants present in empirical path loss models for urban areas at 879 MHz [72]. In [73,74], the GA was used to find the best coefficients and functions to fit the antenna measurements lower than the building and blending scenarios. In [75], a new model using PSO to tune a modified COST-231–WI propagation model was proposed in a 3G network scenario. After modification and verification, the average error gap of the PSO-based model at the test point is −1.96 dBm. In [76], using the PSO to tune the parameters of the COST-231 model improved its ability. From the above references, we can see that IC is typically employed in conjunction with other models for optimizing constants within a model. Table 4 shows the comparison of characteristics for IC. Intelligent computing can be broadly categorized into two types: global search and local search. Genetic algorithms (GA) represent a typical global search algorithm that generates values within a specified range without directionality through "genetic" and "mutation". Global search algorithms are suitable for scenarios with a small optimization range, but they exhibit slower optimization speeds, requiring a greater number of iterations to reach the optimal solution. Particle swarm optimization (PSO) is a typical local search algorithm. After defining the "population size" and "optimization range" for local search, at each iteration, the particle will update its position in the direction of the optimal position. Local optimization algorithms are directional and fast, making them suitable for scenarios with large optimization ranges, but they are prone to getting trapped in local optima. In [77], a comparison was conducted between the GA and PSO algorithms for optimizing the constants of the Hata model in a typical urban scenario. The results indicate that the PSO algorithm outperforms the GA algorithm in terms of prediction accuracy and convergence speed, with an RMSE difference of approximately 1 dB.

Table 4. Comparison of characteristics for IC.

Type	Directionality	Speed	Range	Features
Global search	Yes	Slow	Small	The optimal parameters can be found with enough iterations
Local search	No	Quick	Large	Easy to fall into local optimum

ANNs are widely used in the field of channel modeling due to their excellent self-learning and adaptive capabilities. In [74,78], the authors provided that the model based on ANFIS is more efficient, faster, and more accurate than the physical and empirical methods. In [79], the authors provided that the proposed ANFIS-based model offers desirable advantages in terms of simplicity, high prediction accuracy, and good generalization ability. As long as there are enough measurements, ANNs can be used for modeling any scenario, such as blending [80–82], indoor [83], suburban [84,85], and street scenarios [86]. The Back Propagation NN (BPNN) is used to model the blending [19,87] and cabin scenarios [88]. In particular, Ojo et al. developed two models based on the radial basis function NN (RBFNN) and the multilayer perception NN (MLPNN) by using the measured data as input variables for blending scenarios [89]. In [90], a pre-trained network called VGG-16 is proposed for modeling. They make the necessary modifications and fine-tune it with their training set. In [91], Popoola et al. use an extreme learning machine-LEM algorithm, which can reduce the training time on the premise of ensuring the accuracy of the trained model outdoors. In addition to direct modeling with ANN, it can also be used for auxiliary modeling. In [22,92], CNN extracts environmental information from satellite images to fit the path loss formula for blending and suburb scenarios for blending and suburb.

The above reference indicates that neural networks can achieve modeling tasks in various scenarios as long as there are sufficient training data and appropriate training strategies are selected. Figure 6 shows the steps to establish a channel model using neural networks. The modeling process can be divided into two stages: preparation and modeling. In the preparation stage, an analysis of the modeling scenario is required, as different neural networks exhibit varying performances in different scenarios. In [89], modeling was built using RBFNN and MLPNN in multi-antenna scenarios. Then, the models were compared with generic models such as COST-231, Free Space, Ericson, ITU, and ECC33. The comparison results indicated that the model based on RBFNN exhibited slightly better predictive accuracy than the one based on MLPNN. Both models significantly outperformed generic models in terms of prediction consistency and accuracy, with an average absolute prediction error improvement of over 10 dB. In [93], the modeling performance of ANFIS and GRNN neural networks was compared in a tropical urban scenario. The result shows that the model based on GRNN had a reduced RMSE of 0.59 compared to the one based on ANFIS. Both models achieved approximately a 4.71 dB improvement over a linear regression method composed of COST-231–Hata and COST-231–Walfisch–Ikegami models. The choice of neural network type depends on the modeling scenario. Once the model is determined, the next steps involve analyzing the application scenario, collecting, and preprocessing the feature data. The data collection principle involves selecting differentiating data strongly correlated with radio wave propagation. For example, in flat terrain areas with minimal topographical variations, the influence of terrain factors may not be a significant consideration during model training. Data preprocessing includes removing erroneous data and normalizing data. The aim of normalizing data is to ensure consistent measurement of various feature parameters. Finally, hyperparameters of the neural network are set using optimization algorithms such as the IC algorithm, as detailed above. In the modeling phase, the collected data are proportionally divided into training and testing sets. The training set is used to train the model, while the testing set is used to evaluate the trained model's performance. If the trained model does not meet the requirements, adjustments are made by tuning hyperparameters, changing the model, or modifying feature data.

In addition, there are mixed methods used for modeling. For instance, Sotiroudis et al. proposed a path loss propagation model based on ANN for urban scenes and used the Differential Evolution (DE) algorithm to lay out an optimal ANN for path loss prediction [21]. In [94], a model is proposed based on a fuzzy method and curve-fitting for a suburban scene. Fuzzy sets are used to distinguish between the transmission discontinuities encountered during propagation. Then, the fuzzy sets are used to fit the curve. As another example, Jo et al. first used Principal Component Analysis (PCA), which can reduce the number of features of the dataset and simplify the learning model to assist feature selection. They then built the path-loss model using ANN for the suburb scenario [95]. In [96], PCA extracted relevant features from some selected path profile attributes for NLoS cases. Then, the path-loss based on the polynomial regression method model for urban is built.

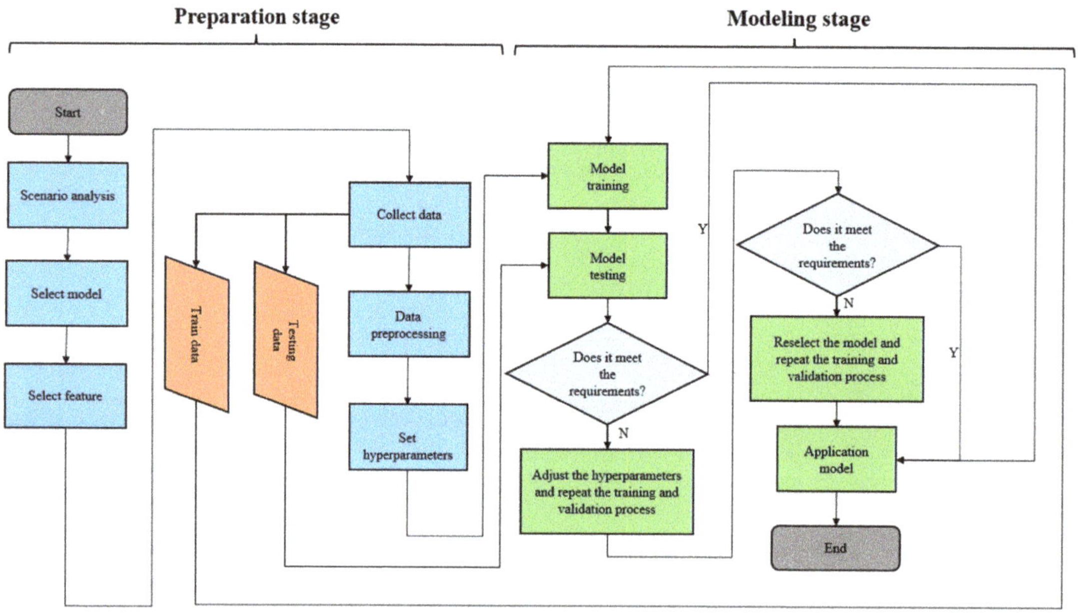

Figure 6. The modeling process of neural network.

2.3. Semi-Empirical Model

The semi-empirical model combines the advantages of the deterministic and the empirical models. This type of model replaces some complex electromagnetic calculations with statistical laws to reduce computational complexity [97–99]. This model's accuracy and calculation complexity are between the two models. Figure 7 shows typical semi-empirical models.

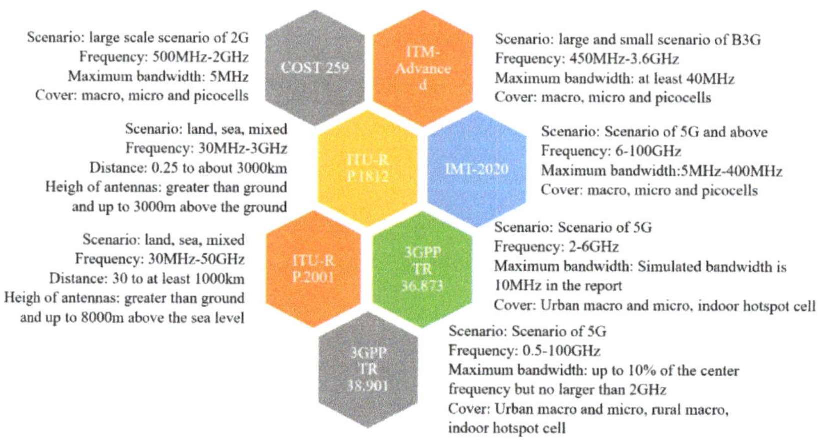

Figure 7. The typical semi-empirical models.

2.3.1. The Typical Model

The random geometric method (RGM), which does not need detailed environmental parameters, is one of the most popular semi-empirical modeling methods. Both the COST-259 and the IMT-Advanced models are the standard models of radio communication based on this method. The COST-259 model is suitable for macrocell radio communication. In order to research the diversity and adaptive antenna system, the directional characteristic modeling of the channel is particularly emphasized during the development of this model [14]. The IMT-Advanced model is the beyond third-generation (B3G) mobile communication system proposed by ITU. Depending on the environment, this model can support low-mobility to high-mobility applications and a variety of data rates. The IMT-Advanced system can also provide high-quality multimedia applications that significantly improve the quality of service under a wide range of services and platforms [15]. The ITU-R P.1812 (below referred to as P.1812) and the ITU-R P.2001 (below referred to as P.2001) models proposed by ITU are also widely used as semi-empirical channel models. The P.1812 model supplements the P.1546 model. It is a propagation prediction method suitable for terrestrial point-to-surface traffic. The height of the antennas of the transmitter and receiver is up to 3000 m above the ground [100]. However, this model is unsuitable for propagation prediction of air–ground or space–ground communication. The model obtains terrain profile information through the digital map. This model considers the end feature, location variability, and building entrance loss. This model also considers the effect of LoS propagation, diffraction, tropospheric scattering, and ducting on path loss [101]. The P.2001 model is a widely used model for terrestrial propagation. It can predict the terrestrial propagation path loss caused by signal enhancement and fading in the effective range of 0% to 100% beyond the annual average. The model's predicted value is obtained by combining the predicted values of four independent sub-models according to the statistical correlation, and the model is particularly suitable for the Monte Carlo method [102].

In addition, with the deployment of 5G and 6G communication systems, many corresponding models emerged one after another. The most important are IMT-2020, 3GPP TR 36.873, and 3GPP TR 38.901. The IMT-2020 model is the new version of the IMT model, which includes the new capabilities of the IMT that go beyond those of the IMT-Advanced. Specifically, the ITM-2020 supports frequencies up to 100 GHz and large bandwidth, three-dimensional (3D) modeling, large antenna array, blockage modeling, spatial consistency, etc. The scenarios of IMT-2020 include enhanced mobile broadband, ultra-reliable and low-latency communications, and massive machine-type communications. It can support low- to high-mobility applications and enhance data rates according to the number of users or the service requirements. IMT-2020 can not only be mobile telecommunication but also a tool for enabling massive connections for a wide range of services. It is a typical model for 5G [103]. The 3GPP TR 36.873 (study on 3D channel model for LTE) is also an important reference. This technical report is suitable for the situation of 3D beamforming and FD-MIMO. The path-loss model in this technical report can be applied in the frequency range of 2–6 GHz and for different antenna heights. The usage scenarios include the LOS and NLOS of urban micro/macro, rural macro, and indoor hotspot cells. Among them, the coverage of the base station can reach at least 3 m and up to 10,000 m [104]. The new channel model proposed in the 3GPP TR 38.901 has, to a large degree, been aligned with earlier channel models for <6 GHz, such as 3GPP TR 36.873 or IMT-Advanced. The frequency range of this model defined in this technical report is about 0.5–100 GHz. The supported scenarios are urban microcell street canyon, urban microcell, indoor office, rural microcell, and indoor factory. The bandwidth of this model is supported up to 10% of the center frequency but no larger than 2 GHz. The base station height is 10–150 m, and the mobile station height is 1–22.5 m [105].

2.3.2. Other Models

Figure 8 shows the principle and the scenario of the semi-empirical models in the references. As mentioned earlier, the semi-empirical method is a method that combines the advantages of empirical and deterministic methods. Specifically, using actual measurements to fit the parameters that are not easy to determine in the deterministic method. Several non-empirical path loss models have been built for a small-scale scenario, such as the one/two-slope, COST-231 multi-wall, RT, floating intercept (FI), and close-in free space reference distance (CI) model. For example, the indoor models were proposed based on the CI and the FI model, with compensation factor fit by the measurement [106,107]. Khatun et al. proposed a model based on the above for indoor and outdoor airports [108]. Zhou et al. proposed a model based on ML and the deterministic model [109]. A model based on the CI and FI was proposed after clustering using the space-alternating generalized expectation-maximization algorithm. The proposed model considers the different cluster characteristics, and it also units the directional and omnidirectional path loss model into the same framework, improving the 5G mmWave channel model to some extent. In [110,111], indoor models were proposed based on the two-slope model with compensation factor fit by the measurements. At the same time, a series of models were proposed based on the COST-231 multi-wall model for indoor [112,113] and home [114]. A log-distance path loss model is the extension of the Friis model and includes the random shadow effect caused by the signal being blocked by hills, trees, buildings, etc. Therefore, it is often used to predict the obstructed scene between the receiver and transmitter, such as smart homes [115], offices [116], and subway tunnels [117]. On the other hand, Bhuvaneshwari et al. proposed a model based on the RT for indoor [118]. The model was built by modifying Fresnel's reflection coefficient with measurement data. Heereman et al. proposed a model based on the one-slope model for a large meeting room. Through measurement, it was found that the single-slope model with a PL index varying between 1.2 and 1.7 can accurately describe PL, and in the case of human existence, the PL index increases to 2 [119]. Diago–Mosquera et al. proposed a model based on Kriging for I2O using 3.5 GHz measurement analysis [120]. This model uses Kriging-aided design to provide a modeling method to reduce cost or limit measurement activities. What is more, it can also model without measured data. In addition, Zhang et al. proposed a model integrating the ML, RT, and COST-231–WI for UAVs [70]. This model used the RT to produce data for training and testing. Random forest and KNN are used to correct the COST-231–WI model. Similarly, the RT is also used to generate data for training and testing purposes [121], and ANN is used to build a model.

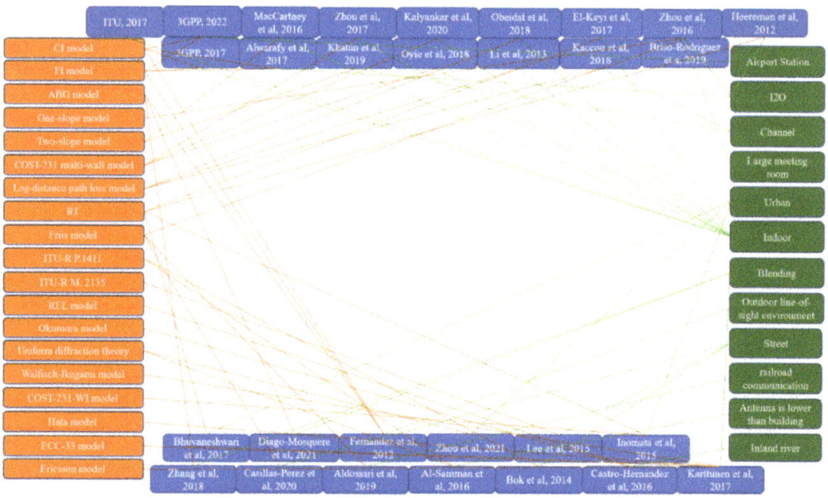

Figure 8. The basic model and the scenario of the semi-empirical models in the references [70,103–130].

In large-scale scenes, the Friis, ITU-R, and 3GPP models are the most used models. Most semi-empirical models are based on these models. For example, Casillas–Perez et al. proposed the WABG model based on the 3GPP ABG model for blending scenarios [122]. This model can use different available datasets for path loss calculation. It overcomes the problem related to unbalanced data described in the literature regarding using weighting policies. Aldossari et al. proposed a path loss model based on the ABG model using ML for blending scenarios [123]. They first used regression techniques to fit the measurement data, which reduces the required number of measurements and the complexity. Then, this model was built using these processed data. Al-Samman et al. use the same method to propose a path loss model for an outdoor LoS environment [124]. In [125,126], path loss models are proposed based on the Friis model and measurement data for street scenarios and high-speed railroad communication. In [127,128], path loss models were developed based on the Uniform Geometrical Theory of Diffraction. In addition, Lee et al. also extended the ITU-R P.1411 (below referred to as P.1411) model to include NLOS, which reasonably explains propagation along street roads where low-height antennas are located [127]. Moreover, Inomata et al. used the measured data to add a correction factor to the P.1411 model to make it suitable for a street scenario [129].

Through analyzing the typical ABG and CI models, Karttunen et al. found that the estimated shadowing variance can be significantly overestimated if the "true" shadowing, the offset variations, and the slope of the straight-line fit between different street canyons are not adequately distinguished [130]. Based on the above analysis, a path loss model was proposed for the street scenario. This model overcomes the disadvantages of the traditional models. The result shows that the proposed model works well in terms of spatial consistency and is better than the existing deterministic and empirical models. Yu et al. proposed a model that uses the Okumura–Hata model to take the place of the equation transformed from the Friis in the Round Earth Loss (REL) model. The model also takes diffraction loss caused by obstacles into account. The result shows that the performance of the proposed model is better than the Okumura–Hata and RLE [131]. Moreover, Bhuvaneshwari et al. used RT to improve the path loss prediction of the COST 231–WI model [132]. In the revised model, RT and the statistically determined empirical measurements are combined. A loss term is computed using the method of images for multiple reflections. As a result, the error is considerably lowered for the proposed hybrid model. In [133], a hybrid model based on k-means and fuzzy logic is presented. The model can choose the best model for prediction from the Friis, Walfisch–Ikegami, Hata, ECC-33, Stanford University Interim, and ERICSSON models.

2.4. Model Comparison and Analysis

The channel model is generally measured by accuracy, complexity, and universality [134,135]. Accuracy reflects the difference between the measurement and the prediction; complexity reflects the calculation quantity and operation time of a model; universality reflects the usability of a model in different scenarios and environments. A perfect model should be the best compromise between accuracy, complexity, and universality. Table 5 is a comparison of three typical models. From Table 5, we can see that:

(1) The principles of deterministic models are intricate, empirical models are straightforward, and semi-empirical models lie in between.
(2) Correspondingly, deterministic models exhibit the lowest execution efficiency, while empirical models demonstrate the highest effectiveness.
(3) The advantage of deterministic models lies in their higher precision, whereas empirical models have the lowest precision, and semi-empirical models fall in between.
(4) Empirical and semi-empirical models have limited application scenarios, and their suitability is relatively constrained by distance and frequency range. In contrast, deterministic models are generally more versatile. However, there are exceptions with some empirical and semi-empirical models, such as ITU-R P.370, P.1546, P.1812, and P.2001, which have broad applicability, cover a wide frequency range, and offer good prediction accuracy. However, their computational complexity is relatively high.

Table 5. Comparison of Typical Models.

Models		Key Performance Indicators					
Type	**Name**	**Principle**	**Scenarios**	**Distance/km**	**Frequency/MHz**	**Efficiency**	**Accuracy**
Deterministic	RT	Complex	Multi-scenario	Short and long	>30	Low	High
	PE	Complex	Multi-scenario	Short and long	full-band	Low	High
Semi-empirical	COST-259	Median	Urban, suburb, rural	<20	150–2000	Median	Median
	IMT-Advanced	Median	Indoor, microcellular, high speed, base coverage urban	0.003–5	450–6000	Median	Median
	IMT-2020	Median	Multi-scenario		1185–2200	Median	Median
	3GPP TR 36.873	Median	Urban, suburb, rural, indoor	0.01–5	2000–6000	Median	Median
	3GPP TR 38.901	Median	Urban, suburb, rural, indoor	0.001–10	500–100,000	Median	Median
	ITU-R P.1812	Median	Multi-scenario	0.25–3000	30~6000	Median	Median
	ITU-R P.2001	Median	Multi-scenario	3–1000	30–50,000	Median	Median
Empirical	Okumura	Simple	Quasi-smooth urban area	1–20	150–1500	High	Low
	Carey	Simple	Flat ground	<130	35–460	High	Low
	Longley–Rice	Median	Ground, sea	1–2000	20–40,000	High	Median
	Lee	Simple	Urban, suburb, rural	>20	450–2000	High	Low
	Okumura–Hata	Simple	Urban, suburb, rural	1–20	150–1500	High	Low
	COST-231–Hata	Simple	Urban, suburb, rural	1–20	1500–2000	High	Low
	COST-231–WI	Simple	Urban, dense urban	0.02–5	800–2000	High	Low
	ITU-R P.370	Median	Multi-scenario	<1000	30–1000	High	Median
	ITU-R P.1546	Median	Multi-scenario	1–1000	30–4000	High	Median

In conclusion, we can see that the empirical model has good complexity but bad accuracy and universality; the deterministic model has good accuracy, high complexity, and high university; the accuracy, complexity, and universality of the semi-empirical model are relatively balanced. Figure 9 shows the radar charts of these three channel models in accuracy, complexity, and universality.

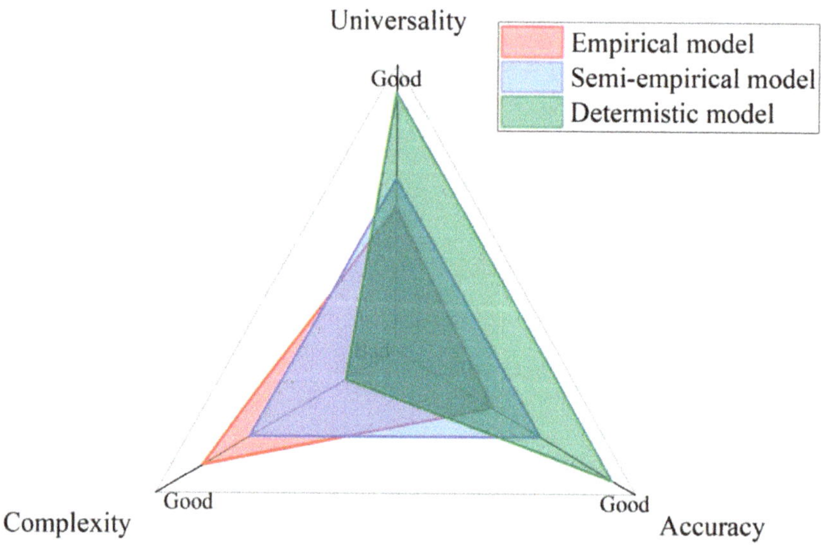

Figure 9. Performance comparison of three types of models.

Figure 10 shows the percentage distribution of modeling methods in the surveyed literature. From the figure, it is evident that empirical methods and semi-empirical methods are currently the predominant modeling approaches, accounting for 62.00% and 32.00% of the surveyed references, respectively. Within empirical methods, the most commonly employed strategy is linear fitting, representing a substantial share of 37.50%. Artificial intelligence algorithms such as ANN, IC, and SML also hold significant proportions, at 28.57%, 14.29%, and 10.71%, respectively. Among semi-empirical methods, modeling based on the IC model, FI model, and RT method occupy the top three positions.

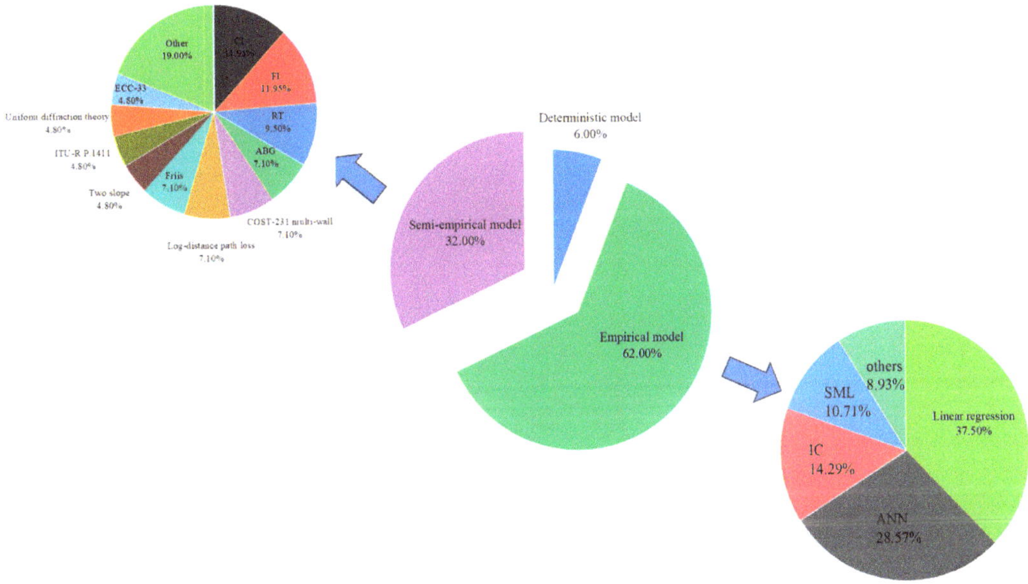

Figure 10. Statistical chart of the proportion of modeling methods in the survey model.

3. Future Prospects of Wireless Communication

3.1. Future Prospects of the Channel Model

With the evolution of the radio communication system, the measurement method, and intelligent technology, the current modeling method has challenged the demand for higher-quality radio communication services. The refinement and intelligence of the current channel modeling method need to be improved. In the future, the large-scale combination model suitable for large-scale scenes, the wide-area generalization model suitable for multi-scenes, and a more efficient intelligent model will be the main development direction of channel modeling. Figure 11 shows the development direction of channel modeling in the future.

3.1.1. Intelligent Modeling Process

It is challenging to balance the predicted efficiency and accuracy of the traditional channel modeling method. Modeling methods driven by artificial intelligence technologies, such as ML and IC, have excellent learning and prediction abilities, strong nonlinear fitting and adaptive abilities, good mining of complex features in high latitude and high redundancy data, and efficient performance. The ML can train channel models offline with measurement or simulation. In addition, due to its high nonlinearity, it is an ideal choice for predicting propagation parameters such as multipath fading. Moreover, it can also be used to model the specific or general model due to its high flexibility [136]. Figure 12 shows the classification and the advantages of intelligent modeling methods.

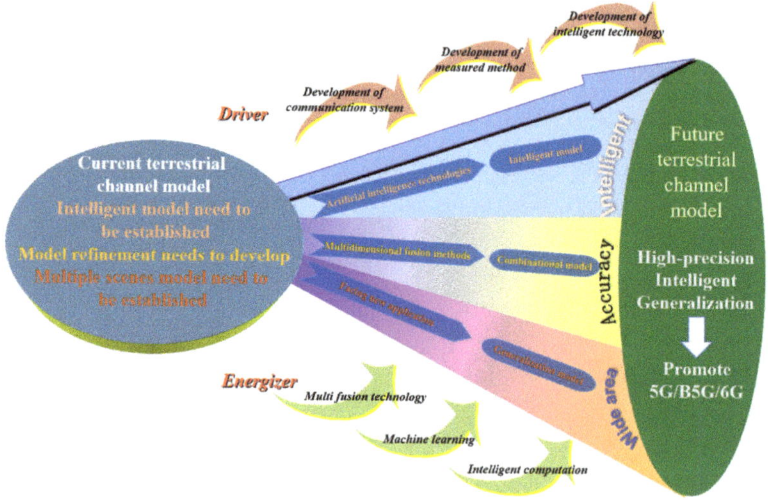

Figure 11. The development direction of channel modeling in the future.

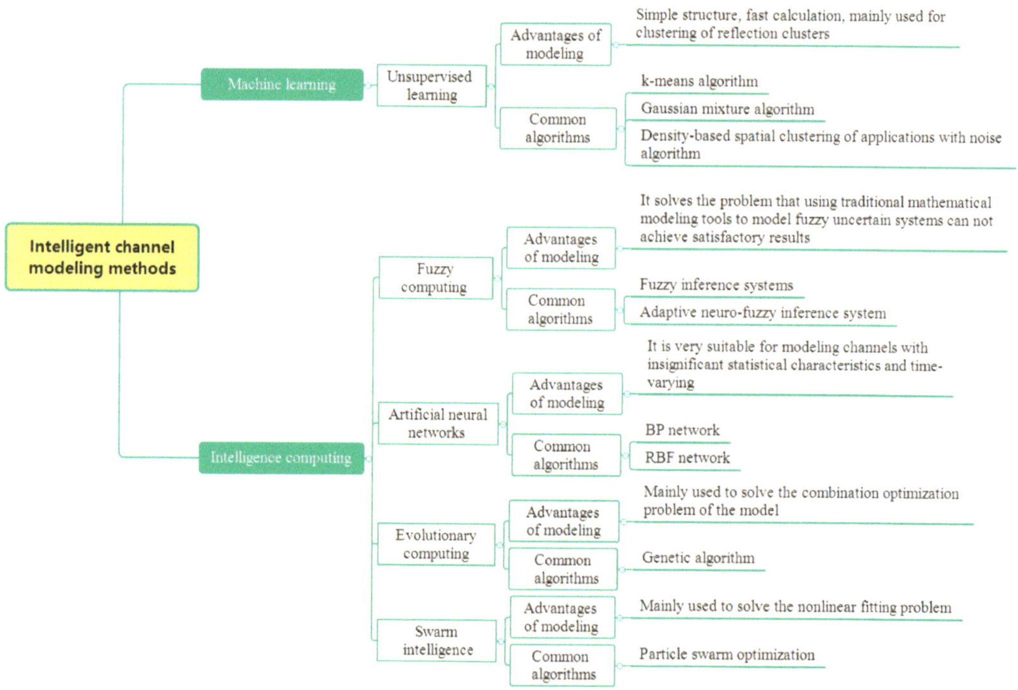

Figure 12. The classification and advantages of intelligent modeling methods.

Modeling Method Based on ML

ML, which can self-learn and self-adapt, is a method that uses computers to build statistical models based on a great deal of data. We can efficiently utilize the built models to predict and analyze anonymous data. An ML algorithm can be supervised, semi-supervised, or unsupervised, depending on whether the data used for training is labeled. ML algorithms have excellent advantages in dealing with simple classification problems because of their simple structure and fast computing speed, making them the main algorithm for channel multipath clustering. In the field of ML, clustering belongs to unsupervised learning. K-means, the Gaussian mixture algorithm (GMA), and the density-based spatial clustering of applications with noise algorithm (DBSCAN) are the common clustering models. The K-means algorithm adopts Euclidean distance. Therefore, the model can easily achieve and converge quickly. However, it cannot capture the propagation characteristics of a multipath channel. The GMM algorithm is based on the statistical distribution of data. The DBSCAN algorithm is density-based. Both of these algorithms account for the statistical characteristics of multipath components (MPC) and can cluster dense datasets of any shape. However, when the cluster distribution is uneven, the DBSCAN algorithm has a poor clustering effect [137]. In [66], a multipath component clustering architecture was proposed based on statistical information and a compact index evaluation criterion of mean and variance. This model used the GMM model and the mean and covariance structure of channel multipath components to achieve multipath component clustering. In [138], He et al. compared a variety of multipath component clustering algorithms, including the KPM algorithm, fuzzy C-means clustering algorithm (FCM), kernel power density-based (KPD) algorithm, and DBSCAN algorithm. The first two algorithms need the number of clusters as an a priori message, and the last two algorithms can automatically generate a more reasonable number of clusters. In addition, to model the non-stationary characteristics of frequency and space through related cluster groups, the paper [139] further clusters based on clusters using the K-means clustering algorithm to obtain related cluster groups. That is, a cluster group is composed of multiple similar clusters. In addition, supervised learning is also widely used in channel modeling. In order to improve the accuracy of frequency band prediction for FM broadcasting in Beijing, Wang et al. established a high-precision model based on support vector machine (SVR) [140] and statistical machine learning algorithms [141], achieving good results and achieving high-precision modeling in Beijing.

Modeling Method Based on IC

At present, IC used in channel modeling mainly includes fuzzy inference system (FIS), ANN, evolutionary algorithm (EC), and swarm intelligence (SI). ANFIS is a kind of adaptive network equivalent to FIS in function [142]. This network solves the problem that traditional mathematical modeling tools cannot obtain satisfactory results in modeling fuzzy uncertain systems. At the same time, compared with the non-interpretability of the general ANN, ANFIS gives weight to the physical meaning of the inference parameters in the fuzzy logic, making the ANN interpretable. In [143], the author inputs the height and distance of the transceiver antenna into the fuzzy system to obtain the path loss in the areas to be predicted. The prediction results show that the path loss of the city is the largest, and the average loss increases by 10 dB every ten kilometers. The model is proven to have better performance by accurately predicting path loss. In [94], a model for predicting path loss based on the fuzzy method is proposed. This model uses free space, vegetation terrain, flat terrain, and rural terrain to establish a fuzzy set. The results make clear that the path loss index for open areas is 2.2, and for low vegetation, small towns, and high vegetation areas, the path loss index is 3.3, 4.1, and 4.7, respectively.

An ANN is composed of a number of interconnected neurons. The input layer, hidden layer, and output layer are the necessary structures of an ANN. It learns and optimizes through BP. Self-learning, self-adaptive, and nonlinear fitting are characteristics of ANNs. The ANN can effectively simulate the radio channel characteristics of real scenes by using massive measured datasets, fully training to determine the relationship between inputs and outputs. Therefore, the ANN is suitable for modeling channels with insignificant and time-varying statistical characteristics. The channel characteristics can be determined and modeled accordingly using ANNs, treating the channel as data. In [142], a model based on the ANN and adaptive evolutionary algorithm for the urban environment was built. This model was compared with the RT. It makes clear that the prediction performance of the proposed model is better than the RT.

The main application of Evolutionary Computing (EC) is to solve combinatorial optimization problems. The process is simulated through program iteration, and the problem to be solved is regarded as an environment. The optimal solution will be found through natural selection among a population of possible solutions. In [73], a prediction model based on the microcellular area's GA is proposed, which is applicable to the frequency within 900 MHz and the distance between 0.1 and 2 km. This model can well explain the influence of building area on path loss. Similarly, in [72], a model based on the GA was proposed. The GA is used to tune the free space mobile information system and the Ericsson model. The results show that the model test data are similar to the actual measurement data.

SI comes from research on the group behavior of social insects such as ants and bees. SI is a distributed, self-organized, and systematic collective behavior, whether natural or artificial, usually composed of a group of simple individuals without central control who interact with each other and the environment [144]. Among them, PSO is the most widely used SI in channel modeling. The PSO is a stochastic optimization algorithm based on the foraging behavior of birds, which is used to solve nonlinear problems [145]. A particle can be searched in space by combining three vectors: the personal best vector (x), the position vector (v), and the fitness value. The PSO algorithm can be applied during either the initialization phase or the iteration phase [146]. At initialization, the velocity and position of each vector are assigned randomly. In the iteration stage, the speed and position can be modified by the equation in [147] to approximate the best solution. Because the PSO does not need detailed propagation environment information, it is widely used in channel modeling. In [145], He et al. created an adaptive radial basis function (RBF) ANN model for path loss estimation using the PSO algorithm. The comparison results show that the performance of the proposed model is better than other RBF–NN-based models. Tahat and Taha applied statistical adjustment technology based on PSO to correct the path loss output of COST-231–WI. Compared with the measured values, this model has a more minor standard deviation than the traditional COST-231–WI model [75].

3.1.2. Hybridization Methods of Multi-Models

Radio channel models are often designed for different scenes, so each model has a different emphasis. There are many different scenes when predicting large-scale areas. Usually, satisfactory results will not be obtained using a single model. Therefore, we can combine different channel models to reach the goal of improving the accuracy of prediction. Multi-model fusion technology is mainly divided into subjective and objective fusion technology. In subjective fusion technology, researchers subjectively divided the predicted area of the independent sub-model according to the characteristics of each model. The typical fusion model is the advanced propagation model (APM) of the US. This model fuses the PE and the geometrical optical (GO) based models [148]. APM selects the Flat Earth (FE) model in the region where the propagation angle is greater than 5° and the propagation distance is less than 2.5 km. Because of the short propagation distance and the large transmitting angle, the earth can be approximately flat. Therefore, there are fewer obstacles in this region, so the FE model is used for prediction. Due to the propagation angle of the PE model varying with frequency, when the frequency is greater than 100 MHz, the maximum propagation angle of PE is 5°. Therefore, APM uses PE for prediction in the low-altitude region, where the propagation angle is suitable for the PE model. Above this region, where the altitude is high and there are almost no obstacles, a relatively simple Extended Optics (EO) model is used to predict. The EO is based on the assumption of parallel rays and is suitable for predicting high-altitude propagation without obstacles. Finally, the RO model is selected to predict propagation loss in other regions. This is because the RO model is suitable for flat surfaces and assumes that the refraction environment is horizontal and uniform. The region's division of the APM is shown in Figure 13. The objective fusion technology does not involve the subjective judgment of experts, and the prediction results are obtained entirely through objective formula calculation. Wang et al. used entropy weight theory and root mean square error as the evaluation standard to establish a combined prediction model by objectively weighting the ITU-R model [149].

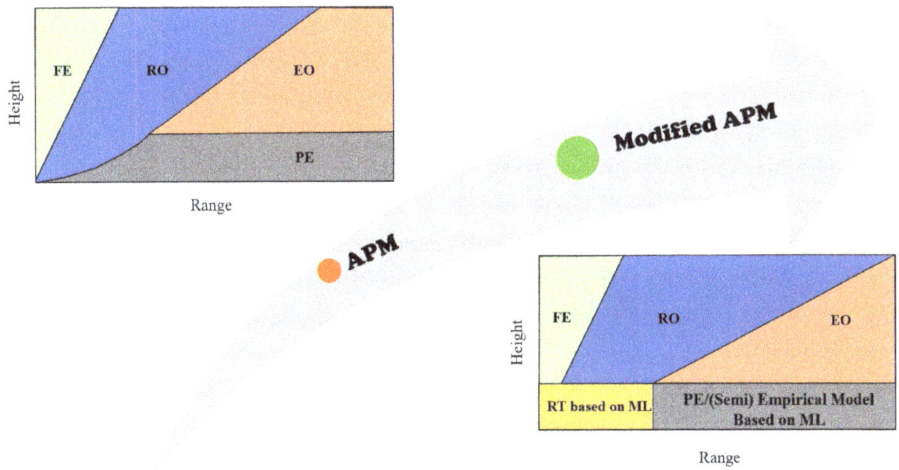

Figure 13. The propagation region's division of APM and future modification.

Based on the previous analysis, the propagation region is re-divided based on APM. Figure 13 gives the new region's division for modifying APM. The region division is changed in the near-ground, where the propagation distance is less than 10 km, and the propagation height is less than 100 m. Dense buildings characterize the region. Therefore, RT can realize better prediction accuracy and operation speed by combining ML. Beyond this region, we can select a mixed method combining PE, semi-empirical, and empirical models, which can also be assisted by machine learning. This region is characterized by sizeable topographic relief and many ground obstacles. Because this region is close to the ground, we can easily obtain the radio environment. Therefore, it is easy to see that ML can effectively combine traditional methods for support modeling. As mentioned earlier, the propagation angle of PE varies with frequency. When the frequency reaches 20 GHz, the maximum propagation angle of PE is only 0.4°. Therefore, as the carrier frequency continues to increase in the future, the PE method will no longer be suitable for prediction in this area.

3.2. Potential Technology for Future Communication

Future radio communication will be a highly heterogeneous four-dimensional "space, air, ground, and sea" communication [150]. The communication enables comprehensive, multi-level, and three-dimensional coverage to meet communication applications in diverse environments and under various requirements. To achieve four-dimensional communication, the current wireless networks require even faster transmission rates and lower latency to support intelligent coordination among various communication systems. Longer transmission distances are necessary to facilitate long-distance transmission scenarios, such as those in air, space, and sea. The degree of sensory integration needs to be further enhanced to empower emerging applications like autonomous driving and digital twinning. Furthermore, there is a need to increase channel capacity and spectrum utilization further to address the explosive growth of wireless services and frequency devices. In response to the requirements mentioned above, new technologies such as terahertz (THz) communication, reconfigurable intelligent surfaces (RIS), integrated communication and sensing

(ICAS), orbital angular momentum (OAM), and ultra-large antenna arrays technology (ULAAs) will emerge as crucial enabling technologies driving the future development of wireless communication. Figure 14 shows the evolution from current communication to future communication.

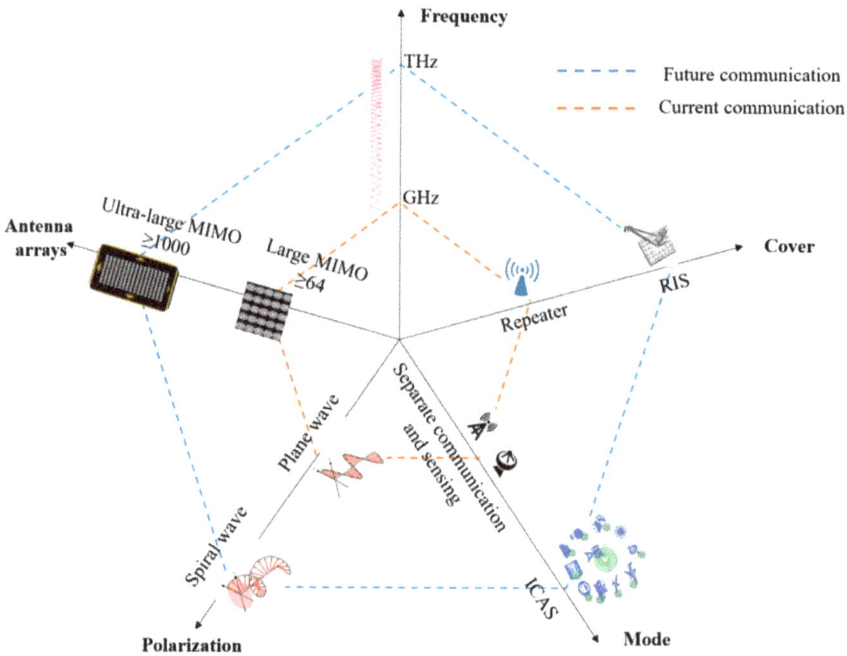

Figure 14. The evolution from current communication to future communication.

3.2.1. The Application Technology of THz

Because millimeter wave has the advantages of continuity and wide bandwidth, it is the critical application of 5G [151]. Due to the increasing data volume and the pursuit of a high data transmission rate, millimeter wave has yet to meet the current development, which causes people to focus on the THz band with higher frequency and wider bandwidth [152]. The THz frequency band, which ranges from 0.1 to 10 THz, has rich bandwidth and spectrum resources. It is one of the key technologies of future communication. Combined with multi-antenna technology, THz is expected to achieve Tbps-level high-speed mobile communication. However, with the increase in frequency, the wavelength is shorter and shorter. The frequency coverage will significantly decrease, which will increase the networking cost. Meanwhile, there may be apparent changes in propagation characteristics. For example, because the wavelength of THz is close to molecules and small particles in the atmosphere (such as water vapor and oxygen), THz's absorption or scattering effect will be more obvious [153,154]. In addition, narrow beams in THz communication are prone to misalignment, leading to a decrease in communication link capacity and even link interruption, which is also a major challenge for THz communication applications [155].

Fortunately, the beamforming technology based on MIMO can improve the transmission distance of THz communication. In [156], Enahoro et al. explored the performance of the least mean square algorithm (LMS) beamforming algorithm. The result shows that the LMS algorithm increases signal quality by eliminating interfering signals and noise by nulling them, while sending maximum signal (beams) to the desired direction. In [157], the prospect of ultra-massive multi-input multi-output (MIMO) technology to combat the distance problem at the THz band is considered. They achieved high gain and directionally adjustable narrow beam transmission based on MIMO technology. It also can significantly improve coverage by using RIS. ICAS can effectively solve the problem of narrow beam misalignment. RIS and ICAS will be discussed in the following text.

3.2.2. Reconfigurable Intelligent Surface Technology

As mentioned earlier, the high attenuation and small coverage of THz communication have become one of the key issues in its deployment. In the past, we usually improved signal coverage by establishing relay stations, but this will greatly increase the cost of deployment [158]. RIS can solve the problems of low coverage and low-frequency efficiency of THz communication. RIS integrates a large number of low-cost passive reflection elements on a planar surface, and each element independently reflects radio waves by controlling their amplitude and phase, thereby achieving the reconfiguration of the propagation environment [159]. Compared with traditional relay stations, RIS-assisted wireless networks can significantly improve communication quality and reduce networking costs. In the future, RIS can play the following role in wireless communication.

Act as a network repeater to increase coverage: By deploying RIS to establish a virtual line-of-sight link and receiver, the coverage range of millimeter wave communication can be greatly improved, and communication quality can be improved [160]. Wu et al. proposed a RIS-aided single-cell wireless system where one RIS is deployed to assist in the communications between a multi-antenna access point (AP) and multiple single-antenna users. The result shows that the received SNR increases quadratically with the number of reflecting elements of the RIS, which is more cost-efficient than the conventional massive MIMO or multi-antenna AF relay [161]. At the same time, Peng et al. utilized RIS to assist in UAV communication; thus, the communication quality and flexibility of the ground-to-air network were improved [162].

Improve the security of wireless communication: The physical layer security is a fundamental issue in wireless communications [163]. RIS can adaptively adjust the phase shifts of its reflecting units to strengthen the desired signal and/or suppress the undesired signal. Yang et al. utilized RIS to maximize the SNR of the legitimate user path, thereby cutting the eavesdropper path. They achieved the aim of improving the security of wireless communication [163]. Cui et al. maximized the secrecy rate of the legitimate communication link by jointly designing the AP's transmit beamforming and the RIS's reflect beamforming [164]. Chen et al. proposed a method of adjusting the reflecting coefficients of the RIS to change the attenuation and scattering of the incident electromagnetic wave so that it can propagate in the desired way toward the intended receiver [165].

In addition, RIS can also be used as the signal reflection hub. In this scenario, it can realize the communication from device to device and achieve communication interference suppression while supporting low-power transmission [159].

3.2.3. Integrated Communication and Sensing Technology

ICAS is also one of the key technologies in future communication. On the one hand, it is because many future application scenarios, such as unmanned driving, digital twins, and virtual reality, require both high-performance sensing and wireless communication [166]. On the other hand, communication and sensing systems have long been studied as two independent systems. However, they have great similarities in spectrum occupation, system architecture, antenna composition, and signal processing. Through joint design, such as signal transmission integration and hardware architecture integration, communication and sensing functions can be achieved simultaneously on the same set of devices and frequency spectrum to alleviate the shortage of spectrum resources, reduce equipment interference, and reduce equipment costs [167].

The integration of communication and perception realizes the integrated design of communication, perception, and computing functions through means such as air interface and protocol joint design, spectrum resource sharing, and software and hardware device sharing, resulting in the fusion and symbiosis of communication and perception functions. While transmitting information, communication and perception integration can also analyze the characteristics of radio waves such as direct, reflected, and scattered waves to locate, measure distance, measure speed, image, detect, identify, reconstruct the environment, and other target or environmental information [168].

As the key technology for future communication, ICAS has been extensively researched by the academic community [169]. Carlos et al. explored the full duplex ICAS system from a signal-processing perspective [170]. In [171,172], the effectiveness of communication and the bounds of the sensing are discussed. In [173,174], design schemes for wireless communications and radar sensing shared waveforms were proposed. In [175,176], new frameworks for ICAS were proposed. Preeti et al. proposed an adaptive preamble design to balance the positioning accuracy and communication rate of the ICAS system [177].

3.2.4. Orbital Angular Momentum Technology

The explosive growth of wireless services and devices is a major issue that needs to be considered in future communications. However, due to the strict limitations of spectrum resources in traditional multiplexing technologies (such as time division multiplexing, frequency division multiplexing, and code division multiplexing), current multiplexing technologies cannot meet the growing demand for ultra-high communication capacity from users. OAM can provide a new degree of multiplexing freedom for wireless communication systems and is recognized as one of the potential key technologies in future communication [178].

Compared with traditional planar waves, the vortex electromagnetic waves carrying OAM have a helical phase structure in the wavefront and a helical phase factor. In addition, different modalities theoretically satisfy strict orthogonality. Thus, in OAM wireless communication, multiple signals can be modulated to multiple OAM modes for transmission. This enables universal data transmission under limited resources and improves channel capacity. Research on OAM wireless communication has begun. Chen et al. proposed a misalignment–robust receiving model by extracting the phases of channel state information and implementing joint phase compensation and signal detection. Compared with the conventional model, the proposed model is more robust [179]. Jing et al. proposed a simple channel–independent beamforming model with fast symbol-wise maximum likelihood detection. Though the bit error rate of the proposed model is the same as that of conventional models, the numbers of complex additions and complex multiplications are much smaller than those of the traditional model [180]. Liang et al. proposed a hybrid orthogonal division multiplexing model using OAM multiplexing and orthogonal frequency division multiplexing in conjunction [181]. Hu et al. proposed an OAM-based independent analog beam selection by using the divergence to solve the problem that the interferences among different users degrade the system performance.

3.2.5. Ultra-Large Antenna Arrays Technology

Radio communication systems have already adopted MIMO and have demonstrated significant performance improvements [182]. With the increased RF, the spacing between antennas is reduced so that a more extensive antenna array can be constructed [183,184]. In 2010, Marzetta first proposed providing large-scale MIMO antenna arrays for the base station to improve users' service [185]. In addition, using 3D antenna arrays can increase system capacity [186]. According to the channel measurement result, it has been verified that the 3D MIMO technology achieves 33% channel capacity gain compared to 2D MIMO technology [187]. With the increase in antennas, 3D MIMO channels become dispersed in the spatial domain, significantly increasing the channel capacity [188]. Therefore, large-scale MIMO technology is a crucial technology of 5G [189]. For future communication, as the frequency increases, ULAAs can be used. However, many problems are still to be solved in modeling channels with the ULAA. For example, the measurement results of using virtual 40×40 planar antenna arrays show that the channel characteristics are non-stationary on the array [190]. Channel models should, therefore, take into account the non-stationarity of antenna arrays. In addition, more antennas will produce narrower beams, so more accurate multipath propagation angles (MPCs) are required to make the radio wave propagation beam point accurately to users. In addition, it is also a challenge to use ULAAs for channel measurement from the perspective of channel measurement. On the one hand, applying ULAAs in HF channel transmitters is difficult, especially in the THz band. On the other hand, with the increase in antenna array size, the amount of measurement data increases exponentially, and more computing costs are required.

4. Conclusions

This paper systematically overviews the existing terrestrial radio channel models. Specially, these models are sorted into deterministic, empirical, and semi-empirical models. Then, we analyzed the features of the three kinds and summarized the semi-empirical model. The results show that the deterministic model has the highest accuracy, does not need a large number of measured data, and is suitable for various scenarios, but its calculation and simulation are time-consuming and not suitable for large-scale scenes. The empirical model has features of high universality, low complexity, simple calculation, poor accuracy, a large amount of measured data required, poor scene applicability, and suitable for small-scale scenes. The semi-empirical model has a moderate amount of computation, adaptation to large-scale scenes, low complexity, and higher accuracy. Namely, the three typical models are relatively balanced in universality, accuracy, and complexity.

At the same time, development trends in the future are also analyzed: the large-scale combination model suitable for large-scale scenes, the wide-area generalization model suitable for multi-scenes, and a more efficient intelligent model will be the main development direction of channel modeling. After that, for the requirement of intelligence and refinement, we propose to use the intelligent method based on the ML or the IC to realize the intelligent modeling process and use the fusion multi-model technology to achieve refined modeling. Finally, we look forward to the potential technologies of future communication, which include the THz application technology, reconfigurable intelligent surface technology, integrated communication and sensing technology, orbital angular momentum technology, and ultra-large antenna array technology. We hope this paper will stimulate more interest in modeling terrestrial radio channels.

Author Contributions: Conceptualization, J.W. and Y.H.; methodology, J.W. and Y.H.; software, J.W., Y.H. and C.Y.; validation, J.W. and Y.H.; formal analysis, J.W., Y.H. and C.Y.; investigation, J.W., Y.H. and C.Y.; resources, J.W. and C.Y.; data curation, J.W. and Y.H.; writing—original draft preparation, J.W. and Y.H.; writing—review and editing, J.W., Y.H. and C.Y.; supervision, J.W., Y.H. and C.Y.; project administration, J.W.; funding acquisition, J.W. All authors have read and agreed to the published version of the manuscript.

Funding: This research was funded by the State Key Laboratory of Complex Electromagnetic Environment Effects on Electronics and Information System (No. CEMEE2022G0201).

Conflicts of Interest: The authors declare no conflict of interest.

References

1. Meng, Y.S.; Lee, Y.H.; Ng, B.C. Empirical near ground path loss modeling in a forest at VHF and UHF bands. *IEEE Trans. Antennas Propag.* **2009**, *57*, 1461–1468. [CrossRef]
2. Lin, L.K.; Zhao, Z.W.; Teng, X.G. Research and Enlightenment of ITU Radio Wave Propagation Standard. *Stand. Res.* **2023**, *9*, 30–36.
3. Hu, R.R.; Zhao, Z.W.; Lin, L.K.; Sun, S.J.; Liu, X.L. The statistical relationship between refractivity gradient in the lowest 1 km near the ground and ground meteorological parameters. *Chin. J. Radio Sci.* **2020**, *35*, 896–901.
4. Zhao, Z.W.; Lin, L.K.; Zhang, Y.S.; Du, P.; Wang, H.G.; Han, J. Progress and prospect of radio meteorology research. *Chin. J. Radio Sci.* **2020**, *35*, 801–812.
5. Wang, J.; Yang, C.; Yan, N.N. Study on digital twin channel for the B5G and 6G communication. *Chin. J. Radio Sci.* **2021**, *36*, 340–348.
6. Yao, F.Q.; Zhang, Y.; Liu, Y.X. Security and control for electromagnetic spectrum. *J. Command Control* **2015**, *1*, 278–283.
7. Zhao, Z.W.; Lin, L.K. Research and prospect of tropospheric radio wave propagation. *Chin. J. Radio Sci.* **2023**, *38*, 591–656.

8. Yang, C.F.; Wu, B.C.; Ko, C.J. A ray-tracing method for modeling indoor wave propagation and penetration. *IEEE Trans. Antennas Propag.* **1998**, *46*, 907–919. [CrossRef]
9. Teti, J.G. Parabolic equation methods for electromagnetic wave propagation [Book Review]. *IEEE Antennas Propag. Mag.* **2001**, *43*, 96–97. [CrossRef]
10. Okumura, Y. Field strength and its variability in VHF and UHF land-mobile radio service. *Rev. Electr. Commun. Lab.* **1968**, *16*, 825–873.
11. Hata, M. Empirical formula for propagation loss in land mobile radio services. *IEEE Trans. Veh. Technol.* **1980**, *29*, 317–325. [CrossRef]
12. *ITU-R P.1546-6*; Method for Point-to-Area Predictions for Terrestrial Services in the Frequency Range 30 MHz to 4000 MHz. ITU-R.: Geneva, Switzerland, 2019.
13. Dalela, C.; Prasad, M.V.S.N.; Dalela, P.K. *Tuning of COST-231 Hata Model for Radio Wave Propagation Predictions*; Academy & Industry Research Collaboration Center: Tamil Nadu, India, 2012.
14. Molisch, A.F.; Asplund, H.; Heddergott, R.; Steinbauer, M.; Zwick, T. The COST259 directional channel model-part I: Overview and methodology. *IEEE Trans. Wirel. Commun.* **2006**, *5*, 3421–3433. [CrossRef]
15. Parkvall, S.; Furuskär, A.; Dahlman, E. Evolution of LTE toward IMT-advanced. *IEEE Commun. Mag.* **2011**, *49*, 84–91. [CrossRef]
16. He, R.; Ding, Z. Applications of machine learning in wireless communications. *Telecommunications* **2019**, *81*, 15–94.
17. Zhang, J.H.; Tang, P.; Yu, L.; Jiang, T.; Tian, L. Channel measurements and models for 6G: Current status and future outlook. *Front. Inf. Technol. Electron. Eng.* **2020**, *21*, 39–61. [CrossRef]
18. Bharti, A.; Adeogun, R.; Pedersen, T. Learning Parameters of Stochastic Radio Channel Models From Summaries. *IEEE Open J. Antennas Propag.* **2020**, *1*, 175–188. [CrossRef]
19. Ayadi, M.; Zineb, A.B.; Tabbane, S. A UHF path loss model using learning machine for heterogeneous networks. *IEEE Trans. Antennas Propag.* **2017**, *65*, 3675–3683. [CrossRef]
20. Eichie, J.O.; Oyedum, O.D.; Ajewole, M.; Aibinu, A.M. Comparative analysis of basic models and artificial neural network based model for path loss prediction. *Prog. Electromagn. Res. M* **2017**, *61*, 133–146. [CrossRef]
21. Sotiroudis, S.P.; Goudos, S.K.; Gotsis, K.A.; Siakavara, K.; Sahalos, J.N. Application of a composite differential evolution algorithm in optimal neural network design for propagation path-loss prediction in mobile communication systems. *IEEE Antennas Wirel. Propag. Lett.* **2013**, *12*, 364–367. [CrossRef]
22. Thrane, J.; Zibar, D.; Christiansen, H.L. Model-aided deep learning method for path loss prediction in mobile communication systems at 2.6 GHz. *IEEE Access* **2020**, *8*, 7925–7936. [CrossRef]
23. Faruk, N.; Abdulrasheed, I.Y.; Surajudeen-Bakinde, N.T.; Adetiba, E.; Oloyede, A.A.; Abdulkarim, A.; Atayero, A.A. Large-scale radio propagation path loss measurements and predictions in the VHF and UHF bands. *Heliyon* **2021**, *7*, e07298. [CrossRef]
24. Romain, C.; Chiehping, L.; Thierry, T.; Derek, C.; Grégory, G.; Jelena, S.; Camillo, G.; Yoann, C.; Jack, C.; Golmie, N. Calibration of Ray-Tracing With Diffuse Scattering Against 28-GHz Directional Urban Channel Measurements. *IEEE Trans. Veh. Technol.* **2020**, *69*, 14264–14276.
25. Bhatia, G.S.; Corre, Y.; Di Renzo, M. Efficient Ray-Tracing Channel Emulation in Industrial Environments: An Analysis of Propagation Model Impact. In Proceedings of the 2023 Joint European Conference on Networks and Communications & 6G Summit (EuCNC/6G Summit), Gothenburg, Sweden, 6–9 June 2023; pp. 180–185.
26. Demetrescu, C.; Constantinou, C.C.; Mehler, M.J. Corner and rooftop diffraction in radiowave propagation prediction tools: A review. In Proceedings of the VTC '98. 48th IEEE Vehicular Technology Conference. Pathway to Global Wireless Revolution (Cat. No.98CH36151), Ottawa, ON, Canada, 21 May 1998; pp. 515–519.
27. Ling, H.; Chou, R.C.; Lee, S.W. Shooting and bouncing rays: Calculating the RCS of an arbitrarily shaped cavity. *IEEE Trans. Antennas Propag.* **1989**, *37*, 194–205. [CrossRef]
28. Zeng, P. Convergence Analysis of Indoor Ray Tracing. Master's Thesis, Beijing University of Posts and Telecommunications, Beijing, China, 2021.
29. Tan, S.Y.; Tan, H.S. A microcellular communications propagation model based on the uniform theory of diffraction and multiple image theory. *IEEE Trans. Antennas Propag.* **1996**, *44*, 1317–1326. [CrossRef]
30. Lentovich, M.A.; Fock, V.A. Solution of propagation of electromagnetic waves along the earth's surface by the method of parabolic equations. *J. Phys. USSR* **1946**, *10*, 13–23.
31. Zelley, C.A.; Constantinou, C.C. A three-dimensional parabolic equation applied to VHF/UHF propagation over irregular terrain. *IEEE Trans. Antennas Propag.* **1999**, *47*, 1586–1596. [CrossRef]
32. Afsharinejad, A.; Davy, A.; Jennings, B.; Rasmann, S.; Brennan, C. A path-loss model incorporating shadowing for THz band propagation in vegetation. In Proceedings of the 2015 IEEE Global Communications Conference, San Diego, CA, USA, 6–10 December 2015; pp. 1–6.
33. Aslam, M.I.; Zekavat, S.R. New channel path loss model for near-ground antenna sensor networks. *IET Wirel. Sens. Syst.* **2012**, *2*, 103–107. [CrossRef]
34. Danufane, F.H.; Di, R.M.; De, R.J.; Tretyakov, S. On the path-loss of reconfigurable intelligent surfaces: An approach based on Green's theorem applied to vector fields. *IEEE Trans. Commun.* **2021**, *69*, 5573–5592. [CrossRef]
35. Li, W.; Hu, X.; Jiang, T. Path loss models for IEEE 802.15. 4 vehicle-to-infrastructure communications in rural areas. *IEEE Internet Things J.* **2018**, *5*, 3865–3875. [CrossRef]

36. Safwat, N.E.D.; Newagy, F.; Hafez, I.M. Air-to-ground channel model for UAVs in dense urban environments. *IET Commun.* **2020**, *14*, 1016–1021. [CrossRef]
37. Haneda, K.; Järveläinen, J.; Karttunen, A.; Kyrö, M.; Putkonen, J. A Statistical Spatio-Temporal Radio Channel Model for Large Indoor Environments at 60 and 70 GHz. *IEEE Trans. Antennas Propag.* **2015**, *63*, 2694–2704. [CrossRef]
38. Järveläinen, J.; Haneda, K.; Karttunen, A. Indoor Propagation Channel Simulations at 60 GHz Using Point Cloud Data. *IEEE Trans. Antennas Propag.* **2016**, *64*, 4457–4467. [CrossRef]
39. Kwakkernaat, M.R.J.A.E.; Herben, M.H.A.J. Diagnostic Analysis of Radio Propagation in UMTS Networks Using High-Resolution Angle-of-Arrival Measurements. *IEEE Antennas Propag. Mag.* **2011**, *53*, 66–75. [CrossRef]
40. Rodríguez-Corbo, F.A.; Azpilicueta, L.; Celaya-Echarri, M.; Lopez-Iturri, P.; Alejos, A.V.; Shubair, R.M.; Falcone, F. Deterministic and Empirical Approach for Millimeter-Wave Complex Outdoor Smart Parking Solution Deployments. *Sensors* **2021**, *21*, 4112. [CrossRef]
41. Kasampalis, S.; Lazaridis, P.I.; Zaharis, Z.D.; Bizopoulos, A.; Paunovska, L.; Zettas, S.; Cosmas, J. Longley-Rice model prediction inaccuracies in the UHF and VHF TV bands in mountainous terrain. In Proceedings of the 2015 IEEE International Symposium on Broadband Multimedia Systems and Broadcasting, Ghent, Belgium, 17–19 June 2015; pp. 1–5.
42. Rivera, A.V.; Unda, A.N.; Ramos, B.; Macías, E.L.S.A.; Suarez, A.; Gomez, J. Lee microcell propagation model: A complex case empirical study. *WSEAS Trans. Commun.* **2015**, *14*, 33–42.
43. Low, K. Comparison of urban propagation models with CW-measurements. In Proceedings of the Vehicular Technology Society 42nd VTS Conference-Frontiers of Technology, Denver, CO, USA, 10–13 May 1992; pp. 936–942.
44. *ITU-R P.370-7*; VHF and UHF Propagation Curves for the Frequency Range from 30 MHz to 1000 MHz, Broadcasting Services. ITU-R.: Geneva, Switzerland, 1953.
45. Han, S.Y.; Abu-Ghazaleh, N.B.; Lee, D. Efficient and consistent path loss model for mobile network simulation. *IEEE/ACM Trans. Netw.* **2015**, *24*, 1774–1786. [CrossRef]
46. Zhu, Q.; Yao, M.; Bai, F.; Chen, X.; Zhong, W.; Hua, B.; Ye, X. A general altitude-dependent path loss model for UAV-to-ground millimeter-wave communications. *Front. Inf. Technol. Electron. Eng.* **2021**, *22*, 767–776. [CrossRef]
47. Zhao, X.; Geng, S.; Coulibaly, B.M. Path-loss model including LOS-NLOS transition regions for indoor corridors at 5 GHz [wireless corner]. *IEEE Antennas Propag. Mag.* **2013**, *55*, 217–223. [CrossRef]
48. Al-Hourani, A.; Kandeepan, S.; Jamalipour, A. Modeling air-to-ground path loss for low altitude platforms in urban environments. In Proceedings of the 2014 IEEE Global Communications Conference, Austin, TX, USA, 8–12 December 2014; pp. 2898–2904.
49. Wang, Y.; Wang, X.L.; Qin, Y.; Liu, Y.; Lu, W.J.; Zhu, H.B. An empirical path loss model in the indoor stairwell at 2.6 GHz. In Proceedings of the 2014 IEEE International Wireless Symposium (IWS 2014), Xi'an, China, 24–26 March 2014; pp. 1–4.
50. Nilsson, M.G.; Gustafson, C.; Abbas, T.; Tufvesson, F. A path loss and shadowing model for multilink vehicle-to-vehicle channels in urban intersections. *Sensors* **2018**, *18*, 4433. [CrossRef]
51. Mardeni, R.; Solahuddin, Y. Path loss model development for indoor signal loss prediction at 2.4 GHz 802.11 n network. In Proceedings of the 2012 International Conference on Microwave and Millimeter Wave Technology (ICMMT), Shenzhen, China, 5–8 May 2012; Volume 2, pp. 1–4.
52. Yu, Y.; Liu, Y.; Lu, W.J.; Zhu, H.B. Path loss model with antenna height dependency under indoor stair environment. *Int. J. Antennas Propag.* **2014**, *2014*, 482615. [CrossRef]
53. Santos, P.M.; Abrudan, T.E.; Aguiar, A.; Barros, J. Impact of position errors on path loss model estimation for device-to-device channels. *IEEE Trans. Wirel. Commun.* **2014**, *13*, 2353–2361.
54. Olasupo, T.O.; Otero, C.E.; Olasupo, K.O.; Kostanic, I. Empirical path loss models for wireless sensor network deployments in short and tall natural grass environments. *IEEE Trans. Antennas Propag.* **2016**, *64*, 4012–4021. [CrossRef]
55. Sulyman, A.I.; Nassar, A.T.; Samimi, M.K.; MacCartney, G.R.; Rappaport, T.S.; Alsanie, A. Radio propagation path loss models for 5G cellular networks in the 28 GHz and 38 GHz millimeter-wave bands. *IEEE Commun. Mag.* **2014**, *52*, 78–86. [CrossRef]
56. Hakim, G.P.; Alaydrus, M.; Bahaweres, R.B. Empirical approach of ad hoc path loss propagation model in realistic forest environments. In Proceedings of the 2016 International Conference on Radar, Antenna, Microwave, Electronics, and Telecommunications (ICRAMET), Jakarta, Indonesia, 3–5 October 2016; pp. 139–143.
57. Sawada, H.; Fujii, K.; Kasamatsu, A.; Ogawa, H.; Ishizu, K.; Kojima, F. Path loss model at 300 GHz for indoor mobile service applications. *IEICE Commun. Express* **2016**, *5*, 424–428. [CrossRef]
58. Genc, A. A new path loss model based on the volumetric occupancy rate for the pine forests at 5G frequency band. *Int. J. Microw. Wirel. Technol.* **2021**, *13*, 144–153. [CrossRef]
59. Keawbunsong, P.; Duangsuwan, S.; Supanakoon, P.; Promwong, S. Quantitative Measurement of Path Loss Model Adaptation Using the Least Squares Method in an Urban DVB-T2 System. *Int. J. Antennas Propag.* **2018**, *2018*, 7219618. [CrossRef]
60. Bilgehan, B.; Ojo, S. Multiplicative based path loss model. *Int. J. Commun. Syst.* **2018**, *31*, e3794. [CrossRef]
61. Wang, Y.; Lu, W.J.; Zhu, H.B. An empirical path-loss model for wireless channels in indoor short-range office environment. *Int. J. Antennas Propag.* **2012**, *2012*, 636349. [CrossRef]
62. Zhao, H.; Wang, Q.; Shi, K. Analysis on human blockage path loss and shadow fading in millimeter-wave band. *Int. J. Antennas Propag.* **2017**, *2017*, 7540202. [CrossRef]

63. He, R.; Zhong, Z.; Ai, B.; Ding, J.; Jiang, W.; Zhang, H.; Li, X. A standardized path loss model for the GSM-railway based high-speed railway communication systems. In Proceedings of the 2014 IEEE 79th Vehicular Technology Conference (VTC Spring), Seoul, Republic of Korea, 18–21 May 2014; pp. 1–5.
64. Nossire, Z.; Gupta, N.; Almazaydeh, L.; Xiong, X. New empirical path loss model for 28 GHz and 38 GHz millimeter wave in indoor urban under various conditions. *Appl. Sci.* **2018**, *8*, 2122. [CrossRef]
65. Alqudah, Y.A. Path loss modeling based on field measurements using deployed 3.5 GHz WiMAX network. *Wirel. Pers. Commun.* **2013**, *69*, 793–803. [CrossRef]
66. Li, Y.; Zhang, J.; Ma, Z. Clustering in wireless propagation channel with a statistics-based framework. In Proceedings of the 2018 IEEE Wireless Communications and Networking Conference (WCNC), Barcelona, Spain, 15–18 April 2018; pp. 1–6.
67. Zhang, Y.; Wen, J.; Yang, G.; He, Z.; Wang, J. Path loss prediction based on machine learning: Principle, method, and data expansion. *Appl. Sci.* **2019**, *9*, 1908. [CrossRef]
68. Wang, A.; Lei, S. Research on modeling of underground path loss prediction based on SA-SVM. *J. Phys. Conf. Ser.* **2022**, *2245*, 012016. [CrossRef]
69. Zhao, X.; Hou, C.; Wang, Q. A new SVM-based modeling method of cabin path loss prediction. *Int. J. Antennas Propag.* **2013**, *2013*, 279070. [CrossRef]
70. Zhang, Y.; Wen, J.; Yang, G.; He, Z.; Luo, X. Air-to-air path loss prediction based on machine learning methods in urban environments. *Wirel. Commun. Mob. Comput.* **2018**, *2018*, 8489326. [CrossRef]
71. Yang, G.; Zhang, Y.; He, Z.; Wen, J.; Ji, Z.; Li, Y. Machine-learning-based prediction methods for path loss and delay spread in air-to-ground millimetre-wave channels. *IET Microw. Antennas Propag.* **2019**, *13*, 1113–1121. [CrossRef]
72. Cavalcanti, B.J.; d'Assunção, A.G.; Mendonça, L.M. Optimizing empirical propagation models for LTE and LTE-A using genetic algorithms at 879 MHz. In Proceedings of the 2017 IEEE-APS Topical Conference on Antennas and Propagation in Wireless Communications (APWC), Verona, Italy, 11–15 September 2017; pp. 312–315.
73. Fernandes, L.C.; Soares, A.J.M. Path loss prediction in microcellular environments at 900 MHz. *AEU-Int. J. Electron. Commun.* **2014**, *68*, 983–989. [CrossRef]
74. Salman, M.A.; Popoola, S.I.; Faruk, N.; Surajudeen-Bakinde, N.T.; Oloyede, A.A.; Olawoyin, L.A. Adaptive Neuro-Fuzzy model for path loss prediction in the VHF band. In Proceedings of the 2017 International Conference on Computing Networking and Informatics (ICCNI), Lagos, Nigeria, 29–31 October 2017; pp. 1–6.
75. Tahat, A.; Taha, M. Statistical tuning of Walfisch-Ikegami propagation model using particle swarm optimization. In Proceedings of the 2012 19th IEEE Symposium on Communications and Vehicular Technology in the Benelux (SCVT), Eindhoven, The Netherlands, 16 November 2012; pp. 1–6.
76. Garah, M.; Oudira, H.; Djouane, L.; Hamdiken, N. Particle swarm optimization for the path loss reduction in suburban and rural area. *Int. J. Electr. Comput. Eng.* **2017**, *7*, 2125. [CrossRef]
77. Banimelhem, O.; Al-Zu'bi, M.M.; Al Salameh, M.S. Hata path loss model tuning for cellular networks in Irbid City. In Proceedings of the 2015 IEEE International Conference on Computer and Information Technology; Ubiquitous Computing and Communications; Dependable, Autonomic and Secure Computing; Pervasive Intelligence and Computing, Liverpool, UK, 26–28 October 2015; pp. 1646–1650.
78. Shoewu, O.O.; Adedoyin, M.A.; Akinyemi, L.A.; Oborkhale, L.I. Fuzzy-logic based path loss models for metropolitan environment. In Proceedings of the 2018 IEEE International Workshop on Signal Processing Systems (SiPS), Cape Town, South Africa, 21–24 October 2018; pp. 71–76.
79. Surajudeen-Bakinde, N.T.; Faruk, N.; Popoola, S.I.; Salman, M.A.; Oloyede, A.A.; Olawoyin, L.A.; Calafate, C.T. Path loss predictions for multi-transmitter radio propagation in VHF bands using Adaptive Neuro-Fuzzy Inference System. *Eng. Sci. Technol. Int. J.* **2018**, *21*, 679–691. [CrossRef]
80. Cheerla, S.; Ratnam, D.V.; Borra, H.S. Neural network-based path loss model for cellular mobile networks at 800 and 1800 MHz bands. *AEU-Int. J. Electron. Commun.* **2018**, *94*, 179–186. [CrossRef]
81. Nguyen, C.; Cheema, A.A. A deep neural network-based multi-frequency path loss prediction model from 0.8 GHz to 70 GHz. *Sensors* **2021**, *21*, 5100. [CrossRef]
82. Benmus, T.A.; Abboud, R.; Shatter, M.K. Neural network approach to model the propagation path loss for great Tripoli area at 900, 1800, and 2100 MHz bands. In Proceedings of the 2015 16th International Conference on Sciences and Techniques of Automatic Control and Computer Engineering (STA), Monastir, Tunisia, 21–23 December 2015; pp. 793–798.
83. Ma, S.; Cheng, H.; Lee, H. A Practical Approach to Indoor Path Loss Modeling Based on Deep Learning. *J. Comput. Sci. Eng.* **2021**, *15*, 84–95. [CrossRef]
84. Cavalcanti, B.J.; Cavalcante, G.A.; Mendonça, L.M.D.; Cantanhede, G.M.; de Oliveira, M.M.; D'Assunção, A.G. A hybrid path loss prediction model based on artificial neural networks using empirical models for LTE and LTE-A at 800 MHz and 2600 MHz. *J. Microw. Optoelectron. Electromagn. Appl.* **2017**, *16*, 708–722. [CrossRef]
85. Duangsuwan, S.; Juengkittikul, P.; Myint Maw, M. Path loss characterization using machine learning models for GS-to-UAV-enabled communication in smart farming scenarios. *Int. J. Antennas Propag.* **2021**, *2021*, 5524709. [CrossRef]
86. Mom, J.M.; Mgbe, C.O.; Igwue, G.A. Application of artificial neural network for path loss prediction in urban macrocellular environment. *Am. J. Eng. Res.* **2014**, *3*, 270–275.

87. Wang, Y.; Liang, M.Y.; Hu, J.; Song, T. A new method for radio wave propagation prediction based on bp-neural network and path loss model. In Proceedings of the 2020 12th International Conference on Knowledge and Smart Technology (KST), Pattaya, Thailand, 9 April 2020; pp. 41–46.
88. Wen, J.; Zhang, Y.; Yang, G.; He, Z.; Zhang, W. Path loss prediction based on machine learning methods for aircraft cabin environments. *IEEE Access* **2019**, *7*, 159251–159261. [CrossRef]
89. Ojo, S.; Imoize, A.; Alienyi, D. Radial basis function neural network path loss prediction model for LTE networks in multitransmitter signal propagation environments. *Int. J. Commun. Syst.* **2021**, *34*, e4680. [CrossRef]
90. Ahmadien, O.; Ates, H.F.; Baykas, T.; Gunturk, B.K. Predicting path loss distribution of an area from satellite images using deep learning. *IEEE Access* **2020**, *8*, 64982–64991. [CrossRef]
91. Popoola, S.I.; Misra, S.; Atayero, A.A. Outdoor path loss predictions based on extreme learning machine. *Wirel. Pers. Commun.* **2018**, *99*, 441–460. [CrossRef]
92. Cheng, H.; Ma, S.; Lee, H.; Cho, M. Millimeter wave path loss modeling for 5G communications using deep learning with dilated convolution and attention. *IEEE Access* **2021**, *9*, 62867–62879. [CrossRef]
93. Abraham, D.C. Soft Computing Techniques for Path Loss Estimation at 1800 MHz in a Tropical Metropolitan Environment: Case Study of Abuja-Nigeria. *Soft Comput.* **2017**, *4*, 8298–8305.
94. Joshi, S.; Sati, G.; Kestwal, M.C. Path loss correction for signal propagation amongst low roof top buildings using fuzzy logic. *Int. J. Comput. Appl.* **2013**, *83*, 43–48. [CrossRef]
95. Jo, H.S.; Park, C.; Lee, E.; Choi, H.K.; Park, J. Path loss prediction based on machine learning techniques: Principal component analysis, artificial neural network, and Gaussian process. *Sensors* **2020**, *20*, 1927. [CrossRef] [PubMed]
96. Juang, R.T. Path loss modelling based on path profile in urban propagation environments. *IET Commun.* **2022**, *16*, 685–694. [CrossRef]
97. Zhu, C.H.; Yao, J.K.; Yang, T.J. Review on radio channel modeling method. *Wirel. Internet Technol.* **2015**, *16*, 26–27.
98. Li, Q.L.; Zhang, J.P.; Zhang, Y.S. Review on sea clutter measurement and modeling. *Chin. J. Radio Sci.* **2023**, *38*, 559–573.
99. Thomas, N.J.; Willis, M.J.; Craig, K.H. The Relative Importance of Different Propagation Mechanisms for Radio Coverage and Interference Prediction in Urban Scenarios at 2.4, 5.8, and 28 GHz. *IEEE Trans. Antennas Propag.* **2006**, *54*, 3918–3920. [CrossRef]
100. Yang, C.; Wang, J.; You, X.; Su, H.; Ji, S.; Zhao, Z. Applicability of ITU-R P.1546 recommendation in typical terrestrial areas of China. *Chin. J. Radio Sci.* **2019**, *34*, 295–301.
101. *ITU-R P.1812-5*; A Path-Specific Propagation Prediction Method for Point-to-Area Terrestrial Services in the VHF and UHF Bands. ITU-R.: Geneva, Switzerland, 2019.
102. *ITU-R P.2001-4*; A General Purpose Wide-Range Terrestrial Propagation Model in the Frequency Range 30 MHz to 50 GHz. ITU-R.: Geneva, Switzerland, 2021.
103. *Report ITU-R M.2412-0*; Guidelines for Evaluation of Radio Interface Technologies for IMT-2020. ITU-R.: Geneva, Switzerland, 2017.
104. *3GPP TR 36.873*; Study on 3D Channel Model for LTE, 3rd Generation Partnership Project. 3GPP: Sophia Antipolis, France, 2017.
105. *3GPP TR 38.901*; Study on Channel Model for Frequencies from 0.5 to 100 GHz. 3GPP: Sophia Antipolis, France, 2022.
106. Alwarafy, A.; Sulyman, A.I.; Alsanie, A.; Alshebeili, S.A.; Behairy, H.M. Path-loss channel models for receiver spatial diversity systems at 2.4 GHz. *Int. J. Antennas Propag.* **2017**, *2017*, 6790504. [CrossRef]
107. MacCartney, G.R.; Deng, S.; Rappaport, T.S. Indoor office plan environment and layout-based mmWave path loss models for 28 GHz and 73 GHz. In Proceedings of the 2016 IEEE 83rd Vehicular Technology Conference (VTC Spring), Nanjing, China, 15–18 May 2016; pp. 1–6.
108. Khatun, M.; Guo, C.; Moro, L.; Matolak, D.; Mehrpouyan, H. Millimeter-wave path loss at 73 GHz in indoor and outdoor airport environments. In Proceedings of the 2019 IEEE 90th Vehicular Technology Conference (VTC2019-Fall), Honolulu, HI, USA, 22–25 September 2019; pp. 1–5.
109. Zhou, L.; Xiao, L.; Yang, Z.; Li, J.; Lian, J.; Zhou, S. Path loss model based on cluster at 28 GHz in the indoor and outdoor environments. *Sci. China Inf. Sci.* **2017**, *60*, 1–11. [CrossRef]
110. Oyie, N.O.; Afullo, T.J.O. A comparative study of dual-slope path loss model in various indoor environments at 14 to 22 GHz. In Proceedings of the 2018 Progress in Electromagnetics Research Symposium (PIERS-Toyama), Toyama, Japan, 1–4 August 2018; pp. 121–128.
111. Kalyankar, S.K.; Lee, Y.H.; Meng, Y.S. Two-slope path loss model for curved-tunnel environment with concept of break point. *IEEE Trans. Intell. Transp. Syst.* **2020**, *22*, 7850–7859. [CrossRef]
112. Li, L.; Ibdah, Y.; Ding, Y.; Eghbali, H.; Muhaidat, S.H.; Ma, X. Indoor multi-wall path loss model at 1.93 GHz. In Proceedings of the MILCOM 2013—2013 IEEE Military Communications Conference, San Diego, CA, USA, 18–20 November 2013; pp. 1233–1237.
113. Obeidat, H.A.; Asif, R.; Ali, N.T.; Dama, Y.A.; Obeidat, O.A.; Jones, S.M.R.; Abd-Alhameed, R.A. An indoor path loss prediction model using wall correction factors for wireless local area network and 5G indoor networks. *Radio Sci.* **2018**, *53*, 544–564. [CrossRef]
114. Kacou, M.; Guillet, V.; El Zein, G.; Zaharia, G. A multi-wall and multi-frequency home environment path loss characterization and modelling. In Proceedings of the 12th European Conference on Antennas and Propagation (EuCAP 2018), London, UK, 9–13 April 2018; pp. 105–110.

115. El-Keyi, A.; Sokun, H.U.; Nguyen, T.N.; Ye, Q.; Zhu, H.J.; Yanikomeroglu, H. A novel probabilistic path loss model for simulating coexistence between 802.11 and 802.15. 4 networks in smart home environments. In Proceedings of the 2017 IEEE 28th Annual International Symposium on Personal, Indoor, and Mobile Radio Communications (PIMRC), Newark, NJ, USA, 8–13 October 2017; pp. 1–5.
116. Zhou, L.; Xiao, L.; Li, J.; Yang, Z.; Lian, J.; Zhou, S. Path loss model based on cluster at 28GHz in the office and corridor environments. In Proceedings of the 2016 IEEE 84th vehicular technology conference (VTC-Fall), Montreal, QC, Canada, 18–21 September 2016; pp. 1–5.
117. Briso-Rodríguez, C.; Fratilescu, P.; Xu, Y. Path loss modeling for train-to-train communications in subway tunnels at 900/2400 MHz. *IEEE Antennas Wirel. Propag. Lett.* **2019**, *18*, 1164–1168. [CrossRef]
118. Bhuvaneshwari, A.; Hemalatha, R.; Satya Savithri, T. Development of an optimized ray tracing path loss model in the indoor environment. *Wirel. Pers. Commun.* **2017**, *96*, 1039–1064. [CrossRef]
119. Heereman, F.; Joseph, W.; Tanghe, E.; Plets, D.; Verloock, L.; Martens, L. Path loss model and prediction of range, power and throughput for 802.11 n in large conference rooms. *AEU-Int. J. Electron. Commun.* **2012**, *66*, 561–568. [CrossRef]
120. Diago-Mosquera, M.E.; Aragón-Zavala, A.; Rodriguez, M. Testing a 5G communication system: Kriging-aided O2I path loss modeling based on 3.5 GHz measurement analysis. *Sensors* **2021**, *21*, 6716. [CrossRef] [PubMed]
121. Fernández Anitzine, I.; Romo Argota, J.A.; Fontán, F.P. Influence of training set selection in artificial neural network-based propagation path loss predictions. *Int. J. Antennas Propag.* **2012**, *2012*, 351487. [CrossRef]
122. Casillas-Perez, D.; Camacho-Gomez, C.; Jimenez-Fernandez, S.; Portilla-Figueras, J.A.; Salcedo-Sanz, S. Weighted ABG: A general framework for optimal combination of ABG path-loss propagation models. *IEEE Access* **2020**, *8*, 101758–101769. [CrossRef]
123. Aldossari, S.; Chen, K.C. Predicting the path loss of wireless channel models using machine learning techniques in mmwave urban communications. In Proceedings of the 2019 22nd International Symposium on Wireless Personal Multimedia Communications (WPMC), Lisbon, Portugal, 24–27 November 2019; pp. 1–6.
124. Al-Samman, A.M.; Rahman, T.A.; Azmi, M.H.; Hindia, M.N. Large-scale path loss models and time dispersion in an outdoor line-of-sight environment for 5G wireless communications. *AEU-Int. J. Electron. Commun.* **2016**, *70*, 1515–1521. [CrossRef]
125. Zhou, F.; Ji, R.; Sun, J.; Cheng, K.; Zhang, X. Analysis on the vehicle-induced path loss for millimetre-wave V2V communication. *IET Microw. Antennas Propag.* **2021**, *15*, 495–502.
126. Bok, J.; Ryu, H.G. Path loss model considering Doppler shift for high speed railroad communication. In Proceedings of the 16th International Conference on Advanced Communication Technology, Pyeong Chang, Republic of Korea, 16–19 February 2014; pp. 1–4.
127. Lee, J.; Kim, M.D.; Chung, H.K.; Kim, J. NLOS path loss model for low-height antenna links in high-rise urban street grid environments. *Int. J. Antennas Propag.* **2015**, *2015*, 651438. [CrossRef]
128. Castro-Hernandez, D.; Paranjape, R. Local tuning of a site-specific propagation path loss model for microcell environments. *Wirel. Pers. Commun.* **2016**, *91*, 709–728. [CrossRef]
129. Inomata, M.; Yamada, W.; Sasaki, M.; Mizoguchi, M.; Kitao, K.; Imai, T. Path loss model for the 2 to 37 GHz band in street microcell environments. *IEICE Commun. Express* **2015**, *4*, 149–154. [CrossRef]
130. Karttunen, A.; Molisch, A.F.; Hur, S.; Park, J.; Zhang, C.J. Spatially consistent street-by-street path loss model for 28-GHz channels in micro cell urban environments. *IEEE Trans. Wirel. Commun.* **2017**, *16*, 7538–7550. [CrossRef]
131. Yu, J.; Chen, W.; Yang, K.; Li, C.; Li, F.; Shui, Y. Path loss channel model for inland river radio propagation at 1.4 GHz. *Int. J. Antennas Propag.* **2017**, *2017*, 5853724. [CrossRef]
132. Bhuvaneshwari, A.; Hemalatha, R.; Satyasavithri, T. Semi deterministic hybrid model for path loss prediction improvement. *Procedia Comput. Sci.* **2016**, *92*, 336–344. [CrossRef]
133. Bhupuak, W.; Tooprakai, S. Minimizing path loss prediction error using k-means clustering and fuzzy logic. *Turk. J. Electr. Eng. Comput. Sci.* **2018**, *26*, 1989–2002. [CrossRef]
134. You, X.; Wang, C.X.; Huang, J.; Gao, X.; Zhang, Z.; Wang, M.; Liang, Y.C. Towards 6G wireless communication networks: Vision, enabling technologies, and new paradigm shifts. *Sci. China Inf. Sci.* **2021**, *64*, 1–74. [CrossRef]
135. Pellegrini, A.; Brizzi, A.; Zhang, L.; Ali, K.; Hao, Y.; Wu, X.; Constantinou, C.C.; Nechayev, Y.; Hall, P.S.; Chahat, N.; et al. Antennas and Propagation for Body-Centric Wireless Communications at Millimeter-Wave Frequencies: A Review [Wireless Corner]. *IEEE Antennas Propag. Mag.* **2013**, *55*, 262–287. [CrossRef]
136. Seretis, A.; Sarris, C.D. An overview of machine learning techniques for radiowave propagation modeling. *IEEE Trans. Antennas Propag.* **2021**, *70*, 3970–3985. [CrossRef]
137. Liu, H.; Fan, Z.L. Survey of application of machine learning in radio channel modelling. *J. Commun.* **2021**, *42*, 134–153.
138. He, R.; Ai, B.; Molisch, A.F.; Stuber, G.L.; Li, Q.; Zhong, Z.; Yu, J. Clustering enabled wireless channel modeling using big data algorithms. *IEEE Commun. Mag.* **2018**, *56*, 177–183. [CrossRef]
139. Huang, Z.; Cheng, X. A general 3D space-time-frequency non-stationary model for 6G channels. *IEEE Trans. Wirel. Commun.* **2020**, *20*, 535–548. [CrossRef]
140. Wang, J.; Wu, Z.L.; Hao, Y.L.; Yang, C. An SVR based Radio Propagation Prediction Model for Terrestrial FM Broadcasing Services in Beijing and Its Surrounding Area. *IEEE Trans. Broadcast.* **2023**. [CrossRef]
141. Wang, J.; Hao, Y.L.; Yang, C. A comprehensive prediction model for VHF radio wave propagation by integrating entropy weight theory and machine learning methods. *IEEE Trans. Antennas Propag.* **2023**, *7*, 6249–6254. [CrossRef]

142. Jang, J.S. ANFIS: Adaptive-network-based fuzzy inference system. *IEEE Trans. Syst. Man Cybern.* **1993**, *23*, 665–685. [CrossRef]
143. Mathew, S.; Shylaja, K.; Jayasri, T. Path loss prediction in wireless communication system using fuzzy logic. *Indian J. Sci. Technol.* **2014**, *7*, 642–647. [CrossRef]
144. Dario, P.; Sandini, G.; Aebischer, P. Swarm intelligence in cellular robotic systems. In *Robots and Biological Systems: Towards a New Bionics?* Springer: Berlin/Heidelberg, Germany, 1993; Volume 38, pp. 703–712.
145. He, T.; Dan, T.; Wei, Y.; Li, H.; Chen, X.; Qin, G. Particle swarm optimization RBF neural network model for internet traffic prediction. In Proceedings of the 2016 International Conference on Intelligent Transportation, Big Data & Smart City (ICITBS), Changsha, China, 17–18 December 2016; pp. 521–524.
146. Saunders, S.R.; Aragón-Zavala, A. *Antennas and Propagation for Wireless Communication Systems*; John Wiley & Sons: Hoboken, NJ, USA, 2007.
147. Duch, W. What is Computational Intelligence and where is it going? In *Challenges for Computational Intelligence*; Springer: Berlin, Germany, 2007; pp. 1–13.
148. Barrios, A.E. Considerations in the development of the advanced propagation model (APM) for US Navy applications. In Proceedings of the 2003 International Conference on Radar, Adelaide, SA, Australia, 3–5 September 2003; pp. 77–82.
149. Wang, J.; Hao, Y.L.; Yang, C. An Entropy Weight-Based Method for Path Loss Predictions. *Radio Sci.* **2023**, *58*, 1–11. [CrossRef]
150. Chen, X.; Sheng, M.; LI, B.; ZHAO, N. Survey on unmanned aerial vehicle communications for 6G. *J. Electron. Inf. Technol.* **2022**, *44*, 781–789.
151. Rappaport, T.S.; Xing, Y.; MacCartney, G.R.; Molisch, A.F.; Mellios, E.; Zhang, J. Overview of millimeter wave communications for fifth-generation (5G) wireless networks—With a focus on propagation models. *IEEE Trans. Antennas Propag.* **2017**, *65*, 6213–6230. [CrossRef]
152. Maviel, L.; Corre, Y.; Gougeon, G.; Amar, D. Analysis of Residential Sub-THz Deployments from Accurate Radio Simulations and Planning Techniques. In Proceedings of the 2022 16th European Conference on Antennas and Propagation (EuCAP), Madrid, Spain, 27 March–1 April 2022; pp. 1–5.
153. Fatemeh, N.; Emidio, M.; Marina, G.; Edward, H.; Costas, C.; Peter, G.; Mikhail, C. Rain Attenuation at Millimeter Wave and Low-THz Frequencies. *IEEE Trans. Antennas Propag.* **2020**, *68*, 421–431.
154. Fatemeh, N.; Rui, D.; Edward, G.H.; Peter, G.; Costas, C.; Mikhail, C.; Marina, G. Low-THz Transmission Through Water-Containing Contaminants on Antenna Radome. *IEEE Trans. Terahertz Sci. Technol.* **2018**, *8*, 63–75.
155. Priebe, S.; Jacob, M.; Kürner, T. Affection of THz indoor communication links by antenna misalignment. In Proceedings of the 2012 6th European Conference on Antennas and Propagation (EUCAP), Prague, Czech Republic, 26–30 March 2012; pp. 483–487.
156. Enahoro, S.; Ekpo, S.C.; Uko, M.C.; Altaf, A.; Ansari, U.E.H.; Zafar, M. Adaptive Beamforming for mmWave 5G MIMO Antennas. In Proceedings of the 2021 IEEE 21st Annual Wireless and Microwave Technology Conference (WAMICON), Sand Key, FL, USA, 28–29 April 2021; pp. 1–5.
157. Sarieddeen, H.; Alouini, M.S.; Al-Naffouri, T.Y. Terahertz-band ultra-massive spatial modulation MIMO. *IEEE J. Sel. Areas Commun.* **2019**, *37*, 2040–2052. [CrossRef]
158. Niu, Y.; Ding, W.; Wu, H.; Li, Y.; Chen, X.; Ai, B.; Zhong, Z. Relay-assisted and QoS aware scheduling to overcome blockage in mmWave backhaul networks. *IEEE Trans. Veh. Technol.* **2019**, *68*, 1733–1744. [CrossRef]
159. Wu, Q.; Zhang, R. Towards smart and reconfigurable environment: Intelligent reflecting surface aided wireless network. *IEEE Commun. Mag.* **2019**, *58*, 106–112. [CrossRef]
160. Björnson, E.; Wymeersch, H.; Matthiesen, B.; Popovski, P.; Sanguinetti, L.; de Carvalho, E. Reconfigurable Intelligent Surfaces: A signal processing perspective with wireless applications. *IEEE Signal Process. Mag.* **2022**, *39*, 135–158. [CrossRef]
161. Wu, Q.Q.; Zhang, R. Intelligent reflecting surface enhanced wireless network via joint active and passive beamforming. *IEEE Trans. Wirel. Commun.* **2019**, *18*, 5394–5409. [CrossRef]
162. Pang, X.; Sheng, M.; Zhao, N.; Tang, J.; Niyato, D.; Wong, K.K. When UAV meets IRS: Expanding air-ground networks via passive reflection. *IEEE Wirel. Commun.* **2021**, *28*, 164–170. [CrossRef]
163. Yang, L.; Yang, J.; Xie, W.; Hasna, M.O.; Tsiftsis, T.; Di Renzo, M. Secrecy performance analysis of RIS-aided wireless communication systems. *IEEE Trans. Veh. Technol.* **2020**, *69*, 12296–12300. [CrossRef]
164. Cui, M.; Zhang, G.; Zhang, R. Secure wireless communication via intelligent reflecting surface. *IEEE Wirel. Commun. Lett.* **2019**, *8*, 1410–1414. [CrossRef]
165. Chen, J.; Liang, Y.C.; Pei, Y.; Guo, H. Intelligent reflecting surface: A programmable wireless environment for physical layer security. *IEEE Access* **2019**, *7*, 82599–82612. [CrossRef]
166. Liu, A.; Huang, Z.; Li, M.; Wan, Y.; Li, W.; Han, T.X.; Chetty, K. A survey on fundamental limits of integrated sensing and communication. *IEEE Commun. Surv. Tutor.* **2022**, *24*, 994–1034. [CrossRef]
167. Xiao, Y.W.; Xiao, Z.Z.; Liang, M. Research and optimization on the sensing algorithm for 6G integrated sensing and communication network. *J. Commun.* **2023**, *44*, 219–230.
168. Liu, Y.; Liao, G.; Xu, J.; Yang, Z.; Zhang, Y. Adaptive OFDM integrated radar and communications waveform design based on information theory. *IEEE Commun. Lett.* **2017**, *21*, 2174–2177. [CrossRef]
169. Liu, F.; Cui, Y.; Masouros, C.; Xu, J.; Han, T.X.; Eldar, Y.C.; Buzzi, S. Integrated sensing and communications: Towards dual-functional wireless networks for 6G and beyond. *IEEE J. Sel. Areas Commun.* **2022**, *40*, 1728–1767. [CrossRef]

170. Barneto, C.B.; Liyanaarachchi, S.D.; Heino, M.; Riihonen, T.; Valkama, M. Full duplex radio/radar technology: The enabler for advanced joint communication and sensing. *IEEE Wirel. Commun.* **2021**, *28*, 82–88. [CrossRef]
171. Shi, C.; Xu, D.; Zhou, Y.; Tu, W. Range-DOA information and scattering information in phased-array radar. In Proceedings of the 2019 IEEE 5th International Conference on Computer and Communications (ICCC), Chengdu, China, 6–9 December 2019; pp. 747–752.
172. Ai, Y.; Yi, W.; Blum, R.S.; Kong, L. Cramer-Rao lower bound for multitarget localization with noncoherent statistical MIMO radar. In Proceedings of the 2015 IEEE Radar Conference (RadarCon), Arlington, VA, USA, 10–15 May 2015; pp. 1497–1502.
173. Sturm, C.; Wiesbeck, W. Waveform design and signal processing aspects for fusion of wireless communications and radar sensing. *Proc. IEEE* **2011**, *99*, 1236–1259. [CrossRef]
174. Liu, F.; Zhou, L.; Masouros, C.; Li, A.; Luo, W.; Petropulu, A. Toward dual-functional radar-communication systems: Optimal waveform design. *IEEE Trans. Signal Process.* **2018**, *66*, 4264–4279. [CrossRef]
175. Rahman, M.L.; Zhang, J.A.; Huang, X.; Guo, Y.J.; Heath, R.W. Framework for a perceptive mobile network using joint communication and radar sensing. *IEEE Trans. Aerosp. Electron. Syst.* **2019**, *56*, 1926–1941. [CrossRef]
176. Zhang, A.; Rahman, M.L.; Huang, X.; Guo, Y.J.; Chen, S.; Heath, R.W. Perceptive mobile networks: Cellular networks with radio vision via joint communication and radar sensing. *IEEE Veh. Technol. Mag.* **2020**, *16*, 20–30. [CrossRef]
177. Kumari, P.; Nguyen, D.H.; Heath, R.W. Performance trade-off in an adaptive IEEE 802.11 ad waveform design for a joint automotive radar and communication system. In Proceedings of the 2017 IEEE International Conference on Acoustics, Speech and Signal Processing (ICASSP), New Orleans, LA, USA, 5–9 March 2017; pp. 4281–4285.
178. Cheng, W.; Zhang, W.; Jing, H.; Gao, S.; Zhang, h. Orbital angular momentum for wireless communications. *IEEE Wirel. Commun.* **2018**, *26*, 100–107. [CrossRef]
179. Chen, R.; Xu, H.; Li, J.; Zhang, Y. Misalignment-robust receiving scheme for UCA-based OAM communication systems. In Proceedings of the 2017 IEEE 85th Vehicular Technology Conference (VTC Spring), Sydney, Australia, 4–7 June 2017; pp. 1–5.
180. Jing, H.; Cheng, W.; Xia, X.G. A simple channel independent beamforming scheme with parallel uniform circular array. *IEEE Commun. Lett.* **2019**, *23*, 414–417. [CrossRef]
181. Liang, L.; Cheng, W.; Zhang, W.; Zhang, H. Joint OAM multiplexing and OFDM in sparse multipath environments. *IEEE Trans. Veh. Technol.* **2020**, *69*, 3864–3878. [CrossRef]
182. Goldsmith, A.; Jafar, S.A.; Jindal, N.; Vishwanath, S. Capacity limits of MIMO channels. *IEEE J. Sel. Areas Commun.* **2003**, *21*, 684–702. [CrossRef]
183. Khan, Z.U.; Alkaraki, S.; Abbasi, Q.H.; Imran, M.A.; Loh, T.H.; Alomainy, A. 3D Printed Slotted Waveguide Antenna Array for Millimeter-wave Communication Systems. In Proceedings of the 2022 16th European Conference on Antennas and Propagation (EuCAP), Madrid, Spain, 27 March–1 April 2022; pp. 1–4.
184. Khan, Z.U.; Loh, T.H.; Belenguer, A.; Alomainy, A. Empty Substrate-Integrated Waveguide-Fed Patch Antenna Array for 5G Millimeter-Wave Communication Systems. *IEEE Antennas Wirel. Propag. Lett.* **2020**, *19*, 776–780. [CrossRef]
185. Liu, G.; Hou, X.; Wang, F.; Jin, J.; Tong, H.; Huang, Y. Achieving 3D-MIMO with massive antennas from theory to practice with evaluation and field trial results. *IEEE Syst. J.* **2016**, *11*, 62–71. [CrossRef]
186. Zhang, J.; Pan, C.; Pei, F.; Liu, G.; Cheng, X. Three-dimensional fading channel models: A survey of elevation angle research. *IEEE Commun. Mag.* **2014**, *52*, 218–226. [CrossRef]
187. Zhang, J.; Tang, P.; Tian, L.; Hu, Z.; Wang, T.; Wang, H. 6–100 GHz research progress and challenges from a channel perspective for fifth generation (5G) and future wireless communication. *Sci. China Inf. Sci.* **2017**, *60*, 1–18. [CrossRef]
188. Zhang, J.; Zheng, Z.; Zhang, Y.; Xi, J.; Zhao, X.; Gui, G. 3D MIMO for 5G NR: Several observations from 32 to massive 256 antennas based on channel measurement. *IEEE Commun. Mag.* **2018**, *56*, 62–70. [CrossRef]
189. Ali, E.; Ismail, M.; Nordin, R.; Abdulah, N.F. Beamforming techniques for massive MIMO systems in 5G: Overview, classification, and trends for future research. *Front. Inf. Technol. Electron. Eng.* **2017**, *18*, 753–772. [CrossRef]
190. Chen, J.; Yin, X.; Wang, S. Measurement-based massive MIMO channel modeling in 13–17 GHz for indoor hall scenarios. In Proceedings of the 2016 IEEE International Conference on Communications (ICC), Kuala Lumpur, Malaysia, 22–27 May 2016; pp. 1–5.

electronics

MDPI

Article

LEO-Assisted Aerial Deployment in Post-Disaster Scenarios Using a Combinatorial Bandit and Genetic Algorithmic Approach

Ehab Mahmoud Mohamed [1,*,†], Sherief Hashima [2,3,†], Kohei Hatano [2,4,†] and Haithem S. Khallaf [5,†]

1 Department of Electrical Engineering, College of Engineering in Wadi Addawasir, Prince Sattam Bin Abdulaziz University, Wadi Addawasir 11991, Saudi Arabia
2 Computational Learning Theory Team, RIKEN-Advanced Intelligence Project (AIP), Fukuoka 819-0395, Japan; sherief.hashima@riken.jp (S.H.); hatano@inf.kyushu-u.ac.jp (K.H.)
3 Engineering Department, NRC, Egyptian Atomic Energy Authority, Cairo 13759, Egypt
4 Department of Informatics, Kyushu University, Fukuoka 819-0395, Japan
5 Reactors Department, Nuclear Research Center, Egyptian Atomic Energy Authority (EAEA), Inshas 13759, Egypt; khallafh@mcmaster.ca
* Correspondence: ehab_mahmoud@aswu.edu.eg
† These authors contributed equally to this work.

Abstract: This paper proposes integrating low earth orbit satellites (LEO-Sats) and multiple aerials to provide rescue services in post-disaster areas. Aerials are distributed to provide wireless connectivity to survivors and rescue workers, while LEO-Sat exhibits backhaul linkages to aerials to connect them with the closest surviving ground base station (GBS). In this context, the aerials' deployment should maximize the total system rate while guaranteeing fairness among the served post-disaster regions within aerials' limited battery budget and LEO-Sat's limited bandwidth resources. Therefore, a combinatorial bandit model with arms fairness and budget constraints (CB-FBC) is proposed to address the aerials' deployment while maintaining fairness in covering post-disaster regions within the aerials' limited battery resources. Additionally, the aerials' transmit communication powers and LEO-Sat's bandwidth resources are optimized according to traffic requests of LEO-aerial linkages using a genetic algorithm (GA). By means of numerical analysis, the proposed GA shows superior performance over other naïve benchmarks.

Keywords: genetic algorithm; LEO satellite; multi-armed bandit; combinatorial bandits; aerials

1. Introduction

Due to their flying, maneuvering, and hovering merits, aerials have been increasingly used as wireless communication platforms, especially for post-disaster aid services. This is due to their ability to quickly reach remote and hard-to-access areas [1–3]. In post-disaster situations, aerials can offer wireless connection for user devices (UDs) owned by survivors and rescue teams where the terrestrial wireless communication networks are completely damaged or malfunctioned [4–7]. Deploying multiple aerials in such scenarios is challenging, requiring multi-objective optimization, including coverage, limited aerial battery life times, heights, and communication performance. Also, the limited transmission coverage of aerials might hinder their ability to establish communication links with the closest ground base station (GBS) or even among themselves, affecting their arrangement and data collection abilities in post-disaster zones.

Therefore, this paper proposes the usage of low earth orbit satellite (LEO-Sat) to assist aerials in rescue missions by creating backhaul links with the nearest survival GBS. Recently, LEO-Sat has garnered significant interest and is currently the subject of ongoing research in diverse wireless communication applications [8–15]. However, limited research studies have explored the potential integration of LEO-Sat and aerial technologies to leverage both benefits and establish more efficient wireless communication platforms [16–18]. In the proposed post-disaster scenario, LEO-Sat effectively assists disseminating/collecting control information and data traffic to/from nearby GBSs via aerials, where only indirect communication links between GBSs and aerials exist. The primary difficulty in this system is how to effectively distribute aerials in the sparsely populated sub-areas, i.e., disaster area clusters/grids. In this scenario, the main objective is to maximize the area's total achievable sum data rates plus maintaining fairness in post-disaster cluster coverage. Also, the bounded aerials' battery capacity and LEO-Sat bandwidth should be anticipated. To ensure fairness in post-disaster cluster coverage, higher user density clusters/grids should be served more frequently than the less-populated ones. This optimization problem has three main challenges. Firstly, coverage fairness should be maintained among post-disaster clusters according to their users' densities, where the wireless infrastructure is entirely malfunctioning, which involves the difficulty of knowing victims' information. Secondly, aerials should conserve their battery energy while maximizing the collected data from the post-disaster zone to prolong their critical rescue operation cycle before recharging. Finally, optimizing aerials' transmit (TX) communication powers as well as LEO-Sat bandwidth resources according to the aerials' traffic reflects another critical challenge. To deal with this highly challenging problem, this work presents the following main contributions:

- We propose to split the LEO-Sat assisted aerials deployment problem into two sub-problems. This is enabled by aerials' TX powers towards the LEO-Sat, and the LEO-Sat bandwidth resources should be allocated based on aerials' traffic needs after deployment. In this regard, the first sub-problem deals with the limited battery budget fair aerials deployment, and the second one deals with optimizing the aerials' TX powers and LEO-Sat bandwidth resources according to the aerials' traffic needs.
- A budget-constrained combinatorial multi-armed bandit (MAB) game with arms' fairness (CB-BFC) [19] will be proposed to address the first sub-problem. In this context, MAB is an advanced online learning tool where a player intends to increase his achievable profit via playing over the bandit's arms. Only through the exploitation–exploration process, without knowing any prior knowledge about the game, the player learns how to always play with the highest-profited arm. In the sub-problem considered, the MAB player will be the nearest survival GBS, the arms will be the post-disaster clusters, and the rewards are the aerials' achievable data rates.
- To ensure fairness in clusters coverage based on users' densities, we propose that GPS localization will be used to pre-estimate users' locations, which will be refined through the aerials' exploration during the proposed combinatorial MAB game.
- In the second sub-problem and after distributing the aerials among clusters, a genetic algorithm (GA) approach [20] will be utilized to optimize the aerials' TX powers and LEO-Sat's bandwidth resources based on the aerials' traffic coming from users' loads in their covered clusters. GA is a well-known optimization algorithm that can effectively address constrained non-linear optimization problems by means of penalty hypothesis.
- Extensive numerical analysis is conducted to confirm the effectiveness of the envisioned approach against benchmark schemes, including basic MAB approaches.

The remainder of this paper is arranged as follows: Section 2 summarizes LEO-Sat aerial-related works. Section 3 previews the envisioned system model of LEO–aerial integration followed by optimization problem formulation. Section 4 figures out the proposed "CB-FBC" algorithm and "GA"-based aerials' powers and LEO-Sat's bandwidth resource allocation. Section 5 discusses the numerical simulations conducted, and Section 6 delivers the concluding remarks of this paper.

2. Related Work

In recent decades, aerial applications in wireless communications have attracted active research due to their maneuvering, flying, and hovering capabilities [2,3]. These functions enable their applications as flying BSs [21], forwarding communication links from GBSs to remote users in non-rural areas [22], providing rescue services in post-disaster areas [5], strengthening communications in hotspot areas using aerial mounted reconfigurable intelligent surfaces (RISs) [6], etc.

On the other side, LEO-Sat has emerged as a promising technology for providing wireless communication services in recent years. These satellites orbit the Earth at a low altitude and can cover a relatively small area, allowing for high-speed, low-latency communication links. LEO-Sat systems are particularly attractive for applications such as sixth-generation (6G) networks, Internet of Things (IoT) connectivity, and other emerging wireless communication technologies. Regarding LEO-enabled wireless communications, in [8], a comprehensive overview of the physical and logical links, architectural and technological components needed to integrate LEO constellations into fifth generation (5G) and beyond 5G (B5G) systems was provided. In [10], a three-dimensional (3D) constellation optimization technique for a high-density LEO-based terrestrial–satellite communication model was proposed to optimize the number of satellites per constellation while attaining the minimum required backhaul per terminal. In [11], a load-balanced routing scheme was introduced to handle the LEO-Sat constellation framework. The work in [12] analyzed the downlink throughput of the LEO-Sat network that uses hopping spot beams. The authors presented upper and lower bounds on the downlink throughput based on the number of spot beams, interference restrictions, and coverage area. In [13], the authors discussed a user-centric handover methodology for ultra-dense LEO-Sat constellations to address the frequent handover problem and provide a higher quality of service. In [14], a low-complex massive multiple input multiple output (mMIMO) transmission technique with full frequency reuse (FFR) in LEO-Sat communication was proposed. The work of [15] investigated the joint optimization in LEO-Sat networks with multi-access edge computing servers to minimize the total delay of ground IoT devices.

Recently, a few researchers focused on integrating LEO-Sat and aerials to expand the coverage range. In [16], a multiplayer deep reinforcement learning (DRL) technique was utilized to optimize millimeter-wave (mmWave) and free-space optical (FSO) links of combined LEO-Sat and aerials. In [17], the authors suggested benefiting from aerials and LEO-Sat for information gathering in 6G networks. The focus was on minimizing the total energy cost while satisfying the data transmission demands, and the issue was solved using algorithms incorporating LEO-Sat-assisted aerial trajectory design. In [18], the authors leveraged LEO-Sat and aerials' caching for content delivery in ground systems to enhance connections and increase user capacity. The joint optimization problem of cache position, resource management, and aerial path was treated using an algorithm based on block coordinate descent and successive convex approximation methods. In [23], the authors of this paper considered the same problem of LEO-assisted aerial distribution but without considering the problem of jointly optimizing the LEO-Sat bandwidth and aerial TX power like the current work.

3. System Model

Figure 1 previews the under-investigation system model of integrating LEO-Sat and aerials to cover post-disaster regions where all nearby GBSs are destroyed or malfunctioned. The management and control (M&C) center orders the nearest survival GBS to the post-disaster area to manage the aerials' deployment assisted by LEO-Sat. To ensure that only one aerial covers a specific post-disaster location, the whole region is virtually divided into a number of clusters. These are depicted as dashed blue lines in Figure 1, with a total number of M clusters. Each cluster comprises K_i UDs, $1 \leq i \leq M$. LEO-Sat facilitates communication between the closest survival GBS and aerials for both traffic and control data. Also, $\mathcal{N}$ refers to the set of aerials, with a total number of N aerials, $N = 5$ in Figure 1, where $1 \leq j \leq N$ and $N < M$, are used to cover the post-disaster area. LEO-Sat continuously covers these aerials and can easily switch between satellites. The effects of this handover on network performance will be the subject of future investigations. At each time step t, the survival GBS determines which clusters the aerials should cover and assigns them accordingly. This is then relayed to the aerials via LEO-Sat backhauling. Relaying and multi-hop transmissions among aerials are out of the scope of this paper and will be the subject of our future investigations. The goal is to properly allocate the aerials to clusters that maximize their attainable data rates while ensuring fair coverage based on UDs' densities in the distributed clusters. This allocation is subject to constraints such as the limited aerial battery budget and the limited LEO-Sat bandwidth resources. In the proposed system model, both rescue workers' and victims' devices are assumed to operate using the same frequency band for simplicity as the main focus is on the LEO-Sat assisted fair aerials distribution. However, we can assume that victims and rescue workers use two different communication bands to mitigate interference. In this case, aerials should be capable of using two different bands' interfaces, one for the workers and the other for the victims, which will complicate the aerials' communication task and put more burdens on their energy consumption. For sure, workers can communicate among themselves using device-to-device (D2D) communications, but they definitely need aerial connectivity to reach the ground monitoring/control center for management and control via LEO-Sat relaying. This interesting point will be a subject of our future investigations by using two different bands, one for workers and the other for victims, including its challenges and possible solutions. Also, this paper proposes a fully centralized scenario for aerial deployment carried out by the M&C center in the survival area, as shown in Figure 1. A decentralized scenario, where rescue workers control aerials deployment, will be the subject of our future investigations due to its numerous challenges, including the limited information on rescue workers, their limited battery capacities that cannot handle the complicated task of aerials deployment, the lack of connectivity to all aerials simultaneously, etc. In the following subsections, the link models of aerial–UD and LEO–aerial are described in detail, and the optimization problem is formulated.

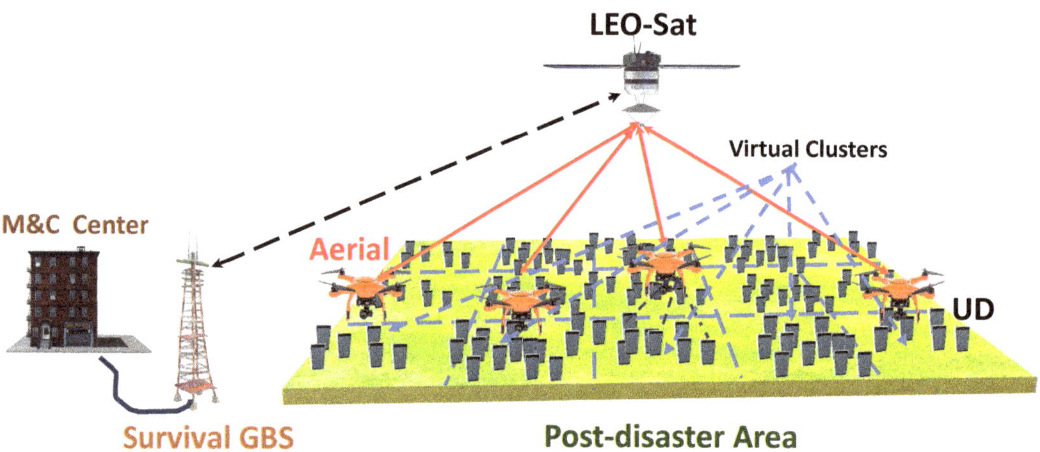

Figure 1. LEO-Sat and aerials integration in post-disaster scenario.

3.1. Aerial-UD Linkage Model

For aerial–UD communication linkage, we leveraged the model of [24], where the total path loss in dB between aerial j and UD k in cluster i can be expressed as follows:

$$\lambda_{jk_i}(d_{jk_i}) = \mathbb{P}^{LoS}\lambda_{jk_i}^{LoS}(d_{jk_i}) + \mathbb{P}^{NLoS}\lambda_{jk_i}^{NLoS}(d_{jk_i}) \tag{1}$$

where

$$\lambda_{jk_i}^{LoS}\left(d_{jk_i}\right) = 20\log\left(\frac{4\pi f_V d_{jk_i}}{c}\right) + \rho^{LoS} \& \lambda_{jk_i}^{NLoS}(d_{jk_i}) = 20\log\left(\frac{4\pi f_V d_{jk_i}}{c}\right) + \rho^{NLoS} \tag{2}$$

Herein, $\lambda_{jk_i}(d_{jk_i})$, is the total path loss. $\lambda_{jk_i}^{LoS}\left(d_{jk_i}\right)$ and $\lambda_{jk_i}^{NLoS}\left(d_{jk_i}\right)$ are the line of sight (LoS), and non-LoS (NLoS) components between aerial j and UD k in cluster i influenced by their related distance d_{jk_i}, respectively. Furthermore, f_V and c are the operating frequency of the aerial–UD link and the speed of light, respectively, where aerial is assumed to use the sub 6GHz wireless communication technology as given in [24]. ρ^{LoS} and ρ^{NLoS} indicate the LoS and NLoS losses in dB, respectively. $\mathbb{P}^{LoS}$ and $\mathbb{P}^{NLoS}$, $\mathbb{P}^{NLoS} = 1 - \mathbb{P}^{LoS}$, define the LoS and NLoS linkage probabilities, mathematically modeled as [24].

$$\mathbb{P}^{LoS} = \left[1 + a\exp\left(-b\left(\varphi_{jk_i} - a\right)\right)\right]^{-1} \tag{3}$$

where a and b are the environmental constant parameters based on the ratio of built-up land area to the total land area, the mean number of buildings per unit area, and the buildings' height distribution. Details of these parameters and how they affect the values of a and b can be found in [25]. φ_{jk_i} is the elevation angle between aerial j and UD k in cluster i, which is equal to $\varphi_{jk_i} = \tan^{-1}\left(\frac{h_j}{d_{Hjk_i}}\right)$, where h_j is the aerial j height and d_{Hjk_i} is the horizontal distance between aerial j and UD k within cluster i. An assumption is made that uplink transmissions from UDs in cluster i to aerial j, where the total data rate between them at time t can be formulated as:

$$\Psi_{j_i}(t) = B_V \sum_{k=1}^{K_i} \log_2\left(1 + \frac{P_{rjk_i}(t)}{\sum_{u=1,\ u\neq k}^{K_i} P_{rju_i}(t) + \sigma^2}\right) \tag{4}$$

where B_V is the total available bandwidth for aerial j. $P_{rjk_i}(t)$ is the received power (RX) at aerial j from UD k in cluster i at time t determined by its TX power P_{tk_ij} and $\lambda_{jk_i}(d_{jk_i}(t))$, where $d_{jk_i}(t)$ is the separation distance between them at time t. $\sum_{u=1,\ u\neq k}^{K_i} P_{rj_iu}(t)$ is the sum of interfering RX powers at aerial j from other UDs in its covered cluster i, while σ^2 is the additive white Gaussian noise (AWGN) power. For further details about these link models and their related parameters, readers are advised to check [24].

3.2. LEO–Aerial Linkage Model

Herein, we leveraged the linkage model of [26], as the RX power at LEO-Sat from aerial j covering cluster i at time t, i.e., $P_{rSj_i}(t)$, is written as follows:

$$P_{rSj_i}(t) = \frac{P_{tj_iS}(t) c G_{tj_i} G_{rS}}{4\pi f_S d_{Sj_i}(t)} \tag{5}$$

Thus, the attainable data rate in bps can be formulated as:

$$\eta_{Sj_i}(t) = B_{Sj_i}(t) \log_2\left(1 + \frac{P_{rSj_i}(t)}{\tau\varepsilon B_{Sj_i}(t)}\right), \tag{6}$$

In (5), $P_{tj_iS}(t)$ is the transmitted power from aerial to LEO-Sat at time t, and G_{tj_i}, G_{rS} are the transmitter and delivered antenna gains from aerial (LEO-Sat) to LEO-Sat (aerial), correspondingly. $d_{Sj_i}(t)$ defines the space between the LEO-Sat and aerial j covering cluster i. Also, f_S and c are the operating frequency of the LEO–aerial link and the speed of light, respectively, and LEO-Sat is assumed to use the sub 6GHz wireless communication technology as given in [26]. In (6), $B_{Sj_i}(t)$ is the allocated bandwidth of the LEO–aerial j link at time t, where it is assumed that the total LEO-Sat bandwidth B_{Smax} is divided among the available LEO-aerial links. Also, τ reflects the noise temperature, and ε is the Boltzmann constant.

3.3. Optimization Problem Formulation

The optimization problem aims to determine the best assignment vector $x_{j_i}(t)$ for aerials-clusters at each time t, intending to maximize their long-term accumulative data rates. This objective should be accomplished while ensuring that cluster coverage is fair over the time horizon, given that $N < M$. The constrained resources of the aerials' batteries and LEO-Sat's bandwidth should also be considered. Mathematically, this optimization problem can be expressed in the following manner:

$$\max_{x_{j_i}(t), P_{tj_iS}(t), B_{Sj_i}(t)} \frac{1}{T} \sum_{t=1}^{T-1} \sum_{j=1}^{N} \sum_{i=1}^{M} x_{j_i}(t) \min\left(\eta_{Sj_i}(t),\ \Psi_{j_i}(t)\right)$$

s.t

$$C1: x_{j_i}(t) \in \{0,1\} \tag{7a}$$

$$C2: \sum_{j=1}^{N} \sum_{i=1}^{M} x_{j_i}(t) \leq N \tag{7b}$$

$$C3: \sum_{i=1}^{M} x_{j_i}(t) = 1\ \forall j \tag{7c}$$

$$C4: x_{j_i}(t)\Gamma_{j_i}(t) \leq \Gamma_{jR}(t) \forall j \tag{7d}$$

$$C5: \frac{1}{T} \sum_{t=0}^{T-1} \sum_{j=1}^{N} x_{j_i}(t) \geq \delta_i, \forall i \tag{7e}$$

$$C6: \sum_{j=1}^{N} \sum_{i=1}^{M} x_{j_i}(t) B_{Sj_i}(t) \leq B_{Smax} \tag{7f}$$

$$C7: P_{tj_iS}(t) \leq P_{Vmax} \tag{7g}$$

where

$$\delta_i = \frac{K_i}{\sum_{i=1}^{M} K_i}, \tag{8}$$

$$\Gamma_{j_i}(t) = P_f T_{f_{j_i}}(t) + (P_h + P_{tj_iS}(t)) T_{h_{j_i}}, \tag{9}$$

where T is the time horizon, $x_{j_i}(t) \in \{0,1\}$ is an assignment binary indicator which equals one if cluster i is chosen to be covered by aerial j at time t otherwise zero. $\eta_{Sj_i}(t)$ given in (6) is the assigned capacity of the LEO–aerial link to cover cluster i by aerial j at time t. Also, $\Psi_{j_i}(t)$ given in (4) is the total data rate between aerial j and cluster i. In (7), the minimum of the aerial-UD link and the aerial–LEO link data rates is taken because UD data will be relayed to LEO-Sat via its coverage aerial. Thus, the total link speed between UD and LEO-Sat will be the minimum between both links as both will exist simultaneously. The constraint C2, which refers to the total number of chosen clusters, is less than or equal to N, i.e., the total number of aerials, as some aerials' batteries may run out of energy during the coverage and need re-charging. The third constraint, C3, reflects that at a time t, each aerial j covers only one cluster i. The fourth constraint, C4, refers to the energy consumption of aerial j, $\Gamma_{j_i}(t)$, needed to assist cluster i at time t should not overcome its remaining full battery capacity $\Gamma_{jR}(t)$ at time t. $\Gamma_{j_i}(t)$ is defined in (9), where it considers flying power P_f, hovering power P_h, and TX communication power $P_{tj_iS}(t)$. Both P_f and P_h are assumed to be constants. However, $P_{tj_iS}(t)$ is a function of time, and it should be optimized based on the current aerial j traffic need. $T_{f_{j_i}}(t)$ indicates the flying time of aerial j, which equals the division of the horizontal distance between its current position and target location in cluster i at time t, over the aerial speed. $T_{h_{j_i}}$ indicates the hovering time of aerial j over cluster i, which is assumed to be constant. The constraint C5 ensures fairness among clusters based on their UDs densities, as given in (8). Constraint C6 means that the sum of the allocated LEO–aerial bandwidths should not exceed the maximum allowable bandwidth of the LEO-Sat, i.e., B_{Smax}. Finally, constraint C7 indicates that the TX communication power of aerial at time t must not exceed its maximum allowable TX power P_{Vmax}.

4. Proposed CB-FBC and Genetic Approach

The prescribed problem in (7) is a non-linear dynamic time sequential programming lacking a closed-form optimal solution. To simplify it, we can take advantage of the fact that both $P_{tj_iS}(t)$ and $B_{Sj_i}(t)$ can be optimized based on the traffic needs of aerials after their deployment among the clusters. This allows us to divide this complex problem into two sub-problems. In the first one, we optimize the values of $x_{j_i}(t)$ while ensuring fairness in cluster coverage and considering the limited remaining energy of aerials. In the second sub-problem, the aerials' total data rates $\Psi_{j_i}(t)$ become fixed based on the optimized $x_{j_i}(t)$ values. Then, using these fixed values, we can optimize the values of $P_{tj_iS}(t)$ and $B_{Sj_i}(t)$. In the following, we propose the "CB-FBC" algorithm to address the first sub-problem, while a "GA" algorithmic approach addresses the second one.

4.1. Optimization of $x_{j_i}(t)$

Herein, the assignment vector $x_{j_i}(t)$ will be optimized under constraints C1 to C5. This sub-problem is a non-linear dynamic time sequential combinatorial optimization problem, where a group of clusters should be selected at each time t under the aforementioned constraints. Additionally, the M&C center has no information about this catastrophic area, including the number of survivors, their distributions, their traffic needs, etc. Such kinds of optimization problems cannot be solved using traditional optimization techniques as there is a lack of full information about the problem. More specifically, the values of $\Psi_{j_i}(t)$ are unknown for the optimizer itself. In these kinds of optimization problems, artificial intelligence by means of online learning provides a sufficient solution. This is due to the exploration and exploitation mechanism of online learning that can arrive at sufficient solutions without knowing the details of the problem. Thus, in this paper, this problem is modeled as a budget constraint combinatorial bandit with arms' fairness [27]. Generally, the MAB hypothesis is an efficient online learning methodology where a player intends to maximize his long-term reward via attempting the bandit's arms [19]. The player has no prior knowledge about arms except his observations about the played arms and their associated rewards. Within the bandit game, the player attempts to compromise between consistently exploiting the arm associated with the most considerable noticeable reward until now or exploring new ones [19]. In some MAB games, a cost should be paid for the played arms, which is subject to the player's limited budget. These types of MAB games are called budget-constrained MAB games [28]. Conversely, combinatorial bandits with arms' fairness given in [29] are bandits where the player chooses a group of arms named the super arm in each time step while guaranteeing long-term fairness in the arms' selection over time horizon. In this section, we will propose the "CB-FBC" algorithm to address the problem of $x_{j_i}(t)$ optimization. In this regard, the player will be the nearest survival GBS, the arms are the post-disaster clusters, and the reward is the sum of the aerials' data rates. The aerials' remaining battery capacity limits this game. Also, fairness should be maintained in post-disaster area clusters coverage based on their users' densities.

Algorithm 1 summarizes the envisioned "CB-FBC" algorithm, which will be implemented in the survival GBS as it is the bandit player. The algorithm inputs are the sets of all available post-disaster clusters and aerials, i.e., $\mathcal{M}$ and $\mathcal{N}$, the initial values of users' densities δ_i, which are pre-estimated using GPS localization, and the design parameter Ω. The algorithm's output is the aerials–clusters assignment vector $x^*_{j_i}(t)$ at time t. For initialization, at $t = 0$, $w_i(t)$ which is the selection vector, $h_i(t)$ which is the number of times cluster i was drawn up to time t, $\hat{\gamma}_i(t)$ which is the mean data rate of cluster i up to time t, and $q_i(t)$ which is the queue of cluster i, are all initiated by zero $\forall i \in \mathcal{M}$. $q_i(t)$ assures fairness among the clusters, as discussed later. For $t = 1$ to T, the upper confidence bound (UCB) for $\forall i \in \mathcal{M}$ are set to $\bar{\gamma}_i(t) = \hat{\gamma}_i(t-1) + \sqrt{(3\log t)/(2h_i(t-1))}$ if $h_i(t-1) > 0$ or $\bar{\gamma}_i(t) = 1, if h_i(t-1) \leq 0$. Then, $q_i(t)$ is estimated for $\forall i \in \mathcal{M}$ as illustrated in Algorithm 1.

At each trial t, the value of $q_i(t)$ is increased by δ_i and decreased by one if cluster i was drawn in time $t-1$. Therefore, if cluster i is not chosen often, its $q_i(t)$ will be doubled by its UD density value, and vice versa. This prioritizes it for being drawn next time step. Hence, after estimating $q_i(t)$ and $\bar{\gamma}_i(t)$ for $\forall i \in \mathcal{M}$, a super arm $A(t) \subset \mathcal{M}$ is chosen according to the following criterion:

$$A(t) = \underset{A(t) \subset \mathcal{M}}{\arg\max} \sum_{i \in A(t)} (q_i(t) + \Omega\bar{\gamma}_i(t)), |A(t)| \leq N, \tag{10}$$

where Ω defines a design parameter that balances between choosing the cluster maximizes the achievable mean data rate or that maximizes fairness upon $q_i(t)$ values. $A(t)$ is evaluated by enumerating the $|A(t)|$ clusters having the highest $(q_i(t) + \Omega\bar{\gamma}_i(t))$. Afterward, the aerials should be distributed among them and attain $x^*_{j_i}(t)$ in a manner that minimizes aerials battery consumption subject to the aerial remaining battery energy at time t, i.e., $\Gamma_{jR}(t)$, as given in constraint C4 in(7). To do that, $\Gamma_{j_i}(t)$ matrix is evaluated for $\forall i \in A(t)$ and $\forall j \in \mathcal{N}$ using (9) by assuming the worst case scenario when $P_{tj_iS}(t) = P_{Vmax}$ for $\forall j \in \mathcal{N}$. If the min $(\Gamma_{j_i}(t)) < \Gamma_{jR}(t)$ for any aerial j, it should fly back for re-charging its battery; otherwise it will fly towards cluster i to cover it and then $x^*_{j_i}(t)$ is set to 1. Afterwards, this cluster will be removed from $A(t)$ by putting $\Gamma_{j_i}(t) = \infty$ for $\forall j \in \mathcal{N}$. Then, $\Gamma_{jR}(t+1)$ are updated for the next time step based on the energy consumed by aerials or that re-charged. Finally, the mean data rates related to the chosen $A(t)$ are noticed, and its associated parameters are updated as stated in Algorithm 1, including updating the value of δ_i based on aerials real observations about users' densities in $A(t)$.

4.2. Optimization of $P_{tj_iS}(t)$ and $B_{Sj_i}(t)$

After obtaining the values of $x^*_{j_i}(t)$ and distributing the aerials among the selected $A(t)$, the values of $\Psi_{j_i}(t)$ become known. Then, in the next step, the values of $P_{tj_iS}(t)$ and $B_{Sj_i}(t)$ should be optimized to accommodate the values of $\Psi_{j_i}(t)$ related to every aerial under constraints C6 and C7 in (7). This optimization problem can be formulated as follows:

$$\max_{P_{tj_iS}(t), B_{Sj_i}(t)} \frac{1}{T} \sum_{t=1}^{T-1} \sum_{j=1}^{N} \min\left(\eta_{Sj_i}(t),\ \Psi_{j_i}(t)\right)$$

$$\text{s.t}$$

$$C6: \sum_{j=1}^{N} B_{Sj_i}(t) \leq B_{Smax} \tag{11a}$$

$$C7: P_{tj_iS}(t) \leq P_{Vmax} \tag{11b}$$

This optimization problem can be further simplified by reformulating it as minimizing the absolute difference between $\eta_{Sj_i}(t)$ and $\Psi_{j_i}(t)$ via optimizing the control parameters $P_{tj_iS}(t)$ and $B_{Sj_i}(t)$, which can be written as follows:

$$\min_{P_{tj_iS}(t), B_{Sj_i}(t)} \sum_{j=1}^{N} \text{abs}\left|\eta_{Sj_i}(t) - \Psi_{j_i}(t)\right|$$

$$\text{s.t}$$

$$C1: \sum_{j=1}^{N} B_{Sj_i}(t) \leq B_{Smax} \tag{12a}$$

$$C2: P_{tj_iS}(t) \leq P_{Vmax} \tag{12b}$$

This problem is non-linear programming due to the absolute function. Actually, various optimization techniques can handle this optimization problem, including GA, sequential quadratic programming (SQP), particle swarm optimization (PSO), interior point optimization, Bayesian optimization, etc. In this paper, we used the GA because it is well-suited for problems with non-linear objective functions like the one given above. The GA operators (crossover and mutation) allow for effective exploration in complex and non-linear spaces. Also, the GA can handle constraints effectively using penalty functions or specialized constraint-handling mechanisms. This is important in the constrained optimization problem like the one given above where $P_{tj_iS}(t)$ must not exceed P_{Vmax} and $B_{Sj_i}(t)$ must not exceed B_{Smax}. Thus, the penalty function provided by the GA is best suited to this scenario. Finally, the GA's population-based approach can help prevent premature convergence to suboptimal solutions. It allows for the simultaneous exploration of multiple candidate solutions, increasing the algorithm's robustness. The comparisons with other candidate optimizers to find the best one are out of the scope of this paper, and they will be the subject of our future investigations. In this paper, a "GA" algorithmic approach is used to address (11b) within its constraints. The inputs to the "GA" algorithm are the values of $\Psi_{j_i}(t)$, B_{Smax}, P_{Vmax} in addition to the adjusting parameters of the "GA" algorithm such as the population size Ps, the maximum number of generations Ge, the percentage of mutation ϵ_m, and the percentage of crossover ϵ_c. The fitness function will be:

$$Fit(t) = \sum_{j=1}^{N} \text{abs}\left|\eta_{Sj_i}(t) - \Psi_{j_i}(t)\right| \tag{13}$$

Algorithm 1: Proposed CB-FBC Algorithm implemented in GBS

Output: $x^*_{j_i}(t)$

Input: $\mathcal{M}, \mathcal{N}$,initiated δ_i using GPS localization, and Ω.

Initialization: At $t = 0$, $w_i(t) = 0$, $h_i(t) = 0$, $\hat{\gamma}_i(t) = 0$, and $q_i(t) = 0 \forall i \in \mathcal{M}$.

for $t = 1, 2, \ldots, T$ **do**
- **for** $i = 1, 2, \ldots, M$ **do**
  - **if** $h_i(t-1) > 0$ **then**
    - $\hat{\gamma}_i(t) = \hat{\gamma}_i(t-1) + \sqrt{(3\log t)/(2h_i(t-1))}$
  - **end**
  - **else**
    - $\hat{\gamma}_i(t) = 1$
  - **end**
- **end**
- $q_i(t) = \max\{q_i(t-1) + \delta_i - w_i(t-1), 0\}$
- • choose the super arm $A(t)$ that achieves: $A(t) = \arg\max_{A(t) \subset \mathcal{M}} \sum_{i \in A(t)} (q_i(t) + \Omega\hat{\gamma}_i(t))$, $|A(t)| \leq N$
- • Select $d^*_{j_i}(t)$ that minimizes aerials energy consumptions as:
- 1. Calculate $\Gamma_{j_i}(t)$ matrix $\forall i \in A(t)$ and $\forall j \in \mathcal{N}$ using (9) assuming $P_{tj_iS}(t) = P_{Vmax}$
- 2. Recharge aerial j or connect it with cluster i as follows:
- **for** $j = 1 : N$ **do**
  - **if** $\min(\Gamma_{j_i}(t)) < \Gamma_{jR}(t)$ **then**
    - aerial j flies back for recharging
  - **end**
  - **else**
    - **if** $\min(\Gamma_{j_i}(t)) \geq \Gamma_{jR}(t)$ **then**
      - a. $\{i, j\} = \arg\min(\Gamma_{j_i}(t))$
      - b. $x^*_{j_i}(t) = 1$
      - c. $\Gamma_{j_i}(t) = \infty \forall j \in \mathcal{N}$
      - d. Update $\Gamma_{jR}(t+1) \forall j \in \mathcal{N}$
    - **end**
  - **end**
- **end**
- • Collect data rates associated with the drawn super arm $A(t)$ then update its corresponding parameters:
  1. $w_i(t) = 1 \ \forall i \in A(t)$.
  2. $h_i(t) = h_i(t-1) + 1 \ \ \forall i \in A(t)$
  3. $\hat{\gamma}_i(t) = \frac{1}{h_i(t)} \sum_{y=1}^{h_i(t)} w_i(y)\Psi_{j_i}(y) \forall i \in A(t)$
  4. Update $\delta_i \forall i \in A(t)$

end

The output of the "GA" algorithm is the sub-optimum values $P^*_{tj_iS}(t)$ and $B^*_{Sj_i}(t)$ as detailed in Algorithm 2. At the beginning of the algorithm, the population "Pop" of $P_{tj_iS}(t)$ and $B_{Sj_i}(t)$ $\forall N$ is constructed randomly in the range of [0,P_{Vmax}] and [0,B_{Smax}], respectively, with a size of Ps. Then, for $1 \leq g \leq G_e$, the fitness function is evaluated for all members in "Pop", and the member characterized with the highest fitness value is selected. The parents, "$Parents$", of size Pa are selected from "Pop" based on the calculated fitness function values. Then, a percentage of parents are crossed over to produce the children, which after mutation produces the new population, "$NewPop$". The generated "$NewPop$" are used to refine the selections of $P^*_{tj_iS}(t)$ and $B^*_{Sj_i}(t)$ in the next iteration as given in Algorithm 2. This algorithm is also implemented in the terrestrial GBS as it is the most energized component in the network. Then, the adjusted values of $P^*_{tj_iS}(t)$ and $B^*_{Sj_i}(t)$ are sent via control signals to LEO-Sat to adjust its bandwidth values and then relay the adjusted $P^*_{tj_iS}(t)$ values to the aerials.

Algorithm 2: Proposed GA implemented in GBS

Output: $P^*_{tj_iS}(t), B^*_{j_iS}(t)$
Input: $\Psi_{j_i}(t), Ps, G_e, \epsilon_m, \epsilon_c, B_{Smax}, P_{Vmax}$, and Fit.
Initialization: Construct the population Pop of size P_s by randomly selecting P_s values of $P_{tj_iS}(t)$ in the range of $[0, P_{Vmax}]$, and $B_{Sj_i}(t)$ in the range of $[0, B_{Smax}]$ for $\forall j \in N$.

for $g = 1, 2, 3 \ldots, Ge$ **do**
- **Evaluate:** $Fit_k(t)$,$k \in P_s$ in Pop using (16)
- **Calculate:** {$P^*_{tj_iS}(t), B^*_{tj_iS}(t)$}= $\max_{\forall k \in P_s}\{Fit_k(t)\}$
- **Select:** the set of $Parents$ of size P_a from PoP using Roulette Wheel, Tournament based on the values of $Fit_k(t)$
- **Set:** $NewPop$={}
- **Crossover:**
- **for** $i = 1, 2 \ldots, \frac{P_a}{2}$ **do**
    - $Parent1$=$Parents(2i-1)$
    - $Parent2 = Parents(2i)$
    - **if** $rand < \epsilon_c$ **then**
        - {$Child1$,$Child2$}=**Crossover**{$Parent1$,$Parent2$}
    - **end**
    - **else**
        - {$Child1$,$Child2$}={$Parent1$,$Parent2$}
    - **end**
- **end**
- $NewPop = \{NewPop, Child1, Child2\}$
- **Mutation:**
- **for** $\forall Child \in NewPop$ **do**
    - **if** $rand < \epsilon_m$ **then**
        - Mutate $Child$ and update it in $NewPop$
    - **end**
- **end**
- Replace Pop by $newPop$

end

5. Numerical Analysis

This section utilizes Monte-Carlo (MC) simulations to demonstrate the efficiency of the envisioned "CB-FBC" and "GA" algorithms approach. The studied post-disaster is 1 Km^2, divided into 36 clusters containing a random number of users/survivors. The LEO-Sat altitude is 550 Km, while the aerial altitude is 100 m. An assumption is made that the nearest terrestrial GBS has full connectivity with the LEO-Sat, and all aerials are within full coverage of the LEO-Sat. The handover process between the LEO-Sat and both GBS and aerials can be conducted smoothly, and its impact on network performance will be investigated in future studies. The total bandwidth allocated for the aerial is 40 MHz, and for the LEO-Sat, it is 100 MHz. For the "GA" algorithm, the number of populations is 20, the maximum number of generations is 50, the percentage of mutation is 0.3, the percentage of crossover is 0.3, and the percentage of creation function is 0.1. Other simulation parameters are provided in Table 1. For performance comparisons, we compare the proposed approach with a basic UCB MAB algorithm given in [19] for cluster selection. Also, the GA is used for the aerials' TX power and LEO-Sat bandwidth optimization for fair comparisons. In the UCB-based clusters selection, the clusters that maximize the upper confidence criterion are selected by the survival GBS for aerial deployment without any fairness guarantee, where the super arm selection is carried out based on the following equation [19]:

$$A(t) = \underset{A(t) \subset \mathcal{M}}{\arg\max} \sum_{i \in A(t)} \left(\bar{\gamma}_i(t) + \sqrt{\frac{2ln(t)}{h_i(t)}} \right), |A(t)| \leq N, \tag{14}$$

where $A(t)$ is the group of selected clusters, $\mathcal{M}$ is the space of all available clusters, $h_i(t)$ is the number of times cluster i was selected just before time t, and $\bar{\gamma}_i(t)$ is the average data rate of cluster i just before time t. Also, naïve benchmarks like nearest and random approaches are involved in the comparisons. In the "Nearest" approach, the aerials always select their nearest clusters without collisions, i.e., no aerials cover the same cluster. If the same cluster is the nearest for two or more aerials, it will be covered by the nearest one of them. The other aerials will cover their second or third nearer cluster, and so on. Also, aerials operate at their maximum TX power values, i.e., P_{Vmax}, while the LEO-Sat bandwidth is equally allocated between the LEO–aerial links. Regarding the "Rand" approach, the clusters are selected randomly, the aerials operate at their maximum TX powers, and the LEO-Sat bandwidth is randomly allocated within the LEO–aerial links.

Table 1. Simulation parameters.

Parameter	Value
B_{Smax}, B_V	100 MHz [26], 40 MHz [24]
P_{tjk_i}, P_{Vmax}	1 Watt, 10 Watt [26]
K_i	Uniformly random in the range [1, 100]
f_V, f_S	2 GHz [24], 2.4 GHz [26]
$\sigma^2(dBm)$	−174 + 10log10(W) + 10 [30]
P_f , P_h	4, 2 Watt [31]
τ and ε	1000 and 1.38×10^{-23} [26]
ρ^{LOS} and ρ^{NLOS}	0.1 dB and 21 dB [24]
UD data load	5 Gbit
T, Ω	1000, 0.01
$G_{tj_i} G_{rS}$	15 dB [26]
a, b	4.88, 0.429 [24]

5.1. *Performance Analysis*

Figure 2 adjusts the number of GA generations, where the average total system rate is evaluated against different numbers of GA generations ranging from 10 to 100. In this simulation, the number of aerials is set to 14 to cover 36 clusters. From this figure, as the number of generations increases from 10 to 50, the average total system rate increases by 28%, while it only increases by 4% when increasing the number of generations from

50 to 100. This means that the average total system rate starts to saturate after 50 number of generations. Considering that the complexity of the GA algorithm increases linearly with an increase in the number of generations, we selected 50 generations to trade off complexity and performance as aforementioned.

Figure 3 displays the average total system rate versus distinct numbers of aerials. As the number of aerials increases, so does the mean total system rate for all compared schemes due to a greater coverage area. However, the proposed approach, named "CB-FBC" and the "GA" algorithms yield a higher mean total system rate than the other benchmark schemes. This is due to their ability to efficiently distribute the aerials across post-disaster clusters while maximizing their sum rates. Moreover, the total system rate of the proposed approach is higher than that of the basic "UCB+GA" MAB algorithm. This is because the proposed CB-FBC algorithm balances between the achievable data rate and the clusters' selection fairness, while UCB only maximizes the confidence of the selected clusters. It is also noteworthy that the "Nearest" algorithm outperforms the "Rand" algorithm in mean sum rate performance because it equally distributes LEO-Sat bandwidth among the aerials, whereas in the "Rand" scheme, they are allocated randomly. Also, Figure 3 illustrates that the proposed approach outperforms the "UCB+GA", "Nearest", and "Rand" methods by 1.168, 1.33 and 1.55 at $N = 2$, correspondingly. At $N = 14$, the proposed approach outperforms them by 1.14, 1.32, and 2 times, respectively.

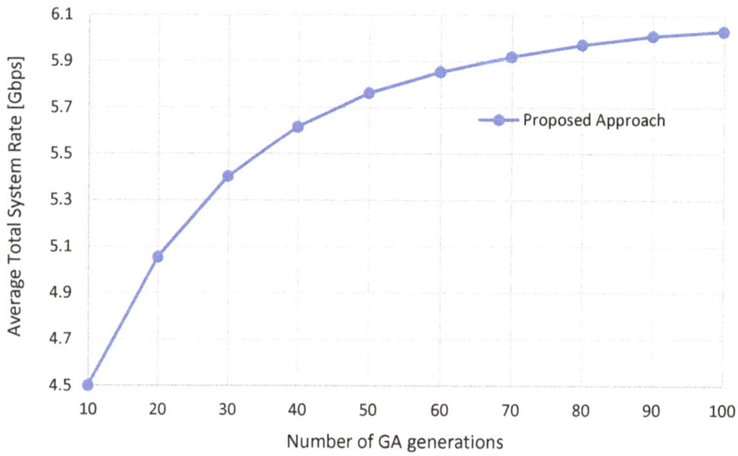

Figure 2. Total system rate against the number of GA generations.

Figure 4 depicts the sum of the aerials' energy consumption in KJ. As the number of aerials increases, so does the sum of the aerials' energy consumption for all compared schemes. However, due to the proposed approach's awareness of the limited aerial battery budget and optimization of aerial TX power, it demonstrates superior performance over other benchmarks regarding aerial energy consumption. The proposed approach has lower aerial energy consumptions than the "UCB+GA" algorithm because it optimizes the flying, hovering, and TX powers of the aerials while UCB only optimizes the aerials' TX powers using the GA algorithm. Moreover, the "Nearest" algorithm shows a better aerial energy consumption achievement than the "Rand" algorithm as it always chooses the clusters nearest to the aerials, which significantly reduces their flying energy consumption. On the other hand, in the "Rand" approach, clusters are randomly selected, which increases their flying energy consumption relative to the "Nearest" algorithm. This is the reason why the "Rand" algorithm has the worst aerial energy consumption performance. Although both the proposed approach and the "Nearest" algorithm almost select the nearest cluster, the proposed approach has lower energy consumption. This is because the proposed approach optimizes the aerials' TX communication powers using "GA" algorithms, while the "Nearest" approach uses fixed maximum TX power allocation. When the number of aerials is low, e.g., $N = 2$, the proposed approach has a lower sum of the aerials' energy consumption than "UCB+GA", the "Nearest" and "Rand" schemes by 5.2%, 42.7%, and 51.88%, respectively. However, for many aerials, e.g., $N = 14$, these values become 19.23%, 40.7%, and 71.97%, respectively. Compared to the scheme proposed in [23], the improvements over "Rand" and "Nearest" schemes were only 7% and 5%, respectively, at $N = 14$ without any improvements at $N = 2$. This superior energy consumption performance of the proposed scheme over that proposed in [23] comes from the aerial TX power optimization mechanism provided by the proposed "GA" algorithm, where the scheme proposed in [23] has no such functionality.

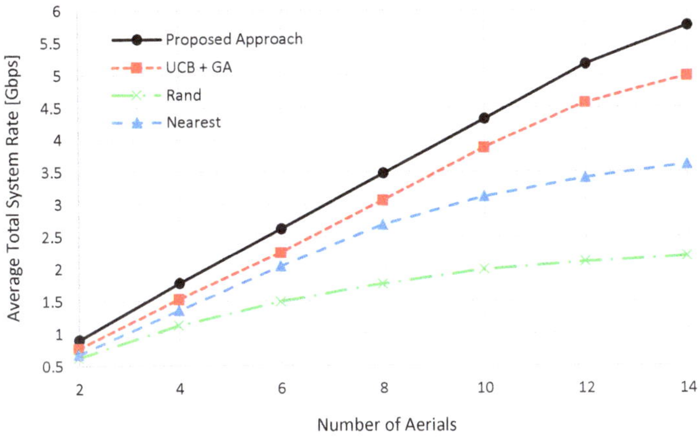

Figure 3. Total system rate against the numbers of aerials.

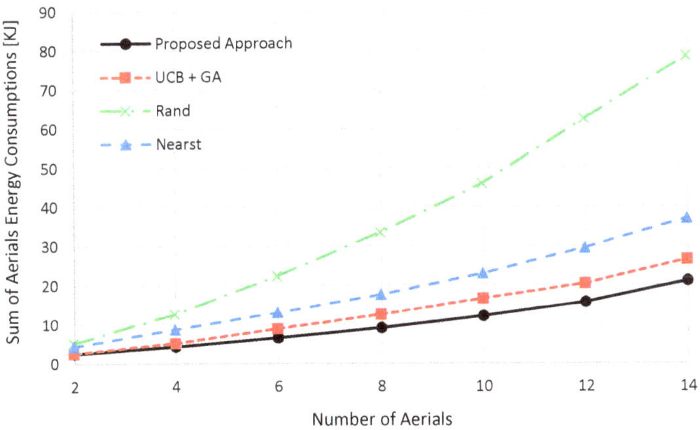

Figure 4. Sum of aerials energy consumption against the number of aerials.

Figure 5 gives the clusters selection fairness index, which is calculated as:

$$\chi = \sum_{i=1}^{M} \text{abs}\left| \frac{h_i(T)}{\sum_{i=1}^{M} h_i(T)} - \delta_i \right| \tag{15}$$

where the term $\frac{h_i(T)}{\sum_{i=1}^{M} h_i(T)}$ defines how many times cluster i was chosen over the time horizon T divided by the total number of clusters selected. If the clusters were drawn upon their user densities, δ_i, χ reaches zero. This means that as the value of χ becomes smaller, better fairness is maintained in the clusters' coverage. From Figure 5, the proposed approach has the lowest χ values at all tested number of aerials. The clusters' fairness index of the "UCB-GA" algorithm is higher than that of the proposed approach because it has no fairness functionality. As the clusters are randomly selected in the "Rand" selection approach, the χ value is almost constant irrespective of the number of aerials used, because the term $\left(\frac{h_i(T)}{\sum_{i=1}^{M} h_i(T)}\right)$ almost has the same value for $\forall i$, which is almost equal to $1/M$. On the other hand, both "UCB+GA" and the "Nearest" approaches have the worst fairness performance among the schemes involved in the comparisons because aerials always select their high confidence or nearest clusters, which causes unfairness in cluster coverage. Moreover, as the number of aerials increases, the χ values of both schemes are decreased, enhancing their fairness performance. This is because many clusters are selected at each time step, which induces better fairness performance over time horizon. As the number of aerials increments, the χ of all schemes tends to approach that of the "Rand" selection. This is due to the limited number of clusters accessible for choice at every trial, and as the number of aerials is equal to the number of clusters; all schemes will merge to the same point of "Rand" selection. From Figure 4, at $N = 2$, the clusters selection fairness index of the proposed scheme is lower than that corresponding to "UAB+GA", "Nearest", and "Rand" selections by 98.5%, 98.57 %, and 95.83%, respectively. However, at $N = 14$, these values become 73.21%,74.64%, and 50.7%, respectively.

Figure 6 displays the ratio of LEO-Sat bandwidth utilization for various schemes. The "Nearest" algorithm fully utilizes the LEO-Sat bandwidth since it distributes it equally among the LEO–aerial links. As shown in Figure 6, its bandwidth utilization ratio equals 1. On the other hand, the "Rand" scheme's bandwidth utilization ratio is nearly equal to 0.5 since the bandwidth of each LEO–aerial link is randomly assigned from the range $[1, B_{S_{max}}/N]$. As the proposed approach and "UCB-GA" algorithms optimize the allocated LEO-Sat bandwidth according to the aerials' traffic requirements using GA, they have the same performance that comes from utilizing the LEO-Sat bandwidth most efficiently. Initially, they achieve a low Sat bandwidth utilization ratio at fewer aerials. However, full Sat bandwidth utilization occurs as the number of aerials increases. This results from aerials' high total traffic demands, which operate the LEO-Sat at its maximum capacity. At $N = 2$, they have a lower SAT bandwidth utilization ratio than the "Nearest" and "Rand" approaches by 82%.

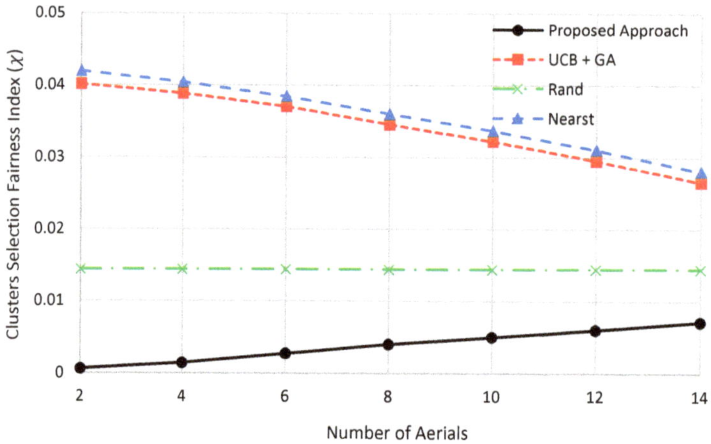

Figure 5. Clusters selection fairness index against the number of aerials.

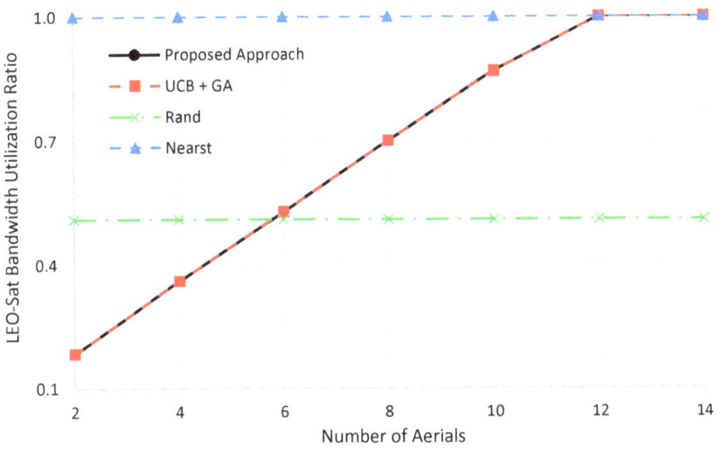

Figure 6. LEO-SAT bandwidth utilization ratio against the numbers of aerials.

5.2. Computational Complexity Analysis

The main computational complexity of the proposed approach issues from the proposed "CB-FBC" and the "GA" algorithms. The computational complexity of the envisioned "CB-FBC" algorithm is equal to $\mathcal{O}(N^2) + \mathcal{O}(N+1)$, where $\mathcal{O}(N^2)$ comes from minimizing energy consumption and $\mathcal{O}(N+1)$ from the sorting operation and updating the parameters of the selected super arm. The computational complexity of a "GA" algorithm is influenced by several factors, such as the population size, the number of generations, the objective function complexity, and "GA" algorithm functions such as mutation, crossover, and selection. Generally, this can be expressed as follows [32]:

$$\text{Complexity of GA} = \mathcal{O}(nGP(C + UOS)) \tag{16}$$

where n is the number of decision variables that is equal to $2N$ in the proposed "GA" algorithm. G and P are the number of maximum generations and the population size, which are set to 50 and 20, respectively. C is the computational complexity of the fitness function given in (16) which is equal to $\mathcal{O}(2N)$. U, O, and S are the computational complexities of the mutation, crossover, and selection functions used by the "GA" algorithm. For simplicity, we can ignore the effect of "GA" algorithm functions as they typically have constant complexity and are lower than the fitness function complexity. Thus, the computational complexity of the envisioned "GA" algorithm equals $\mathcal{O}(4000N^2)$. To sum up, the total computational complexity of the proposed approach, i.e., "CB-FP" + "GA" algorithms, will be in the range of $\mathcal{O}(N^2)$.

Thus, the improved performance of the proposed approach over the naïve algorithms and that proposed in [23] comes with a slight increase in its computational complexity. However, as the algorithm will be run on the power supply feed terrestrial GBS, this will not be a big issue for the proposed approach.

6. Conclusions

In summary, this study suggests a new approach for aerials' deployment in post-disaster areas using LEO-Sat assistance, aiming to maximize the accessible sum rate of aerials simultaneously with achieving fairness in cluster coverage based on UDs' densities and considering the finite existing resources of both LEO-Sat and aerials. The approach utilizes the "CB-FBC" algorithm to distribute aerials eventually while considering fairness in the clusters' coverage and the aerials' battery capacity. A "GA" was also used to optimize the allocation of aerials' TX powers and LEO-Sat's bandwidths. The study conducted extensive numerical analysis, which shows that the proposed approach outperforms naïve benchmark schemes like nearest- and random-based selections regarding total system rate, aerials' batteries' consumption, aerials' TX powers allocation, and LEO-Sat bandwidth utilization. Overall, this study opens opportunities for using LEO–aerial integration in different 6G network scenarios.

Author Contributions: All authors contributed equally to this paper. All authors have read and agreed to the published version of the manuscript.

Funding: The authors extend their appreciation to Prince Sattam Bin Abdulaziz University for funding this research work through the project number (PSAU/2023/01/24766).

Data Availability Statement: Data are available upon request to the authors.

Conflicts of Interest: The authors declare no conflict of interest.

References

1. Hashesh, A.O.; Hashima, S.; Zaki, R.M.; Fouda, M.M.; Hatano, K.; Eldien, A.S.T. AI-Enabled UAV Communications: Challenges and Future Directions. *IEEE Access* **2022**, *10*, 92048–92066. [CrossRef]
2. Mozaffari, M.; Saad, W.; Bennis, M.; Nam, Y.H.; Debbah, M. A Tutorial on UAVs for Wireless Networks: Applications, Challenges, and Open Problems. *IEEE Commun. Surv. Tutor.* **2019**, *21*, 2334–2360. [CrossRef]
3. Nomikos, N.; Gkonis, P.K.; Bithas, P.S.; Trakadas, P. A Survey on UAV-Aided Maritime Communications: Deployment Considerations, Applications, and Future Challenges. *IEEE Open J. Commun. Soc.* **2023**, *4*, 56–78. [CrossRef]
4. Zhang, S.; Liu, J. Analysis and Optimization of Multiple Unmanned Aerial Vehicle-Assisted Communications in Post-Disaster Areas. *IEEE Trans. Veh. Technol.* **2018**, *67*, 12049–12060. [CrossRef]
5. Hashima, S.; Hatano, K.; Mohamed, E.M. Multiagent Multi-Armed Bandit Schemes for Gateway Selection in UAV Networks. In Proceedings of the 2020 IEEE Globecom Workshops (GC Workshops), Taipei, Taiwan, 7–11 December 2020; pp. 1–6.
6. Mohamed, E.M.; Hashima, S.; Hatano, K. Energy Aware Multiarmed Bandit for Millimeter Wave-Based UAV Mounted RIS Networks. *IEEE Wirel. Commun. Lett.* **2022**, *11*, 1293–1297. [CrossRef]
7. Guo, F.; Lu, L.; Zang, Z.; Shikh-Bahaei, M. Machine Learning for Predictive Deployment of UAVs With Multiple Access. *IEEE Open J. Commun. Soc.* **2023**, *4*, 908–921. [CrossRef]
8. Leyva-Mayorga, I.; Soret, B.; Röper, M.; Wübben, D.; Matthiesen, B.; Dekorsy, A.; Popovski, P. LEO Small-Satellite Constellations for 5G and Beyond-5G Communications. *IEEE Access* **2020**, *8*, 184955–184964. [CrossRef]
9. Elhalawany, B.M.; Gamal, C.; Elsayed, A.; Elsherbini, M.M.; Fouda, M.M.; Ali, N. Outage Analysis of Coordinated NOMA Transmission for LEO Satellite Constellations. *IEEE Open J. Commun. Soc.* **2022**, *3*, 2195–2202. [CrossRef]
10. Deng, R.; Di, B.; Zhang, H.; Kuang, L.; Song, L. Ultra-Dense LEO Satellite Constellations: How Many LEO Satellites Do We Need? *IEEE Trans. Wirel. Commun.* **2021**, *20*, 4843–4857. [CrossRef]
11. Dong, C.; Xu, X.; Liu, A.; Liang, X. Load balancing routing algorithm based on extended link states in LEO constellation network. *China Commun.* **2022**, *19*, 247–260. [CrossRef]
12. Mokhtar, A.; Azizoglu, M. On the downlink throughput of a broadband LEO satellite network with hopping beams. *IEEE Commun. Lett.* **2000**, *4*, 390–393. [CrossRef]
13. Li, J.; Xue, K.; Liu, J.; Zhang, Y. A User-Centric Handover Scheme for Ultra-Dense LEO Satellite Networks. *IEEE Wirel. Commun. Lett.* **2020**, *9*, 1904–1908. [CrossRef]
14. You, L.; Li, K.X.; Wang, J.; Gao, X.; Xia, X.G.; Ottersten, B. Massive MIMO Transmission for LEO Satellite Communications. *IEEE J. Sel. Areas Commun.* **2020**, *38*, 1851–1865. [CrossRef]
15. Hao, Y.; Song, Z.; Zheng, Z.; Zhang, Q.; Miao, Z. Joint Communication, Computing, and Caching Resource Allocation in LEO Satellite MEC Networks. *IEEE Access* **2023**, *11*, 6708–6716. [CrossRef]
16. Singya, P.K.; Alouini, M.S. Performance of UAV-Assisted Multiuser Terrestrial-Satellite Communication System over Mixed FSO/RF Channels. *IEEE Trans. Aerosp. Electron. Syst.* **2022**, *58*, 781–796. [CrossRef]
17. Ma, T.; Zhou, H.; Qian, B.; Cheng, N.; Shen, X.; Chen, X.; Bai, B. UAV-LEO Integrated Backbone: A Ubiquitous Data Collection Approach for B5G Internet of Remote Things Networks. *IEEE J. Sel. Areas Commun.* **2021**, *39*, 3491–3505. [CrossRef]
18. Tran, D.H.; Chatzinotas, S.; Ottersten, B. Satellite- and Cache-Assisted UAV: A Joint Cache Placement, Resource Allocation, and Trajectory Optimization for 6G Aerial Networks. *IEEE Open J. Veh. Technol.* **2022**, *3*, 40–54. [CrossRef]
19. Auer, P.; Cesa-Bianchi, N.; Fischer, P. Finite-time Analysis of the Multiarmed Bandit Problem. *Mach. Learn.* **2002**, *47*, 235–256. [CrossRef]
20. Holland, J.H. Genetic Algorithms. *Sci. Am.* **1992**, *267*, 66–73. [CrossRef]
21. Huang, H.; Savkin, A.V. Deployment of Heterogeneous UAV Base Stations for Optimal Quality of Coverage. *IEEE Internet Things J.* **2022**, *9*, 16429–16437. [CrossRef]
22. Ahmed, S.; Chowdhury, M.Z.; Jang, Y.M. Energy-Efficient UAV Relaying Communications to Serve Ground Nodes. *IEEE Commun. Lett.* **2020**, *24*, 849–852. [CrossRef]

23. Mohamed, E.M. LEO satellite assisted UAV distribution using combinatorial bandit with fairness and budget constraints. *PLoS ONE* **2023**, *18*, e0290432. [CrossRef]
24. Sabzehali, J.; Shah, V.K.; Fan, Q.; Choudhury, B.; Liu, L.; Reed, J.H. Optimizing Number, Placement, and Backhaul Connectivity of Multi-UAV Networks. *IEEE Internet Things J.* **2022**, *9*, 21548–21560. [CrossRef]
25. Al-Hourani, A.; Kandeepan, S.; Lardner, S. Optimal LAP Altitude for Maximum Coverage. *IEEE Wirel. Commun. Lett.* **2014**, *3*, 569–572. [CrossRef]
26. Jia, Z.; Sheng, M.; Li, J.; Niyato, D.; Han, Z. LEO-Satellite-Assisted UAV: Joint Trajectory and Data Collection for Internet of Remote Things in 6G Aerial Access Networks. *IEEE Internet Things J.* **2021**, *8*, 9814–9826. [CrossRef]
27. Chen, W.; Wang, Y.; Yuan, Y. Combinatorial Multi-Armed Bandit: General Framework and Applications. In Proceedings of the 30th International Conference on Machine Learning, Atlanta, GA, USA, 17–19 June 2013; Dasgupta, S., McAllester, D., Eds.; Proceedings of Machine Learning: Pittsburgh, PA, USA; Volume 28, pp. 151–159.
28. Sinha, D.; Sankararama, K.A.; Kazerouni, A.; Avadhanula, V. Multi-Armed Bandits with Cost Subsidy. In Proceedings of the AISTATS, Virtual, 18 September 2021.
29. Li, F.; Liu, J.; Ji, B. Combinatorial Sleeping Bandits With Fairness Constraints. *IEEE Trans. Netw. Sci. Eng.* **2020**, *7*, 1799–1813. [CrossRef]
30. Mohamed, E.M.; Hashima, S.; Hatano, K.; Kasban, H.; Rihan, M. Millimeter-Wave Concurrent Beamforming: A Multi-Player Multi-Armed Bandit Approach. *Comput. Mater. Contin.* **2020**, *65*, 1987–2007. [CrossRef]
31. Mohamed, E.M.; Hashima, S.; Aldosary, A.; Hatano, K.; Abdelghany, M.A. Gateway Selection in Millimeter Wave UAV Wireless Networks Using Multi-Player Multi-Armed Bandit. *Sensors* **2020**, *20*, 3947. [CrossRef]
32. Beyer, H.G.; Schwefel, H.P.; Wegener, I. How to analyze evolutionary algorithms. *Theor. Comput. Sci.* **2002**, *287*, 101–130. [CrossRef]

MDPI AG
Grosspeteranlage 5
4052 Basel
Switzerland
Tel.: +41 61 683 77 34

Electronics Editorial Office
E-mail: electronics@mdpi.com
www.mdpi.com/journal/electronics